Lecture Notes
in Control and Information Sciences 273

Editors: M. Thoma · M. Morari

Springer

Berlin
Heidelberg
New York
Barcelona
Hong Kong
London
Milan
Paris
Tokyo

Fritz Colonius, Lars Grüne (Eds)

Dynamics, Bifurcations, and Control

With 85 Figures

Springer

Editors

Fritz Colonius
Universität Augsburg
Institut für Mathematik
Universitätsstraße
86150 Augsburg
Germany

Lars Grüne
J.W. Goethe-Universität
Fachbereich Mathematik
Postfach 11 19 32
60054 Frankfurt am Main
Germany

Cataloging-in-Publication Data applied for
Die Deutsche Bibliothek – CIP-Einheitsaufnahme
Dynamics, Bifurcations, and Control / Fritz Colonius, Lars Grüne (eds)
Berlin; Heidelberg; New York; Barcelona; Hong Kong; London; Milano; Paris; Tokyo: Springer, 2002

(Lecture Notes in control and information sciences; 273)
(Engineering online library)
ISBN 3-540-42890-9

ISBN 3-540-42890-9 Springer-Verlag Berlin Heidelberg New York

Springer-Verlag Berlin Heidelberg New York
a member of BertelsmannSpringer Science + Business Media GmbH

http://www.springer.de

Typesetting: Digital data supplied by author. Data-conversion by PTP-Berlin, Stefan Sossna
Cover-Design: design & production GmbH, Heidelberg
Printed on acid-free paper SPIN 10791459 62/3020Rw - 5 4 3 2 1 0

Preface

This volume originates from the Third Nonlinear Control Workshop "Dynamics, Bifurcations and Control", held in Kloster Irsee, April 1-3 2001. As the preceding workshops held in Paris (2000) and in Ghent (1999), it was organized within the framework of Nonlinear Control Network funded by the European Union (http://www.supelec.fr/lss/NCN).

The papers in this volume center around those control problems where phenomena and methods from dynamical systems theory play a dominant role. Despite the large variety of techniques and methods present in the contributions, a rough subdivision can be given into three areas: Bifurcation problems, stabilization and robustness, and global dynamics of control systems.

A large part of the fascination in nonlinear control stems from the fact that is deeply rooted in engineering and mathematics alike. The contributions to this volume reflect this double nature of nonlinear control.

We would like to take this opportunity to thank all the contributors and the referees for their careful work. Furthermore, it is our pleasure to thank Françoise Lamnabhi-Lagarrigue, the coordinator of our network, for her support in organizing the workshop and the proceedings and for the tremendous efforts she puts into this network bringing the cooperation between the different groups to a new level. In particular, the exchange and the active participation of young scientists, also reflected in the Pedagogical Schools within the Network, is an asset for the field of nonlinear control.

We, as all participants, enjoyed the pleasant atmosphere created by the Schwäbisches Bildungszentrum Kloster Irsee and its staff during the workshop. Last but not least, we appreciate the financial support from the European Union which made it all possible.

Augsburg, Frankfurt a.M.,
September 2001

Fritz Colonius
Lars Grüne

Contents

Part I

Bifurcation Problems

Controlling an Inverted Pendulum with Bounded Controls

Diego M. Alonso[1,3], Eduardo E. Paolini[1], and Jorge L. Moiola[2,3]

[1] Dep. Ing. Eléctrica, Universidad Nacional del Sur, Av. Alem 1253, 8000 Bahía Blanca, Argentina
[2] Mathematical Institute, University of Cologne, 50931 Cologne, Germany
[3] CONICET, Argentina

Abstract. The dynamical behaviour of a simple underactuated mechanical system with a bounded continuous control law is analyzed. The system consists of a pendulum with an inertia disk mounted on its free extreme. It is driven applying torques to the inertia disk by means of a DC motor. The closed-loop system exhibits a rich and complex dynamic when a control parameter is varied. A numerical analysis reveals Hopf, fold and homoclinic bifurcations as the main phenomena. It is shown that the pendulum can be stabilized in its inverted position with zero velocity of the disk if the controller's gains are appropriately chosen.

1 Introduction

In practical applications, physical limitations of actuators impose an important constraint to the design of a control strategy. A control law derived neglecting these limitations will surely result in a global behaviour completely different to that expected. For example, actuator's saturation may cause the appearance of new equilibria, limit cycles, or even chaos. Bifurcation theory plays an active role in this topic in order to analyze these complex behaviours (see for example [2], [3] and [11]), or even to propose other nonconventional techniques in handling appropriately the rich nature of nonlinear systems [16].

In this paper the problem of controlling an oscillatory mechanical system with a restricted control action is addressed. The system is an underactuated pendulum with an inertia disk mounted in its free extreme. The "almost" global stabilization problem at the inverted position with continuous state feedback as solved in [12] and [13] requires a control action capable of dominating the gravity torque. This hard constraint is currently bypassed using a switched or hybrid-type strategy: a nonlinear controller for swinging-up the pendulum, and a linear one to stabilize it in the inverted position. This approach has been successfully applied to many different classes of inverted pendula (some of these can be found, for example, in [4], [5], [9], [14], [15], [17]). For the particular case of the pendulum treated in this paper, the swinging-up approach has been performed using methods from bifurcation theory and energy-type considerations [1].

F. Colonius, L. Grüne (Eds.): Dynamics, Bifurcations, and Control, LNCIS 273, pp. 3–16, 2002.

Although the switched strategy is effective dealing with actuators with insufficient torque, the stabilization with a bounded continuous state feedback is still a challenging problem. Towards this end, a saturated control law is proposed in this paper, and conditions over controllers gains ensuring the stability of the inverted position are derived using classical tools. The dynamical behaviour of the closed loop system is analyzed as a function of a control (bifurcation) parameter. The closed loop system shows a rich dynamics, including Hopf and fold bifurcations, as well as heteroclinic connections of saddle points and homoclinic orbits for certain values of the control parameter. Despite this complex dynamical behavior, it is shown that there is a range of values of the control parameter where the pendulum is swung-up from the downward position to the inverted position.

2 Description of the system

The mechanical system considered is shown in Fig. 1. It is a conventional pendulum with an inertia disk in its free extreme. The system is actuated by means of torques applied to the disk by a DC motor mounted on the pendulum's arm. The system is underactuated since the pendulum can rotate freely around its pivot point.

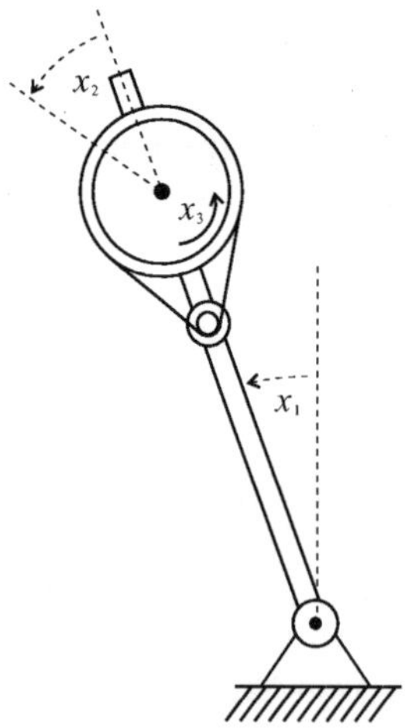

Fig. 1. The underactuated pendulum.

The dynamical model of this system is given by

$$
\begin{aligned}
\dot{x}_1 &= x_2, \\
\dot{x}_2 &= q_1 \sin x_1 + q_2 x_3 - q_3 u, \\
\dot{x}_3 &= -q_1 \sin x_1 - q_2 (1+\rho) x_3 + q_3 (1+\rho) u,
\end{aligned}
$$

where x_1 is the angle of the arm ($x_1 = 0$ at the upright position), x_2 is the angular velocity of the arm, x_3 is the angular velocity of the disk with respect to the arm, u is the control input (voltage applied to the motor), and q_1, q_2, q_3, $\rho > 0$ are constant coefficients derived from physical parameters. The disk position is not considered as a state variable because it is irrelevant for the stabilization of the pendulum in the inverted position, and in order to make the bifurcation analysis simpler.

3 Bounded control law

The control objective is the stabilization of the pendulum in its inverted position with zero velocity of the disk. In other words, the objective is to drive the state (x_1, x_2, x_3) to the target point $(2k\pi, 0, 0)$. Toward this end, consider first an auxiliary control $v = k_1 \sin x_1 + k_2 x_2 + k_3 x_3$, where k_1, k_2 and k_3 are control gains. From a linear analysis, since the system is controllable at $(2k\pi, 0, 0)$, there exist gains k_1, k_2 and k_3 such that v stabilizes the system (at least locally) at the desired point. If $u_{\max}$ is the maximum control effort available from the actuator (or admissible by the system), this bound is considered in the model by means of the saturated version of v given by

$$u = u_{\max} \tanh\left(k_1 \sin x_1 + k_2 x_2 + k_3 x_3\right).$$

3.1 Equilibrium points

The equilibrium points of the closed loop system are $x_e = (k\pi, 0, x_{3e})$, where $k \in \mathbf{Z}$ and x_{3e} are the solutions of

$$q_2 x_{3e} - q_3 u_{\max} \tanh\left(k_3 x_{3e}\right) = 0. \tag{1}$$

As the system has only two meaningful rest positions (namely $x_1 = 0$ and $x_1 = \pm\pi$), the values of the state variable x_1 corresponding to $x_1 = 2k\pi$ ($x_1 = (2k+1)\pi$) will be noted as x_{10} ($x_{1\pi}$), to avoid considering infinitely many equilibrium points.

The trivial solution, $x_{3e} = 0$, of (1) yields the equilibria

$$x_{00} := (x_{10}, 0, 0), \quad x_{\pi 0} := (x_{1\pi}, 0, 0).$$

Both equilibria undergo a supercritical pitchfork bifurcation for $k_3 = k_3^* := q_2 / (q_3 u_{\max})$, arising the following equilibrium points

$$x_{0+} := (x_{10}, 0, \hat{x}_{3e}), \quad x_{0-} := (x_{10}, 0, -\hat{x}_{3e}),$$

and

$$x_{\pi+} := (x_{1\pi}, 0, \hat{x}_{3e}), \quad x_{\pi-} := (x_{1\pi}, 0, -\hat{x}_{3e}),$$

where $\pm\hat{x}_{3e}$ are the non-trivial solutions of (1). The additional equilibria x_{0+} and x_{0-} ($x_{\pi+}$ and $x_{\pi-}$, respectively) correspond to the situation where the pendulum is in the inverted (rest) position and the disk rotates with constant speed $\hat{x}_{3e}$ or $-\hat{x}_{3e}$.

3.2 Stability of the equilibria x_{00}

As stated before, the control objective is the stabilization of the system at the equilibrium point x_{00}. Towards this end, conditions over the controller's gains are derived next.

The characteristic polynomial of the linearized closed-loop system at x_{00} is $P(\lambda) = \lambda^3 + a_2\lambda^2 + a_1\lambda + a_0$, where $a_0 = \rho q_1 (u_{\max} k_3 q_3 - q_2)$, $a_1 = -q_1 + u_{\max} k_1 q_3$ and $a_2 = q_2(1+\rho) + u_{\max} q_3 (k_2 - k_3(1+\rho))$. Hence, this equilibrium point is stable if $a_2 > 0$, $a_2 a_1 - a_0 > 0$, and $a_0 > 0$, or in terms of the controller's gains if

$$\begin{aligned} &k_3 > k_3^*, \\ &k_2 > (1+\rho)(k_3 - k_3^*) =: k_2^*, \\ &k_1 > \frac{q_1}{q_3 u_{\max}(1+\rho)} \frac{k_2(1+\rho) - k_2^*}{k_2 - k_2^*} =: k_1^*. \end{aligned} \tag{2}$$

3.3 Stability of the remaining equilibria

Assuming that $k_2 > k_2^*$ and $k_3 > k_3^*$ the stability of the remaining equilibria will be analyzed in terms of the control parameter k_1.

Equilibrium $x_{\pi 0}$

The closed loop characteristic polynomial of the linearized system at this point is $P(\lambda) = \lambda^3 + a_2\lambda^2 - a_1\lambda - a_0$. It is easy to verify that the equilibrium $x_{\pi 0}$ is unstable (a saddle, with one real, positive eigenvalue) for all $k_1 \in \mathbf{R}$.

Equilibria x_{0+} and x_{0-}

Providing that $k_3 \gg k_3^*$ the non-trivial solutions of (1) can be approximated by $\hat{x}_{3e} \approx \pm 1/k_3^*$. Hence, the closed loop characteristic polynomial at these points is $P(\lambda) = \lambda^3 + \bar{a}_2\lambda^2 + \bar{a}_1\lambda + \bar{a}_0$, with $\bar{a}_0 = \rho q_1 (\bar{u}_{\max} k_3 q_3 - q_2)$, $\bar{a}_1 = -q_1 + \bar{u}_{\max} k_1 q_3$, $\bar{a}_2 = q_2(1+\rho) + \bar{u}_{\max} q_3 (k_2 - k_3(1+\rho))$ and $\bar{u}_{\max} = u_{\max} \operatorname{sech}^2(k_3/k_3^*)$. The analysis of this polynomial reveals that these equilibrium points are unstable (each one is a saddle with one real, positive eigenvalue) for all $k_1 \in \mathbf{R}$.

Equilibria $x_{\pi+}$ and $x_{\pi-}$

Once again, assuming $k_3 \gg k_3^*$ so that $\hat{x}_{3e} \approx \pm 1/k_3^*$, the corresponding characteristic polynomial is $P(\lambda) = \lambda^3 + \bar{a}_2\lambda^2 - \bar{a}_1\lambda - \bar{a}_0$. Then, these equilibrium points are stable (unstable) if $k_1 < \bar{k}_1^*$ ($k_1 > \bar{k}_1^*$), where

$$\bar{k}_1^* = \frac{q_1}{q_3 \bar{u}_{\max}(1+\rho)} \frac{k_2(1+\rho) - \bar{k}_2^*}{k_2 - \bar{k}_2^*},$$

where $\bar{k}_2^* := (1+\rho)(k_3 - q_2/(q_3 \bar{u}_{\max}))$.

4 Local nonlinear analysis

Three Hopf bifurcation points are detected as the control gain k_1 is varied (with $k_2 > k_2^*$ and $k_3 > k_3^*$): one at the equilibrium point x_{00} and other two at each of the antisymmetric points $x_{\pi+}$ and $x_{\pi-}$.

4.1 Hopf bifurcation at the equilibrium point x_{00}

The defining conditions of a Hopf bifurcation are fulfilled when $k_1 = k_1^* =: k_{1\mathrm{HB}1}$. For this critical value of k_1, the system has a pair of complex conjugate eigenvalues with zero real part and frequency $\omega_c = (-q_1 + k_{1\mathrm{HB}1} q_3 u_{\max})^{1/2}$. From (2), the equilibrium x_{00} is unstable if $k_1 < k_{1\mathrm{HB}1}$ and stable if $k_1 > k_{1\mathrm{HB}1}$.

The stability of the emerging limit cycle is analyzed computing the *curvature coefficient* or *curvature index* in the frequency domain [10]. A convenient feedback realization to apply the frequency domain approach is given by

$$\begin{aligned} \dot{\mathbf{x}} &= \mathbf{A}(k_1)\,\mathbf{x} + \mathbf{B}\mathbf{g}(\mathbf{C}\mathbf{x}; k_1), \\ \mathbf{y} &= \mathbf{C}\mathbf{x}, \end{aligned}$$

with

$$\mathbf{A} = \begin{pmatrix} 0 & 1 & 0 \\ \bar{q}_1 & 0 & q_2 \\ \rho q_1 - \bar{q}_1 (1+\rho) & 0 & -q_2(1+\rho) \end{pmatrix}, \ \mathbf{B} = \begin{pmatrix} 0\,0 \\ 1\,0 \\ 0\,1 \end{pmatrix}, \ \mathbf{C} = \begin{pmatrix} 1\,0\,0 \\ 0\,1\,0 \\ 0\,0\,1 \end{pmatrix},$$

$$\mathbf{g}(\mathbf{x}) = \begin{pmatrix} -\bar{q}_1 x_1 + q_1 \sin x_1 - q_3 u_b \\ (-\rho q_1 + \bar{q}_1 (1+\rho))\, x_1 - q_1 \sin x_1 + q_3 (1+\rho) u_b \end{pmatrix},$$

where $\bar{q}_1 := q_1 - u_{\max} k_1 q_3$ and $u_b := u_{\max} \tanh(k_1 \sin x_1 + k_2 x_2 + k_3 x_3)$.

Then using the aforementioned approach, the curvature coefficient at $k_1 = k_{1\mathrm{HB}1}$ is given by

$$\sigma_{1c} = \frac{q_1^2}{8 q_3 u_{\max} \Delta_2} \, \frac{q_2^2 \rho \Delta_2 k_2^2 + q_1 \Delta_3 \left(\Delta_2 + (1+\rho)\,\Delta_3\right)^2}{\Delta_2 \Delta_3 + q_1 \rho \left(\Delta_3^2 + k_2^2\right)} \cdot \frac{\Delta_2^2 + \rho \Delta_2 \left(\Delta_3 - k_3^*\right) - \rho k_2^* k_3^*}{q_3^2 u_{\max}^2 \Delta_2^3 + q_1 \rho \Delta_3},$$

where $\Delta_2 := k_2 - k_2^* > 0$, and $\Delta_3 := k_3 - k_3^* > 0$. The sign of this coefficient depends on the chosen value for k_2 (satisfying $k_2 > k_2^*$), vanishing (degenerate Hopf bifurcation) when

$$k_2 = k_{2\mathrm{d}} := \frac{1}{2}\left(2\Delta_3 + k_3 \rho + \sqrt{\rho \left(4 \Delta_3 k_3^* + k_3^2 \rho\right)}\right).$$

Thus, once k_1 moves away $k_{1\mathrm{HB}1}$ (such that $k_1 > k_{1\mathrm{HB}1}$ but near $k_{1\mathrm{HB}1}$) the emerging limit cycle is stable ($\sigma_{1c} < 0$) if $k_2^* < k_2 < k_{2\mathrm{d}}$, or unstable ($\sigma_{1c} > 0$) if $k_2 > k_{2\mathrm{d}}$.

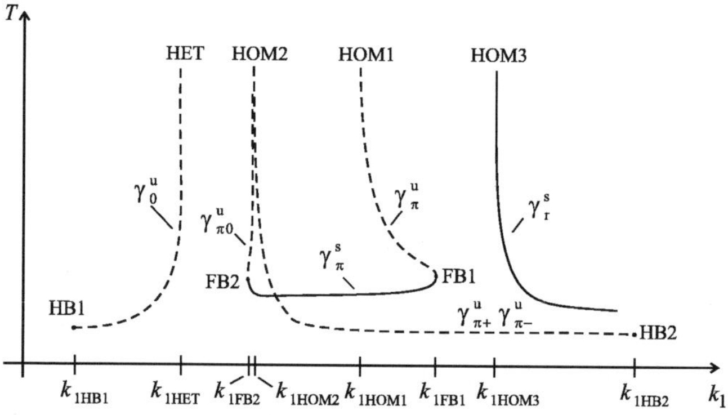

Fig. 2. Qualitative global bifurcation diagram: (*solid lines*) stable periodic solutions; (*dashed lines*) unstable periodic solutions.

4.2 Hopf bifurcation at the equilibrium points $x_{\pi+}$ and $x_{\pi-}$

Both equilibrium points exhibit a subcritical Hopf bifurcation when $k_1 = \bar{k}_1^* =: k_{1\text{HB2}}$. In this case, the stability of the emerging cycle will be determined in the following section by means of numerical methods because the derivation of the curvature index is cumbersome.

5 Numerical analysis of the global dynamical behavior

The global dynamical behavior of the closed loop system is analyzed by means of numerical continuation of limit cycles and simulation of temporal responses using control gain k_1 as the main bifurcation parameter. The numerical continuation has been performed using the well-known AUTO package.

To make the computations useful, the system parameters adopted for the numerical study are $q_1 = 30$, $q_2 = 0.0245$, $q_3 = 0.0393$ and $\rho = 250$, corresponding to those of an experimental prototype built in our laboratory. In this experimental setup the maximum control effort available makes reasonable to choose $u_{\max} = 60$ (in arbitrary units). The control gain k_3 is chosen large enough such that the approximate non-trivial solutions of (1) are close to the exact ones, thus $k_3 = 0.1 \approx 10k_3^*$. The gain k_2 is fixed arbitrarily at $k_2 = 50$.

It is convenient to begin the analysis from the Hopf bifurcations points detected analytically. Numerical continuations of the emerging periodic solutions allow us to find global phenomena such as homoclinic and heteroclinic bifurcations. A qualitative bifurcation diagram is shown in Fig. 2, where the period of the cycles is plotted versus control gain k_1.

The Hopf bifurcation at the equilibrium x_{00} is analyzed first. This bifurcation occurs for $k_1 = k_{1\text{HB1}} \approx 23.084$. For values of k_1 greater than $k_{1\text{HB1}}$ an unstable limit cycle (named γ_0^{u}) emerges; the computation of the stability index gives $\sigma_1 = 16.8 \times 10^{-6} > 0$. The numerical continuation of the amplitude

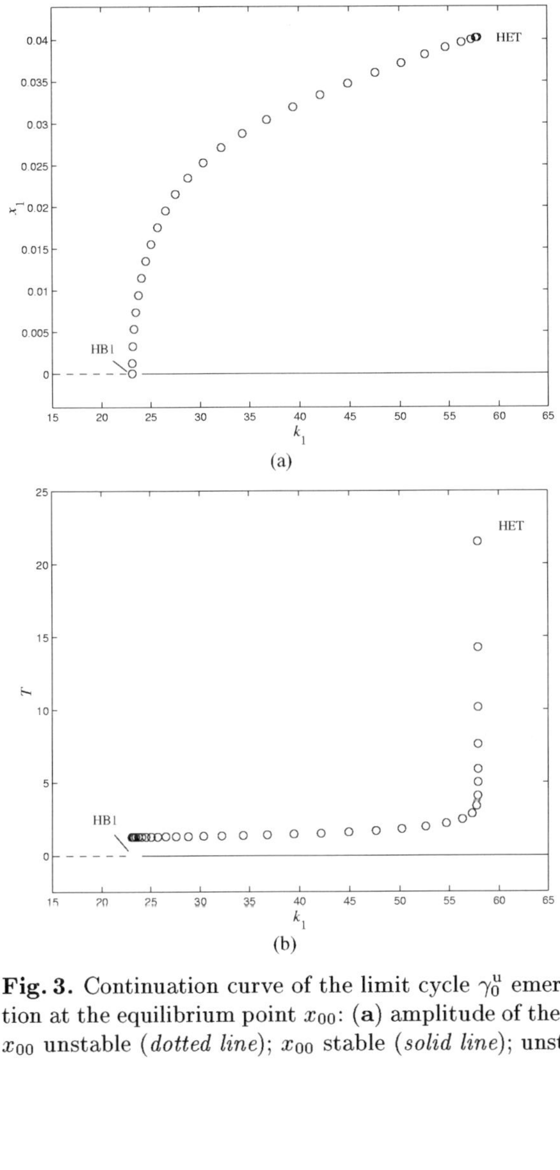

Fig. 3. Continuation curve of the limit cycle γ_0^u emerging from HB1 Hopf bifurcation at the equilibrium point x_{00}: (**a**) amplitude of the state variable x_1; (**b**) period. x_{00} unstable (*dotted line*); x_{00} stable (*solid line*); unstable limit cycles (*circles*).

of x_1 as gain k_1 is varied is shown in Fig. 3-a. For $k_1 = k_{1\text{HET}} \approx 57.88$ this cycle collides with the saddle node equilibrium points x_{0+} and x_{0-} simultaneously, and two heteroclinic trajectories connecting the stable and unstable manifolds of x_{0+} and x_{0-} do arise. In other words, a homoclinic cycle formed of heteroclinic trajectories appears at this bifurcation point (see [7]). This phenomenon is corroborated by the continuation curve of the period of the cycle in terms of the bifurcation parameter. As shown in Fig. 3-b, the period grows to infinity as k_1 approaches $k_{1\text{HET}}$. The periodic trajectory in the state

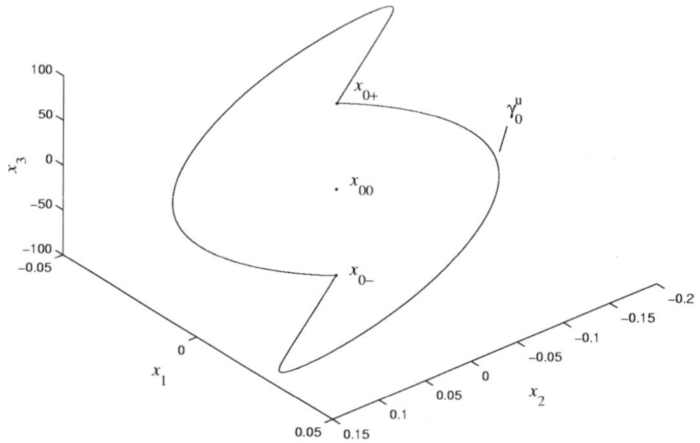

Fig. 4. Limit cycle γ_0^{u} when $k_1 = 57.88$. Note that it is near the special homoclinic cycle formed by heteroclinic trajectories.

space when $k_1 = 57.88$ (close to $k_{1\mathrm{HET}}$) is shown in Fig. 4. Notice that this is a very small amplitude oscillation in the $x_1 - x_2$ plane.

Now the attention is focused on the cycles emerging from the HB2 Hopf bifurcation at $x_{\pi+}$ and $x_{\pi-}$. These singular points undergo a subcritical Hopf bifurcation for $k_1 = k_{1\mathrm{HB2}} \approx 2.9 \times 10^6$. Two antisymmetric unstable cycles, named $\gamma_{\pi+}^{\mathrm{u}}$ and $\gamma_{\pi-}^{\mathrm{u}}$ corresponding to $x_{\pi+}$ and $x_{\pi-}$ respectively, do emerge for $k_1 < k_{1\mathrm{HB2}}$. The evolution of these cycles is investigated for decreasing values of k_1 (see Fig. 5), and the maximum amplitude of the state variable x_1 is depicted in the range $95 < k_1 < 300$ of the main bifurcation parameter. Note that cycles $\gamma_{\pi+}^{\mathrm{u}}$ and $\gamma_{\pi-}^{\mathrm{u}}$ are indistinguishable in this plot since they are antisymmetric and their maximum amplitudes are identical.

For $k_1 = k_{1\mathrm{FB1}} \approx 233.3$ a fold bifurcation of periodic solutions occurs. A stable limit cycle and an unstable one (γ_π^{s} and γ_π^{u}, respectively) appear in addition to the existing pair $\gamma_{\pi+}^{\mathrm{u}}$ and $\gamma_{\pi-}^{\mathrm{u}}$. This situation is depicted in Fig. 6, where the trajectories in the state space are shown for $k_1 = 233$. It provides a typical and unique configuration of four limit cycles for this system. Notice the different order of the amplitudes of the symmetric and antisymmetric cycles. As k_1 decreases below $k_{1\mathrm{FB1}}$, the amplitude of γ_π^{u} increases and the amplitude of γ_π^{s} decreases. The unstable cycle γ_π^{u} is shown in Fig. 7 for $k_1 = 189$, where the other cycles $\gamma_{\pi+}^{\mathrm{u}}$, $\gamma_{\pi-}^{\mathrm{u}}$ and γ_π^{s} are not shown to simplify the figure since they are slightly distorted versions of the ones appeared in Fig. 6. For $k_1 = k_{1\mathrm{HOM1}} \approx 187$ the periodic trajectory γ_π^{u} collides with the saddle node equilibria x_{0+} and x_{0-}. Although in Fig. 7 these points do appear repeatedly at $x_1 = 0$ and at $x_1 = 2\pi$, they are physically the same as stated in Sect. 3.1. Thus, at $k_1 = k_{1\mathrm{HOM1}}$ two homoclinic orbits to x_{0+} and x_{0-} do arise.

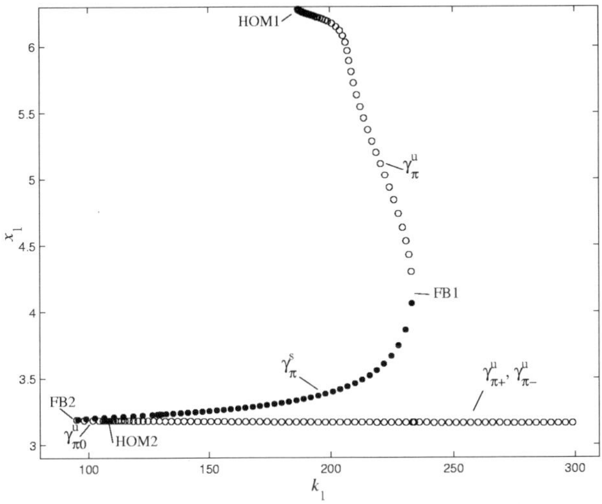

Fig. 5. Continuation curve of the limit cycles emerging from HB2 Hopf bifurcation at the equilibrium points $x_{\pi+}$ and $x_{\pi-}$: (*circles*) unstable periodic solutions; (*dots*) stable ones.

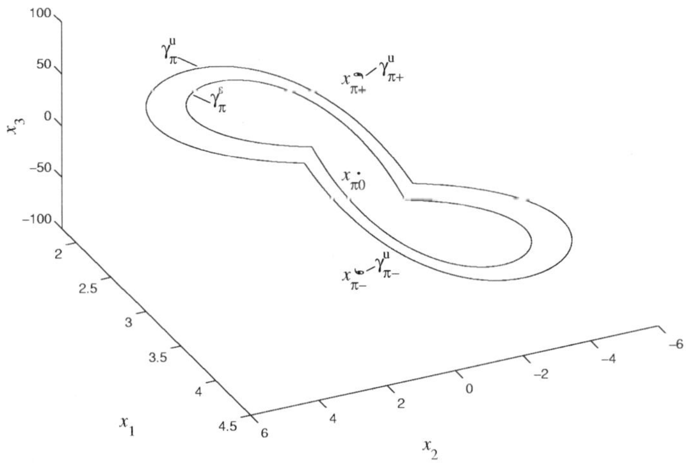

Fig. 6. Coexisting limit cycles after FB1 fold bifurcation ($k_1 = 233$).

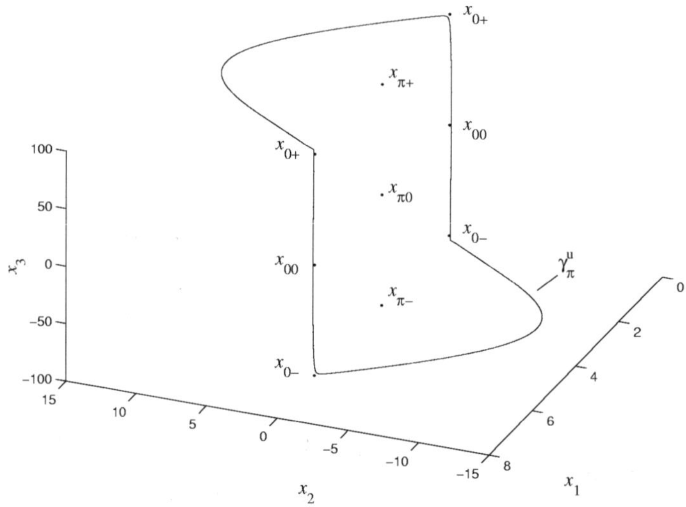

Fig. 7. Unstable limit cycle γ_π^{u} very close to the HOM1 homoclinic bifurcation ($k_1 = 189$).

In addition, heteroclinic trajectories connecting x_{0+} with x_{0-} also appear, giving rise to a special homoclinic orbit.

Let us return now to the description of the pair of antisymmetric cycles $\gamma_{\pi+}^{\mathrm{u}}$ and $\gamma_{\pi-}^{\mathrm{u}}$ introduced in Fig. 5. For decreasing values of k_1 both cycles exhibit increasing amplitudes, until they collide each other at the saddle node point $x_{\pi 0}$. This homoclinic bifurcation occurs for $k_1 = k_{1\mathrm{HOM2}} \approx 107$ causing a double saddle loop. In order to visualize this phenomenon, Fig. 8 shows the trajectories of the system in the state-space, revealing the change in the shape of these cycles when k_1 is varied. Note that once this homoclinic bifurcation occurs, an unstable cycle named $\gamma_{\pi 0}^{\mathrm{u}}$ emerges for $k_1 < k_{1\mathrm{HOM2}}$ (the antisymmetric cycles disappeared in the homoclinic cycle, or they coalesced to create $\gamma_{\pi 0}^{\mathrm{u}}$). Then, having in mind the global continuation of periodic solutions depicted in Fig. 5, it is easy to see that there is a small parameter range (between $k_{1\mathrm{FB2}}$ and $k_{1\mathrm{HOM2}}$) in which two periodic solutions coexist. In this regard, Fig. 9 shows the cycles $\gamma_{\pi 0}^{\mathrm{u}}$ and γ_π^{s} for $k_1 = 96$. Finally, both cycles collide in a fold bifurcation for $k_1 = k_{1\mathrm{FB2}} \approx 94.99$ as it has been shown schematically in Fig. 2 and numerically in Fig. 5.

Another global bifurcation takes place for $k_1 = k_{1\mathrm{HOM3}} \approx 300.5$. This is a homoclinic bifurcation, and a stable periodic motion arises for $k_1 > k_{1\mathrm{HOM3}}$ (considering x_1 modulo 2π). This periodic motion corresponds to rotations of the pendulum around the pivot point rather than oscillations (see Fig. 10).

Although other bifurcation phenomena have not been detected in the explored range of parameter k_1, the complex dynamical behavior that a system

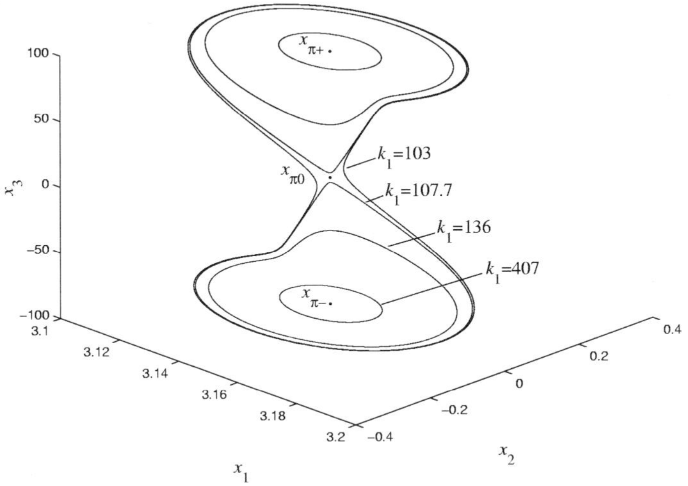

Fig. 8. Evolution of the unstable antisymmetric limit cycles $\gamma^{u}_{\pi+}$ and $\gamma^{u}_{\pi-}$ for different values of k_1.

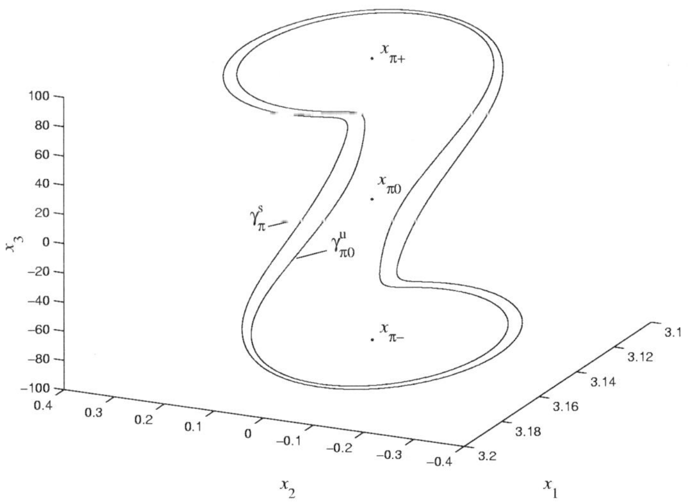

Fig. 9. Limit cycles γ^{s}_{π} and $\gamma^{u}_{\pi 0}$ before FB2 fold bifurcation ($k_1 = 96$).

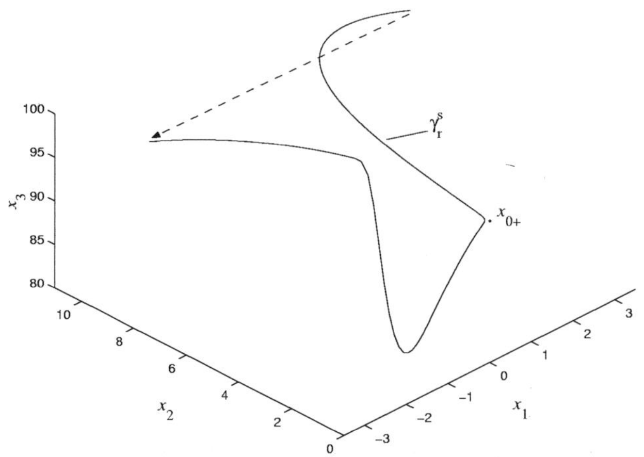

Fig. 10. Trajectory for $k_1 = 301 > k_{1\mathrm{HOM3}}$, considering x_1 modulo 2π.

may exhibit in the neighborhood of an homoclinic bifurcation ([8]) deserves a deeper analysis in the future.

6 Desired operating behaviour

One of the main motivation of the present work was to obtain a smooth continuous (non switched) control law that semi-globally stabilizes the pendulum in its upright position. Towards this end, appropriate values of k_1 giving the desired behavior are pursued next. Two intervals within the range of values of parameter k_1 shown in Fig. 2 appear suitable for accomplishing this specific purpose. The first interval comprises k_1 in the range $k_{1\mathrm{HET}} < k_1 < k_{1\mathrm{FB2}}$, where the target point x_{00} is stable and do not exist any periodic orbit. However, $x_{\pi+}$ and $x_{\pi-}$ are also stable, so a careful examination of the basins of attraction should be performed since there exist a strong competition between stable equilibria.

The second one is defined by $k_{1\mathrm{FB1}} < k_1 < k_{1\mathrm{HOM3}}$. In this case the situation is similar with respect to the equilibrium competition, but now unstable limit cycles do exist around $x_{\pi+}$ and $x_{\pi-}$. These unstable cycles (namely $\gamma^{\mathrm{u}}_{\pi+}$ and $\gamma^{\mathrm{u}}_{\pi-}$) play an important role in securing the desired behaviour. As obtained from simulations performed for k_1 within this range, the unstable cycles restrict the basins of attraction of $x_{\pi+}$ and $x_{\pi-}$. For example with $k_1 = 275$ the pendulum is driven from an initial condition near the rest position to the inverted one, where it is stabilized with zero velocity of the disk as shown in Fig. 11. A more complete analysis of the actual basin of attraction of x_{00} is under progress, and will be reported elsewhere.

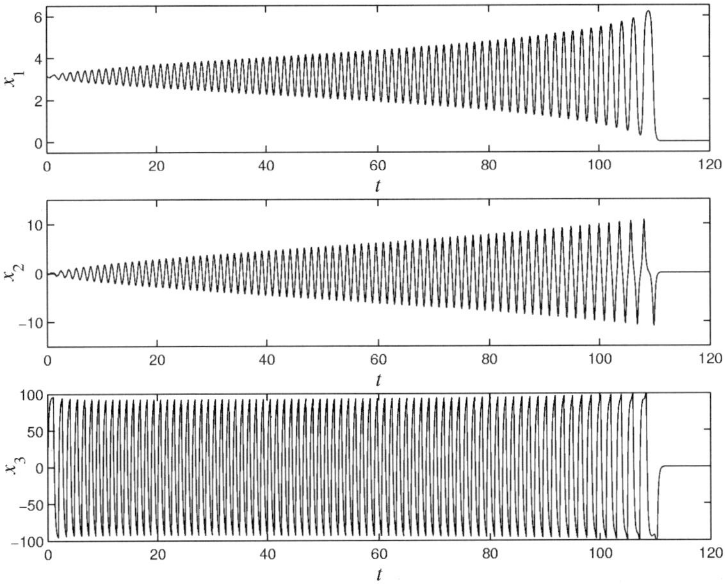

Fig. 11. Swinging-up and stabilization at the inverted position $(x_1, x_2, x_3) = (0, 0, 0)$. Initial condition $(\pi, 0, 20)$. Feedback gains: $k_1 = 275$, $k_2 = 50$ and $k_3 = 0.1$.

7 Conclusions

A bounded continuous control law for the pendulum with an inertia disk has been proposed. Conditions on the controller's gains ensuring the stability of the upright equilibrium point have been derived. Hopf bifurcations have been detected analytically when varying one of the control parameters. The global dynamical behaviour has been studied by means of numerical continuation of the limit cycles appearing from Hopf bifurcations. This numerical analysis reveals the existence of fold bifurcations, as well as heteroclinic and homoclinic cycles. Finally, a range of values of the control parameter where the controller swings-up the pendulum to the (desired) inverted position has been provided. This *ideal* control law will be validated in the prototype built in our laboratory in order to provide new insights between the mathematical model and the experimental apparatus. In the actual system, one may expect the appearance of imperfect homoclinic bifurcations ([6]), among other phenomena, since it is impossible to have perfect symmetry in any physical system.

Acknowledgments

J.L.M. acknowledges the Alexander von Humboldt Foundation and the University of Cologne for their support.

References

1. D. M. Alonso, E. E. Paolini, and J. L. Moiola, "On anticontrol of Hopf bifurcations." *Proc. 39th IEEE Conference on Decision and Control*, Sydney, Australia, pp. 2072-2077, Dec. 12-15, 2000.
2. J. Aracil, K. J. Åström and D. Pagano, "Global bifurcations in the Furuta pendulum." *Proc. IFAC Nonlinear Control System Design*, Enschede, The Netherlands, pp. 37-41, 1998.
3. J. Aracil, E. Ponce and T. Álamo, "A frequency-domain approach to bifurcations in control systems with saturations. " *Int. J. of Systems Science* **31**, pp. 1261-1271, 2000.
4. K. J. Åström and K. Furuta, "Swinging up a pendulum by energy control." *Proc. 13th IFAC World Congress*, San Francisco, USA, 1996.
5. K. Furuta, M. Yamakita, and S. Kobayashi, "Swing-up control of inverted pendulum using pseudo-state feedback." *Proc. Instn. Mech. Engrs.* **206**, pp. 263-269, 1992.
6. P. Glendinning, J. Abshagen and T. Mullin, "Imperfect homoclinic bifurcations." Preprint, available on line at: http://arxiv.org/abs/nlin/0103054, March 2001.
7. J. Guckenheimer and P. Holmes, *Nonlinear Oscillations, Dynamical Systems, and Bifurcations of Vector Fields.* Applied Mathematical Sciences **42**, Springer Verlag, New York, 1983.
8. E. A. Jackson, *Perspectives of Nonlinear Dynamics* **2**. Cambridge University Press, 1990.
9. R. Lozano, I. Fantoni and D. Block, "Stabilization of the inverted pendulum around its homoclinic orbit." *Systems & Control Letters* **40**, pp. 197-204, 2000.
10. J. L. Moiola and G. Chen, *Hopf Bifurcation Analysis. A Frequency Domain Approach.* World Scientific Pub., Singapore, 1996.
11. M. G. Ortega, J. Aracil, F. Gordillo and F. Rubio, "Bifurcation analysis of a feedback system with dead zone and saturation." *IEEE Control Systems Magazine* **20**, pp. 91-101, Aug. 2000.
12. R. Ortega and M. Spong, "Stabilization of underactuated mechanical systems via interconnection and damping assignment." *IFAC Workshop on Lagrangian and Hamiltonian Methods in Nonlinear Systems*, Princeton, NJ, March 16-18, 2000.
13. L. Praly, R. Ortega and G. Kaliora, "Stabilization of nonlinear systems via forwarding $mod\{L_gV\}$." *2nd NCN Workshop*, Paris, June 5-9, 2000.
14. A. Shiriaev, A. Pogromsky, H. Ludvigsen and O. Egeland, "On global properties of passivity-based control of an inverted pendulum." *Int. J. Robust Nonlinear Control* **10**, pp. 283-300, 2000.
15. M. W. Spong, "The swing up control problem for the acrobot." *IEEE Control Systems Magazine* **15**, pp. 49-55, Feb. 1995.
16. T. L. Vincent, "Utilizing chaos in control system design." In: G. Chen (Ed.) *Controlling Chaos and Bifurcations in Engineering Systems*, ch. 5, CRC Press, 1999.
17. Q. Wei, W. P. Dayawansa, and W. S. Levine, "Nonlinear controller for an inverted pendulum having restricted travel." *Automatica* **31** 6, pp. 841-850, 1995.

Bifurcations of Neural Networks with Almost Symmetric Interconnection Matrices

Mauro Di Marco[1], Mauro Forti[1], and Alberto Tesi[2]

[1] Dipartimento di Ingegneria dell'Informazione,
Università di Siena, Italy
[2] Dipartimento di Sistemi e Informatica,
Università di Firenze, Italy

Abstract. The symmetry of the neuron interconnection matrix ensures that additive neural networks are completely stable, i.e., each trajectory converges towards some equilibrium point. However, the crucial point is that in any practical realization it is not possible to implement perfectly symmetric interconnections, and therefore robustness of complete stability with respect to small perturbations of the nominal symmetric neuron interconnections is an extremely important issue. This paper addresses such a basic problem showing that symmetry of the interconnection matrix is not sufficient to ensure robustness of complete stability. More precisely, some third-order neural network configurations are discussed through which the existence of non-convergent dynamics as close to the symmetry condition as one pleases is proved. These include stable limit cycles generated via local Hopf bifurcations or global heteroclinic bifurcations. Moreover, numerical simulations show the presence of period-doubling bifurcations which originate more complicated attractors.

1 Introduction

By complete stability of a neural network it is meant that each trajectory of the differential equations converges towards some equilibrium point (a stationary state), possibly within a set of many equilibria [1]. It is well known that such a peculiar dynamical property has important potential applications to several signal processing problems. Indeed, completely stable neural networks have proven well suited for implementing content addressable memories, solving combinatorial optimization problems, and performing feature extraction in image processing [2–7].

For example, in an associative memory, the neural network is provided with initial conditions representing partial information or information corrupted by noise. The process of retrieval of the whole information is accomplished during the transient motion towards the stable equilibrium point to which the initial condition belongs to [2,4]. Since a completely stable neural network usually has an associated Lyapunov function which decreases along the trajectories, another natural and typical application concerns real time optimization problems. In this case the main goal is to design the neural network interconnecting structure and the nonlinear neuron activation functions

F. Colonius, L. Grüne (Eds.): Dynamics, Bifurcations, and Control, LNCIS 273, pp. 17–35, 2002.

so that the Lyapunov function matches the cost function to be minimized [2,8].

The most basic results on complete stability have been established for neural networks where the neuron interconnection matrix is symmetric. This implies that the neural network is described by gradient-like systems of differential equations [2–5]. Conversely, it is known that in the general case neural networks with largely non-symmetric interconnection matrices may exhibit complex behaviours, which include limit cycles and even more complicated attractors [1,6,9].

Clearly, being a symmetric matrix is not a generic property. Moreover, in any hardware implementation, it is impossible to realize perfectly symmetric neuron interconnections. Therefore, the fundamental issue naturally arises of the robustness of complete stability with respect to small perturbations of the nominal symmetric interconnections [10,11]. It is stated in [10] that "since such networks require symmetry in the interconnecting structure, robustness issues in the qualitative analysis ... are of paramount importance." An analogous concept is expressed in [11], where it is stated that "it is still an open question as to whether a neural network with *nearly* symmetric interconnections can exhibit limit cycles, and if so, under what conditions."

In view of the practical electronic implementation it is thus mandatory to understand whether complex dynamics, which are certainly present *far* from symmetry for many classes of neural networks, can also occur for parameter sets *arbitrarily* close to the symmetry condition. In this paper, an answer to this basic questions is given. The main result is that non-convergent dynamics can in fact be found even as close as one pleases to the nominal case of symmetric interconnection matrices. Said another way, this proves that symmetry of the interconnection matrix is not sufficient to guarantee robustness of complete stability in the general case.

The structure of the paper is briefly outlined as follows. Section 2 presents the neural network model of interest and the needed background material. Section 3 and Sect. 4 discuss some nearly symmetric neural networks for which it is possible to show the presence of limit cycles. Section 5 shows the existence of more complicated attractors close to symmetry. Finally, Sect. 6 contains some concluding remarks.

Notation

$\mathbb{R}^n$: real n-space
$x = (x_1, \cdots, x_n)^t \in \mathbb{R}^n$: column vector
$A = [A_{ij}] \in \mathbb{R}^{n \times n}$: square matrix
A^t : transpose of A
A^{-1} : inverse of A
$\ker A$: kernel of A
$\sigma\{A\}$: spectrum of A
$\|A\|$: matrix norm of A

$E_n \in \mathbb{R}^{n\times n}$: identity matrix
$0_{n\times m} \in \mathbb{R}^{n\times m}$: null matrix
$D = \text{diag}(d_1, \cdots, d_n) \in \mathbb{R}^{n\times n}$: diagonal matrix with $D_{ii} = d_i$, $i = 1, \cdots, n$
$\text{span}(u, v)$: linear subspace generated by $u, v \in \mathbb{R}^n$

2 Neural network model and preliminaries

Let us consider a *nominal* neural network which obeys the system of differential equations

$$\dot{x} = -Dx + A_0 G(x) + I, \tag{N}$$

where $x \in \mathbb{R}^n$ is the vector of the neuron state variables, and $G(x) = (g_1(x_1), \cdots, g_n(x_n))^t : \mathbb{R}^n \to \mathbb{R}^n$ has components g_i modeling the nonlinear input-output activations of the neurons. Matrix $A_0 \in \mathbb{R}^{n\times n}$ describes the neuron interconnections, while $D = \text{diag}(d_1, \cdots, d_n) \in \mathbb{R}^{n\times n}$ has diagonal entries $d_i > 0$, $i = 1, \cdots, n$, corresponding to the neuron self-inhibitions. Finally, vector $I \in \mathbb{R}^n$ has components which represent constant inputs to the neurons.

Model (N) describes the dynamics of the general class of neural networks which possess an additive neuron interconnecting structure. It includes the Hopfield model [2], the Cellular Neural Networks [5], and several other classes of neural architectures widely employed in practice [1,12].

We shall consider two basic models for the neuron nonlinearities g_i. The first one is the standard unit-gain piecewise-linear function defined by

$$g_i(x_i) = g_{pwl}(x_i) = \begin{cases} 1 & \text{for } x_i \geq 1 \\ x_i & \text{for } x_i \in (-1, 1) \\ -1 & \text{for } x_i \leq -1 \end{cases} . \tag{1}$$

Then, we consider the smooth sigmoidal nonlinearity

$$g_i(x_i) = g_{apwl}(x_i) = \frac{1}{2p} \log \frac{\cosh(p(x_i + 1))}{\cosh(p(x_i - 1))}. \tag{2}$$

Nonlinearity g_{apwl} has been proposed in [13] to approximate, as p increases, the ideal piecewise-linear function g_{pwl}, see Fig. 1. We may think of g_{apwl} as a realistic model for the electronic (e.g., VLSI) implementation of the nonlinearity g_{pwl}. We remark that the neuron activation g_{pwl} is widely employed in neural network models [5,14], the most important case being that of the standard Cellular Neural Networks [5] [1]. We finally note that the

[1] It is observed that the best choice of p for g_{apwl} is actually related to the technology used in the neuron implementation. For example, values of p of some units give a good approximation for the implementation of g_{pwl} in MOS technology, see, e.g., the experimental results in [15]. Instead, in the realization of g_{pwl} using op amps, higher values of p are more realistic, since in this case the implemented nonlinearity possesses almost straight segments and sharp corners [16].

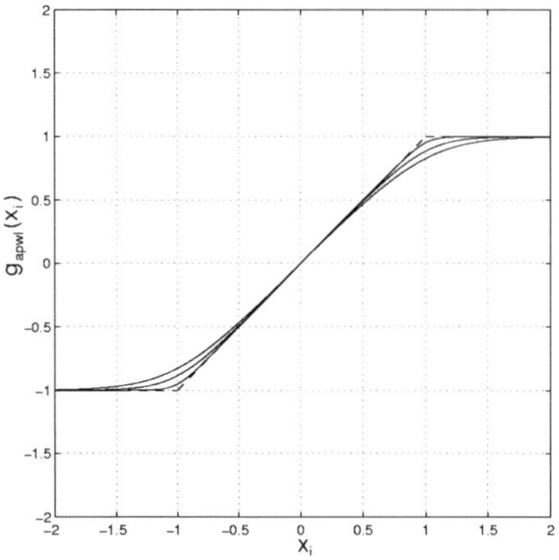

Fig. 1. Sigmoidal nonlinearity g_{apwl}, $p = 2, 3, 7$ (solid lines), compared with the ideal piecewise-linear nonlinearity g_{pwl} (dashed line)

results in this paper can easily be extended to other neuron nonlinearities, such as the high-gain sigmoidal functions used in the Hopfield model [2].

Henceforth, we assume that the *nominal* interconnection matrix A_0 satisfies the next fundamental hypothesis.

Assumption 1 *The interconnection matrix* A_0 *is* symmetric, *i.e.*, $A_0^t = A_0$.

In addition to (N), we consider the *perturbation* model defined as

$$\dot{x} = -Dx + AG(x) + I, \tag{P}$$

where

$$A = A_0 + \Delta A, \tag{3}$$

being $\Delta A \in \mathbb{R}^{n \times n}$ in general non-symmetric. Note that (P) is obtained from (N) by modifying the interconnection matrix, only.

The main goal is to investigate the loss of complete stability of neural network (P). To this end, we first recall the following definitions and results.

Definition 1. The neural network (N) (or (P)) is said to be completely stable if each trajectory converges towards an equilibrium point, as $t \to \infty$.

Definition 2. The neural network (N) (or (P)) is said to be almost completely stable if there is at most a set of zero measure of initial conditions originating non-convergent trajectories.

Note that though slightly weaker than complete stability, the property of almost complete stability is actually equivalent to complete stability for the applications, since such non-convergent trajectories cannot be observed in physical experiments. Obviously, there is the loss of almost complete stability when there exists an open set of initial conditions for which the corresponding trajectories of the neural network do not converge towards an equilibrium point.

The principal result on complete stability when $g_i(x_i) = g_{pwl}(x_i)$ is that established in [5], and is reported in the next theorem [2].

Theorem 1 ([5,18]). *Suppose that the nominal neural network (N) satisfies Assumption 1, and $g_i(x_i) = g_{pwl}(x_i)$, $i = 1, \cdots, n$. Then, (N) is completely stable.*

An analogous result is valid in the sigmoidal case.

Theorem 2 ([1,2]). *Suppose that the nominal neural network (N) satisfies Assumption 1, and $g_i(x_i) = g_{apwl}(x_i)$, $i = 1, \cdots, n$. If the equilibrium points are isolated, then (N) is completely stable.*

When the neuron interconnection matrix is *non-symmetric*, complete stability is no longer guaranteed. In fact, it has been found that in the general case largely non-symmetric neural networks may exhibit non-convergent dynamics, including stable limit cycles and chaotic attractors [19–22]. For completeness, it is also worth to mention that there are special classes of non-symmetric neural networks which are completely stable in the generic case. One important example is that of cooperative neural networks, which are defined by the property that $A_{ij} \geq 0$, for all $i \neq j$ [1]. The second example concerns neural networks possessing a unique equilibrium point which is globally asymptotically stable [8]. We refer the reader to [1,12] for a more comprehensive review on existing results on complete stability of neural networks.

The crucial point to note is that the symmetry condition required in Theorem 1 and Theorem 2 makes the electronic implementation of the neural network quite a delicate task. Indeed, in any practical neural network realization, it is not possible to obtain perfectly symmetric neuron interconnections, since this requires the exact equality between pairs of physically distinct conductances. To better illustrate this point, consider Fig. 2 which represents a possible electronic scheme to implement an additive neural network [2]. The neuron interconnections A_{12} and A_{21} are actually given by

$$A_{12} = \frac{1}{R_{12}}; \qquad A_{21} = \frac{1}{R_{21}},$$

[2] It is of importance to remark the following facts. The proof of complete stability in [5] (see Theorem 1) was found to contain some gaps [17]. In a recent work [18], a new method has been introduced to analyze complete stability of the standard Cellular Neural Networks [5]. In that paper it is rigorously shown that the symmetry of A_0 indeed implies complete stability of (N). The result in [18] is general and it includes, among other cases, also degenerate situations where (N) possesses infinitely many (non-isolated) equilibrium points.

where R_{12} and R_{21} are the resistances of the two resistors encircled in the figure. Due to tolerances, it will never happen that $R_{12} = R_{21}$.

The previous considerations make it evident that robustness of complete stability with respect to small perturbations of the nominal symmetric neuron interconnections is an extremely important issue. The fundamental question we address in this paper is indeed to discover whether there are critical situations for which (P) is not almost completely stable, even for perturbations ΔA of a nominal symmetric matrix A_0 that are arbitrarily small (i.e., $\|\Delta A\| \to 0$).

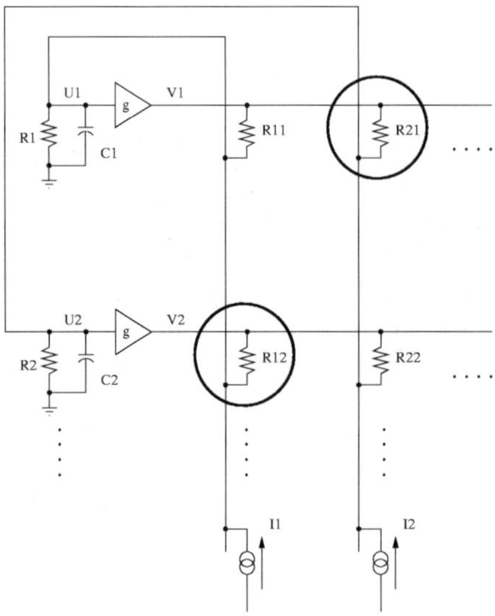

Fig. 2. Schematic of an electronic implementation of an additive neural network

As noted before, it is already known that complex dynamical behaviours exist when the neural network interconnections are *largely* non-symmetric [20,22]. The main problem is thus to discover whether such non-convergent dynamics can be found even *arbitrarily close* to the nominal case of a symmetric interconnection matrix A_0. To this end, some basic third-order neural networks are discussed. The main result obtained is that for such neural networks it is possible to analytically prove the existence of stable limit cycles as close as one pleases to symmetry. These can be originated by local Hopf-like bifurcations of an equilibrium point, or by global heteroclinic bifurcations. Moreover, there is the experimental evidence of the possible occurrence of more complex bifurcations close to symmetry, such as period-doubling bifurcations, leading to the birth of complicated attractors.

3 Limit cycles in a competitive neural network

Consider the third-order neural network

$$\begin{cases} \dot{x}_1 = -x_1 - \alpha g_{pwl}(x_2) - \beta g_{pwl}(x_3) \\ \dot{x}_2 = -x_2 - \beta g_{pwl}(x_1) - \alpha g_{pwl}(x_3) \\ \dot{x}_1 = -x_3 - \alpha g_{pwl}(x_1) - \beta g_{pwl}(x_2) \end{cases} \tag{4}$$

where the parameters α, β are assumed to be positive. This means that there are inhibitory interconnections between distinct neurons, i.e., the neural network is competitive.

It is straightforward to check that (4) can be written in the form (P), where

$$A_0 = -\begin{bmatrix} 0 & 1 & 1 \\ 1 & 0 & 1 \\ 1 & 1 & 0 \end{bmatrix},$$

$$\Delta A = -\begin{bmatrix} 0 & \alpha - 1 & \beta - 1 \\ \beta - 1 & 0 & \alpha - 1 \\ \alpha - 1 & \beta - 1 & 0 \end{bmatrix},$$

$$D = E_3,$$

$$G(x) = (g_{pwl}(x_1), g_{pwl}(x_2), g_{pwl}(x_3))^t,$$

and

$$I = (0, 0, 0)^t.$$

Note that the origin is always an equilibrium point of (4).

First, we note that $A_0 + \Delta A$ is symmetric if $\alpha = \beta$. Therefore, according to Theorem 1, (4) is completely stable for (α, β) belonging to the straight half-line

$$C_s = \{(\alpha, \beta) : \alpha - \beta = 0, \alpha > 0, \beta > 0\}. \tag{5}$$

In accordance with our problem of discovering situations where the perturbation model (P) is not completely stable, even for $\|\Delta A\| \to 0$, we discuss the dynamical behaviours exhibited by (4) when the bifurcation parameters (α, β) are varied in a neighborhood of $(1, 1)$.

These results are summarized in the bifurcation diagram depicted in Fig. 3. More details on the dynamics of (4), and the proof of the results here discussed, can be found in [23].

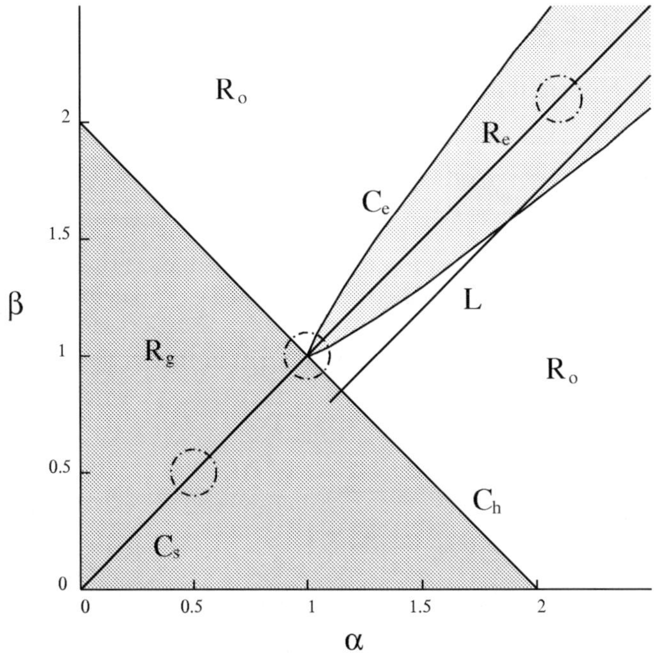

Fig. 3. Bifurcation diagram for the third-order competitive neural network (4)

The diagram is specified by C_s and the two bifurcation curves

$$C_h = \{(\alpha,\beta) : \alpha + \beta - 2 = 0, \alpha > 0, \beta > 0\}$$

and

$$C_e = \{(\alpha,\beta) : (\alpha + \alpha\beta - \beta^2 - 1)(\beta + \alpha\beta - \alpha^2 - 1) = 0, \alpha \geq 1, \beta \geq 1\}.$$

These two curves identify the three open parameter regions R_g, R_o, and R_e shown in the figure.

The dynamics of (4) for $\alpha, \beta > 0$ are summarized as follows. For $(\alpha,\beta) \in R_g$, the origin of (4) is the unique equilibrium point, and it is globally asymptotically stable. Within region R_e, (4) is completely stable, but now it has multiple asymptotically stable equilibrium points. Finally, for $(\alpha,\beta) \in R_o$, the neural network has a stable limit cycle which attracts almost all trajectories, i.e., R_o is an oscillatory region for (4).

It is of interest to analyze in some detail the bifurcations related to the curves C_h, C_e. Consider the straight half-line

$$L = \{(\alpha,\beta) : (\alpha,\beta) = (\tilde{\alpha} + \epsilon, \tilde{\beta} + \epsilon), \epsilon > -\delta\}$$

where $(\tilde{\alpha},\tilde{\beta}) \in C_h \backslash (1,1)$ $(\tilde{\alpha} > \tilde{\beta})$, and $\delta > 0$ is sufficiently small, see Fig. 3.

As already mentioned, for $\epsilon \in (-\delta, 0)$, i.e., $(\alpha, \beta) \in R_g$, the origin of (4) is globally asymptotically stable. When $\epsilon = 0$, i.e., $(\alpha, \beta) \in C_h \backslash (1,1)$, it is seen that the origin of (4) is still the unique equilibrium point, however, it is now a globally attractive center. For sufficiently small $\epsilon > 0$, i.e., $(\alpha, \beta) \in R_o$, the origin continues to be the unique equilibrium point of (4). Now, however, it is unstable and is surrounded by a stable limit cycle which is globally attractive (Fig. 4).

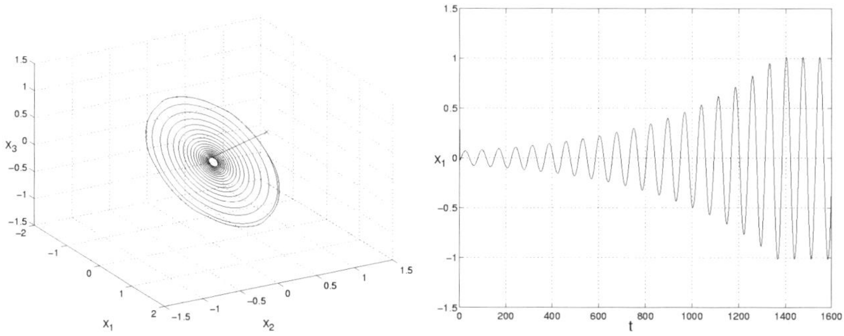

(a) (b)

Fig. 4. Trajectory converging towards a stable limit cycle. System parameters: $\alpha = 1.052$, $\beta = 0.952$

Therefore, the crossing of the curve C_h leads to a bifurcation resembling the Hopf one at the origin [24]. This is confirmed by the fact that a pair of eigenvalues of the Jacobian of (4) at the origin crosses the imaginary axis as ϵ crosses 0. Despite this similarity, we remark that, differently from a standard Hopf bifurcation (see next section), the size of the cycle does not tend to 0 as ϵ tends to 0. This is related to the lack of smoothness of the vector field defining (4) [24].

As ϵ is further increased, the amplitude and the period of the limit cycle increase. When the point along L eventually reaches C_e, the cycle disappears. More precisely, on C_e a global bifurcation occurs which leads to the birth of multiple equilibrium points (6 double equilibrium points, in addition to the origin). Moreover, a heteroclinic orbit is seen to connect the double equilibrium points, see Fig. 5. The appearance of such an orbit indeed causes the death of the limit cycle. If we finally enter region R_e, the 6 double equilibrium points bifurcate and give rise to 6 asymptotically stable equilibrium points, and 6 unstable ones (the origin is still an unstable equilibrium point).

In summary, if we consider any circle in the parameter space centered at $(\alpha, \beta) = (1,1)$ (see Fig. 3), all the bifurcations and dynamics previously illustrated can be found within the circle, and this is true no matter how small the circle is chosen. This implies that there are open parameter sets (α, β) arbitrarily close to the symmetric matrix corresponding to $(\alpha, \beta) = (1,1)$,

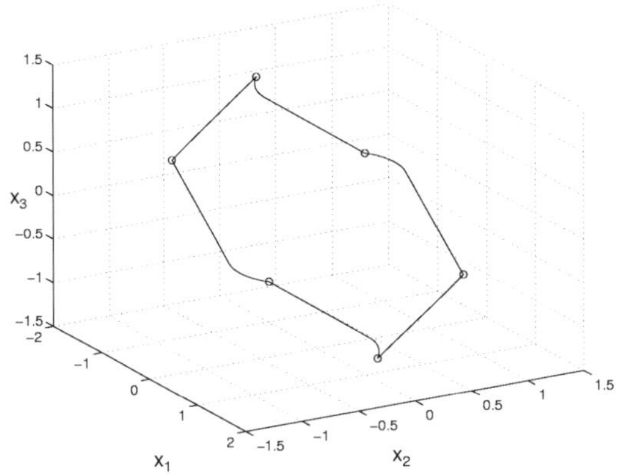

Fig. 5. Heteroclinic connection between 6 double equilibrium points. System parameters: $\alpha = 1.2222$, $\beta = 1.1222$

for which almost all trajectories of (4) persistently oscillate. Note that this is not true for any other point $(\alpha, \beta) \in C_s \setminus (1, 1)$. Indeed, for such a point it is possible to find a circle around it where the complete stability property is preserved (see Fig. 3).

4 Hopf bifurcations in sigmoidal neural networks

In the previous section we have shown that complete stability of (4) is not robust with respect to arbitrarily small perturbation of the nominal symmetric interconnection matrix, only if the nominal system parameters are $\alpha = 1$, $\beta = 1$. Indeed, such a peculiarity of the point $(\alpha, \beta) = (1, 1)$ can be justified by computing the eigenvalues of the Jacobian $J_0(0)$ at the origin of (4), when $\Delta A = 0$. The nominal Jacobian is given by $J_0(0) = -D + A_0 G'(0)$, where $G'(0) = \mathrm{diag}(g'_{pwl}(0), \ldots, g'_{pwl}(0))$, and a simple calculation yields (see Fig. 6.a)

$$\sigma\{J_0(0)\} = \sigma\{-D + A_0 G'(0)\} = \{0, 0, -3\}.$$

The presence of two zero eigenvalues suggests the existence of a suitable one-parameter family of small perturbations $\Delta A(\mu)$, $\mu \in \mathbb{R}$, such that the eigenvalues of $J_0(0) + \Delta A(\mu) G'(0)$ display a crossing of the imaginary axis (see Fig. 6.b), thus making possible the birth of a Hopf bifurcation under suitable smoothness assumptions on the nonlinear map $G(x)$.

In this section, this idea is exploited to ensure the existence of limit cycles arbitrarily close to the symmetry condition for a class of inputless neural networks of order n which obey the system of differential equations

$$\dot{x} = -Dx + (A_0 + \Delta A) G(x), \tag{6}$$

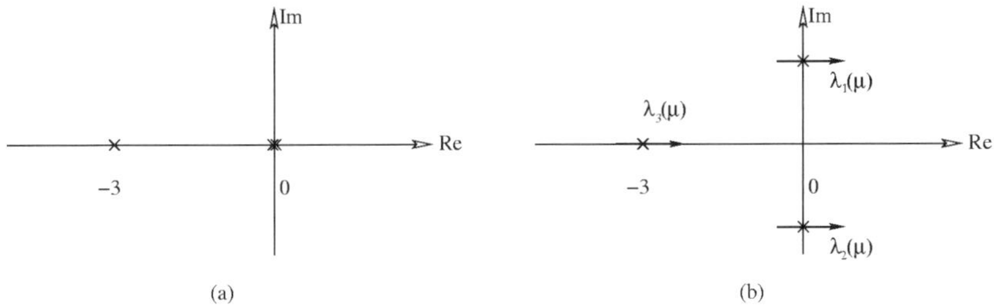

Fig. 6. (**a**) Eigenvalues of $J_0(0) = -D + A_0 G'(0)$; (**b**) Eigenvalues of $J_0(0) + \Delta A(\mu) G'(0)$ as a function of parameter μ

where $x \in \mathbb{R}^n$ and $G(x) = (g_{apwl}(x_1), \ldots, g_{apwl}(x_n))^t \in \mathbb{R}^n$, being $g_{apwl}(x_i)$ the smooth sigmoidal nonlinearity defined in (2). The diagonal matrix D is assumed as

$$D = G'(0) = g'_{apwl}(0) E_n,$$

while the nominal interconnection matrix $A_0 \in \mathbb{R}^{n \times n}$ and the related perturbation $\Delta A \in \mathbb{R}^{n \times n}$ satisfy the two assumptions given below.

Assumption 2 *The nominal interconnection matrix A_0 has the form*

$$A_0 = U \left(E_n + \begin{bmatrix} 0_{2\times 2} & 0_{2\times(n-2)} \\ 0_{(n-2)\times 2} & A_1 \end{bmatrix} \right) U^t, \tag{7}$$

where $A_1 \in \mathbb{R}^{(n-2)\times(n-2)}$ is a symmetric matrix with real negative eigenvalues, and $U \in \mathbb{R}^{n\times n}$ is a unitary matrix.

It is straightforward to verify that Assumption 2 implies that the nominal Jacobian $J_0(0)$ at the origin of (6) has the following property.

Property 1 *Let $J_0(0) = -D + A_0 G'(0)$ and let $u, v \in \mathbb{R}^n$ be the first two columns of U, i.e., $U = [u, v, W]$, $W \in \mathbb{R}^{n\times(n-2)}$. Then,*

$$\sigma\{J_0(0)\} = \{0, 0\} \cup \sigma\{g'_{apwl}(0) A_1\} \tag{8}$$

and

$$\mathrm{span}(u, v) = \ker J_0(0). \tag{9}$$

Note that $J_0(0)$ has two zero eigenvalues and the others are real negative, i.e., the situation is like that depicted in Fig. 6.a.

The next assumption defines the structure of the perturbation ΔA in (6).

Assumption 3 *Let $u, v \in \mathbb{R}^n$ be the first two columns of U and let $\omega \in \mathbb{R}$, $\omega > 0$, be given. Then, ΔA is defined as the one-parameter family of perturbations*

$$\Delta A(\mu) = \frac{\mu}{g'_{apwl}(0)} E_n + \frac{\omega}{g'_{apwl}(0)} (uv^t - vu^t), \tag{10}$$

where $\mu \in \mathbb{R}$ is a variable parameter.

According to the above assumptions, system (6) can be rewritten as the following one-parameter family of differential equations

$$\dot{x} = -G'(0)x + [A_0 + \Delta A(\mu)]G(x). \tag{11}$$

Our aim is to show that system (11) possesses a stable limit cycle (i.e., it is no longer completely stable), even if $||\Delta A(\mu)|| \to 0$. More specifically, it can be shown that the origin of (11) undergoes a supercritical Hopf bifurcation at $\mu = 0$. To this end, let $J(0, \mu)$ denote the Jacobian at the origin of (11) for a given value of the Hopf parameter μ. A simple computation yields

$$J(0, \mu) = g'_{apwl}(0)[A_0 - E_n] + \mu E_n + \omega(uv^t - vu^t),$$

and, exploiting Property 1, it turns out that

$$\sigma\{J(0, \mu)\} = \{\mu + j\omega, \mu - j\omega, \mu + g'_{apwl}(0)\lambda_3, \ldots, \mu + g'_{apwl}(0)\lambda_n\},$$

where $\lambda_3, \ldots, \lambda_n$ are the real negative eigenvalues of matrix A_1 in (7). Therefore, we are exactly in a situation like the one depicted if Fig. 6.b and it is easy to check that all the hypotheses of the classical Hopf theorem are satisfied (see [24]), thus ensuring the existence of limit cycles for small positive values of μ.

To determine whether the limit cycle is stable (supercritical bifurcation), the stability coefficient β_2 has to be evaluated. An algorithm to do this is proposed in [25] (see also [26]). If $\beta_2 < 0$ the Hopf-generated limit cycle is stable.

The quantity β_2 is also related to the amplitude ϵ of the limit cycle. Indeed, for small values of μ, it can be shown that the following approximate relation holds [24]

$$\epsilon^2 = \left| \frac{2}{\beta_2} \right| \mu.$$

Exploiting the structure of model (11), the following result can be proven (see [27] for details and generalizations).

Proposition 1. *Suppose that Assumption 2 and Assumption 3 hold. Then, the equilibrium at the origin of system (11) undergoes a supercritical Hopf bifurcation at $\mu = 0$. Furthermore, for small positive values of μ, the amplitude*

ϵ and the period T of the stable Hopf-generated limit cycle are approximately given by

$$\epsilon = \sqrt{\frac{8}{\sum_{i=1}^{n}(u_i^2+v_i^2)^2}\frac{\cosh(p)}{p^2}\mu}, \tag{12}$$

$$T = \frac{2\pi}{\omega}. \tag{13}$$

It is important to note that the amplitude ϵ in (12) does not depend on ω. Hence, it is evident from (10) that stable oscillations are present for arbitrarily small perturbations, i.e., even if $||\Delta A(\mu)|| \to 0$. Equation (12) makes clear the influence of the neuron nonlinearity g_{apwl} (via the parameter p), and of $u = (u_1, \ldots, u_n)^t$, $v = (v_1, \ldots, v_n)^t$, on the limit cycle amplitude. On the contrary, the limit cycle period also depends on ω, and therefore on the size of perturbation (10).

As an application example, we consider the third-order system ($n = 3$) where the interconnection matrix A_0 is defined according to (7) by

$$U = \frac{1}{\sqrt{6}}\begin{bmatrix} 1 & \sqrt{3} & \sqrt{2} \\ 1 & -\sqrt{3} & \sqrt{2} \\ 2 & 0 & -\sqrt{2} \end{bmatrix}$$

and

$$A_1 = -3.$$

Therefore, it results

$$A_0 = \begin{bmatrix} 0 & -1 & 1 \\ -1 & 0 & 1 \\ 1 & 1 & 0 \end{bmatrix}$$

and

$$\Delta A(\mu) = \frac{1}{\sqrt{3}g'_{apwl}(0)}\begin{bmatrix} \mu\sqrt{3} & -\omega & -\omega \\ \omega & \mu\sqrt{3} & \omega \\ \omega & -\omega & \mu\sqrt{3} \end{bmatrix}.$$

Moreover, since $\sum_{i=1}^{3}(u_i^2+v_i^2)^2 = 4/3$, relation (12) reduces to

$$\epsilon = \sqrt{6\frac{\cosh^2(p)}{p^2}\mu}. \tag{14}$$

Several simulations have been performed for different values of p and ω. As an illustrative example, we briefly consider the case $p = 2$ and $\omega = 0.0167$.

μ	ϵ	Amplitude	T	Period
10^{-5}	0.0146	0.0146	376.2962	376.3697
10^{-4}	0.0461	0.0468	376.2962	376.4280
10^{-3}	0.1457	0.1477	376.2962	376.4465
10^{-2}	0.4608	0.4504	376.2962	380.2767
10^{-1}	1.4571	1.1462	376.2962	443.5171

Table 1. Amplitude (estimated and true) and period (estimated and true) of the periodic solution for different values of parameter μ

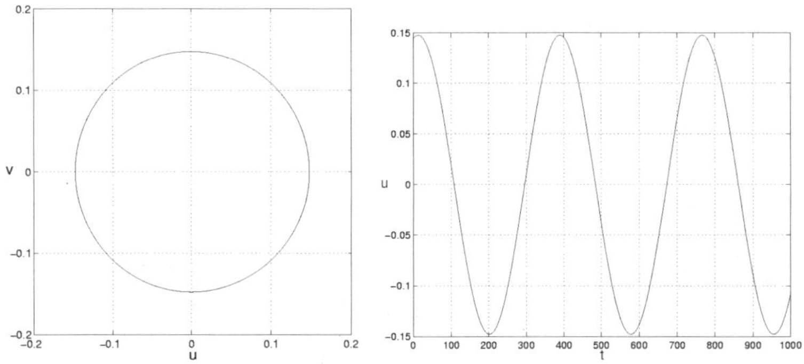

Fig. 7. Simulation results for $\mu = 0.001$: (**a**) limit cycle in the (u, v)-plane; (**b**) limit cycle time evolution along the u-axis

The simulation results are summarized in Table 1. From these results it can be verified that, as expected, (14) is well approximated for small μ. Also, the limit cycle period is quite close to its approximation given in (13).

Figure 7 report the trajectory in the (u, v) plane and the time evolution along the u-axis of the limit cycle obtained for $\mu = 0.001$ (third row of Table 1). Note that this cycle has a practically sinusoidal behaviour. On the contrary, Fig. 8 reports the limit cycle for $\mu = 0.1$ (last row of Table 1), which is far from being sinusoidal. For larger values of μ, the numerical simulations show convergence towards some equilibrium point.

5 Period-doubling bifurcations in a third-order neural network

In the previous sections we have seen that the existence of two zero eigenvalues of the Jacobian at the equilibrium point was the key condition for ensuring the birth of oscillatory behaviours for suitably small perturbations of the nominal symmetric interconnection matrix. This was related to the effect of such perturbations on the eigenvalues at the equilibrium point, as depicted in Fig. 6. Clearly, if we suppose that the nominal Jacobian at the origin has three zero eigenvalues, it should be possible to add perturbations

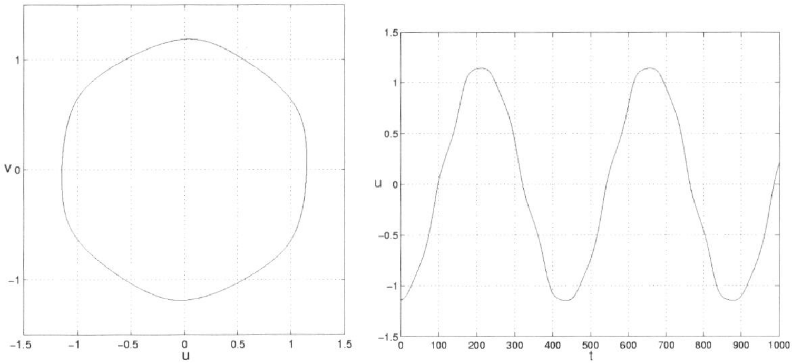

Fig. 8. Simulation results for $\mu = 0.1$: (**a**) limit cycle in the (u,v)-plane; (**b**) limit cycle time evolution along the u-axis

able to generate complicated patterns of eigenvalues, thus making it possible to give rise to bifurcations much more complex than the Hopf ones.

Consider a third-order cooperative-competitive neural network which in the form (P) is described by

$$A_0 = \begin{bmatrix} 1\,0\,0 \\ 0\,1\,0 \\ 0\,0\,1 \end{bmatrix},$$

$$\Delta A = \begin{bmatrix} -0.000810 & 0.000720 & 0.005333 \\ -0.001203 & \mu & 0.005333 \\ -0.002732 & -0.005333 & \mu \end{bmatrix},$$

$$D = E_3,$$

$$G(x) = (g_{pwl}(x_1), g_{pwl}(x_2), g_{pwl}(x_3))^t,$$

$$I = \left(0.003787,\ 0.003709,\ -0.000550\right)^t,$$

and g_{pwl} as defined in (1). The neural network thus corresponds to a one-parameter family of differential equations, μ being the bifurcation parameter. Note that the nominal Jacobian at the origin is $J_0(0) = 0_{3\times 3}$, i.e., it has three zero eigenvalues.

We have simulated the dynamical behaviour of the neural network by varying parameter μ, and some basic results are reported in Fig. 9. It is seen that the neural network displays a cascade of period-doubling bifurcations which lead to the birth of a complicated attractor. The cycles and the complex attractors have a large size, which is comparable with the saturation levels of

g_{pwl}. In fact, in all cases the trajectory $x(t)$ spans a large part of the linear region, and there are time intervals where $x(t)$ slightly enters regions with one saturated variable. It is worth to note that for all values of μ considered in the figure, the maximum absolute value of the entries of ΔA is less than 6×10^{-3}, i.e., the deviations are quite small and well within the typical errors expected in a practical neural network implementation.

The complex attractors for $\mu = 0.00042$ and $\mu = 0.00055$ seem to have some similarities with the Rössler-type attractors present in Chua's circuit [28]. However, one basic difference is that the complex attractors of Fig. 9 involve the saturation of three different nonlinearities, while Chua's circuit is characterized by one nonlinearity, only. Most importantly, to the authors knowledge, complex attractors in Chua's circuit have been observed only far from reciprocity (symmetry) [29]. The same is true of complex attractors already found for neural networks [22].

We finally note that the neural network in this section have been designed through an extensive random search in the system parameter space. Other complicated attractors, with a shape different from that reported in Fig. 9, have been discovered for interconnection matrices close to symmetry. They are not reported here due to space limitations.

6 Conclusion

In this paper, situations where complete stability of neural networks is not robust with respect to small perturbations of the nominal symmetric neuron interconnections have been investigated. It has been shown that unexpectedly rich complex dynamics exist for interconnection matrix which are almost symmetric, or even arbitrarily close to some symmetric one. These dynamics range from stable limit cycles to complex attractors, and have the common feature that the Jacobian at the related equilibrium point of the nominal neural network has two zero eigenvalues at least. Moreover, such limit cycles and complex attractors have usually a large size, even if the perturbation is well within the tolerances typically encountered in the electronic implementation of neural network chips. This drastically impairs the applicability of the neural network to practical situations that require completely stable dynamics.

The main contribution of the paper makes it clear that the symmetry of the nominal interconnection matrix is not sufficient to guarantee robustness of complete stability. Therefore, the development of a theory of robustness of complete stability for neural networks is now a fundamental task to investigate.

References

1. M. Hirsch, "Convergent activation dynamics in continuous time networks," *Neural Networks*, vol.2, pp. 331-349, 1989.

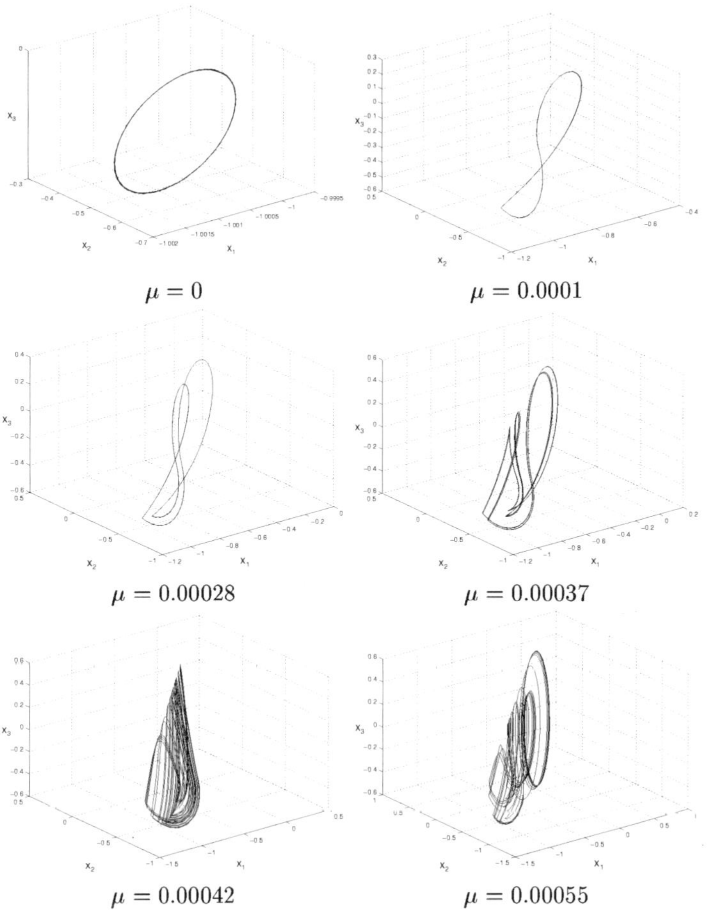

Fig. 9. Cascade of period-doubling bifurcations originating a complex attractor in a third-order cooperative-competitive neural network

2. J. J. Hopfield, "Neurons with graded response have collective computational properties like those of two-state neurons," *Proc. Nat. Acad. Sci.*, vol. 81, pp. 3088-3092, 1984.
3. M. A. Cohen and S. Grossberg, "Absolute stability of global pattern formation and parallel memory storage by competitive neural networks," *IEEE Trans. Syst. Man Cyber.*, vol. 13, pp. 815–825, 1983.
4. J. H. Li, A. N. Michel, and W. Porod, "Qualitative analysis and synthesis of a class of neural networks," *IEEE Trans. Circuits Syst.*, vol. 35, pp. 976–986, 1988.
5. L. O. Chua and L. Yang, "Cellular neural networks: Theory," *IEEE Trans. Circuits Syst. I*, vol. 35, pp. 1257-1272, 1988.
6. L. O. Chua, "CNN: A vision of complexity," *Int. J. Bifurcation and Chaos*, vol. 7, pp. 2219–2425, 1997.

7. R. Fantacci, M. Forti, M. Marini, and L. Pancani, "Cellular neural network approach to a class of communication problems," *IEEE Trans. Circuits Syst. I*, vol. 45, pp. 1457-1467, 1999.
8. M. Forti and A. Tesi, "New conditions for global stability of neural networks with application to linear and quadratic programming problems," *IEEE Trans. Circuits Syst. I*, vol. 42, pp. 354–366, 1995.
9. P. Thiran, *Dynamics and Self-Organization of Locally Coupled Neural Networks.* Lausanne, Switzerland: Presses Polytechniques et Universitaires Romandes, 1997.
10. K. Wang and A. N. Michel, "Robustness and perturbation analysis of a class of nonlinear systems with applications to neural networks," *IEEE Trans. Circuits Syst. I*, vol. 41, pp. 24-32, 1994.
11. M. Vidyasagar, "Location and stability of the high-gain equilibria of nonlinear neural networks," *IEEE Trans. Neural Networks*, vol. 4, pp. 660-672, 1993.
12. S. Grossberg, "Nonlinear neural networks: Principles, mechanisms, and architectures, *Neural Networks*, vol. 1, pp. 17-61, 1988.
13. M. Hänggi, S. Moser, E. Pfaffhauser, and G. S. Moschytz, "Simulation and visualization of CNN dynamics," *Int. J. Bifurcation and Chaos*, vol. 9, pp. 1237-1262, 1999.
14. J. Li, A. N. Michel, and W. Porod, "Analysis and synthesis of a class of neural networks: Linear systems operating on a closed hypercube,"*IEEE Trans. Circuits Syst.*, vol. 36, pp. 1405-1422, 1989.
15. J. M. Cruz and L. O. Chua, "A CNN chip for connected component detection," *IEEE Trans. Circuits Syst.*, vol. 38, pp. 812–817, 1991.
16. L. O. Chua, C. A. Desoer, and E. S. Kuh, *Linear and Nonlinear Circuits.* New York: McGraw-Hill, 1987.
17. C. W. Wu and L. O. Chua, "A more rigorous proof of complete stability of cellular neural networks," *IEEE Trans. Circuits Syst. I*, vol. 44, pp. 370-371, 1997.
18. M. Forti and A. Tesi, "A new method to analyze complete stability of PWL cellular neural networks," *Int. J. Bifurcation and Chaos*, vol. 11, pp. 655-676, 2001.
19. L. O. Chua, *CNN: A Paradigm for Complexity*, World Scientific, Singapore, 1998.
20. F. Zou and J. A. Nossek, "Bifurcation and chaos in cellular neural networks," *IEEE Trans. Circuits Syst. I*, vol. 40, pp. 166-173, 1993.
21. G. Setti, P. Thiran, and C. Serpico, "An approach to information propagation in 1-D cellular neural networks-Part II: Global propagation," *IEEE Trans. Circuits Syst. I*, vol. 45, pp. 790-811, 1998.
22. F. Zou, A. Katérle, and J. A. Nossek, "Homoclinic and heteroclinic orbits of the three-cell cellular neural networks," *IEEE Trans. Circuits Syst. I*, vol. 40, pp. 843-848, 1993.
23. M. Di Marco, M. Forti, and A. Tesi, "Bifurcations and oscillatory behavior in a class of competitive cellular neural networks," *Int. J. Bifurcation and Chaos*, vol. 10, pp. 1267-1293, 2000.
24. M. Farkas, *Periodic Motions*, Springer-Verlag, New York, 1994.
25. L. N. Howard, *Nonlinear Oscillations*, in: Nonlinear Oscillations in Biology, F. C. Hoppensteadt (ed.), American Mathematical Society, Providence, RI, pp. 1-67, 1979.

26. J.-H. Fu and E. H. Abed, "Families of Lyapunov functions for nonlinear systems in critical cases," *IEEE Trans. Automat. Contr.*, vol. 38, pp. 3-16, 1993.
27. M. Di Marco, M. Forti, and A. Tesi, "Hopf bifurcations in neural networks with nearly symmetric interconnections," Tech. Rep. no. 34, Dipartimento di Ingegneria dell'Informazione, Università di Siena, 53100-Siena, Italy, 2000.
28. L. O. Chua, M. Komuro, and T. Matsumoto, "The double scroll family," *IEEE Trans. Circuits Syst.*, vol. 33, pp. 1073-1118, 1986.
29. L. O. Chua and G.-N. Lin, "Canonical realization of Chua's circuit family," *IEEE Trans. Circuits Syst.*, vol. 37, pp. 885-902, 1990.

Bifurcations in Systems with a Rate Limiter

Francisco Gordillo, Ismael Alcalá, and Javier Aracil

Dept. Ingeniería de Sistemas y Automática, Universidad de Sevilla.
Escuela Superior de Ingenieros, Camino de los Descubrimientos, s/n. 41092
Sevilla, Spain

Abstract. Limit cycles analysis of feedback systems with rate limiters in the actuator can be implemented by a classical method in the frequency domain, the harmonic balance method. In this paper, the rate limiter describing function is obtained and applied to the search for limit cycles in such control systems. Three examples with three different bifurcations (saddle-node bifurcation of limit cycles, subcritical Hopf bifurcation at infinity and supercritical Hopf-like bifurcation) are included. The method is approximate but its main advantage is that intuition is gained into a difficult problem.

1 Introduction

Nonlinear control of nonlinear plants leads to nonlinear dynamical systems. The last class of systems can display complex dynamical behaviours, and, what is more important, after (even small) changes in parameters or in their system structure, one can observe qualitative changes in their behaviour modes. This is the realm of bifurcation theory [7,9], which supplies tools to study the points where these changes are produced and the archetypical forms of the state portrait changes in these points. These archetypes are of a great value for the control systems designer as they supply a unified and global perspective on the behaviour modes of the system (he can see what is expected to happen for all the parameter values involved).

Bifurcation theory is a valuable tool for understanding the behaviour richness of nonlinear systems. Roughly speaking, when by moving system parameters one observes a qualitative change in the system response (to be deduced from the state portrait, for instance) it is said that the system undergoes a bifurcation phenomenon. These phenomena can lead to a different number of stationary solutions (equilibrium points), to the appearance of oscillations, or even more complex behaviours (chaos, for instance). After a bifurcation analysis it is possible to split the parameter space into several regions with different asymptotic dynamics [10]. Furthermore, from a certain point of view, bifurcations are related to robustness issues, since only far from the bifurcation points the system displays behaviours that are structurally stable.

In this paper, the presence of a concrete nonlinearity, the rate limiter, in the actuator of control systems is considered. The study of the rate limiter is of a great importance due to fact that this nonlinearity exists in a large type of actuators in which the speed of response is bounded. Furthermore,

F. Colonius, L. Grüne (Eds.): Dynamics, Bifurcations, and Control, LNCIS 273, pp. 37–50, 2002.

the presence of rate limiters in plants that are controlled with PID gives rise to difficult problems, mainly with unstable plants [16,17]. Anti-windup techniques give good solutions for systems with saturation, but they do not work well in systems with rate limiters. In this paper, the rate limiter is studied with the describing function method [3,5,6,8] which helps to analyze the stability of the system, the appearance of limit cycles and other phenomena such as bifurcations. The method is approximate, but the fact that it allows to analyze not only the local stability but also the global one [1,2,11,15] together with the simplicity and facility of the use of the method, justifies the current study. Furthermore, the analysis leads to detect the occurrence of stable and unstable limit cycles that are organized by the archetypes supplied by several types of bifurcations. The describing function method has been previously applied to bifurcation analysis [4,11,14].

In this paper, three examples of systems with rate limiters are presented. These examples show three different kinds of bifurcations: saddle-node bifurcation of limit cycles, subcritical Hopf bifurcation at infinity and supercritical Hopf-like bifurcation. In this way an overall perspective of the behaviour modes displayed by the systems can be reached. This perspective is of a great interest for the control system designer.

In Section 2, the dynamical response of rate limiters is described and the different behaviour modes when the input is sinusoidal are characterized. This analysis is used in Section 3 to obtain the analytical expression of the describing function of rate limiters. In Section 4 this describing function is used to analyze the existence of limit cycles, while in Section 5 the three examples of bifurcations are presented. Finally, some conclusions are drawn in Section 6.

2 Behaviour of rate limiters

In this study the rate-limiter nonlinearity is considered as the block of Fig. 1 with one input and one output. The output attempts to follow its input but with the constraint that the slope of the output is bounded in $\pm m$.

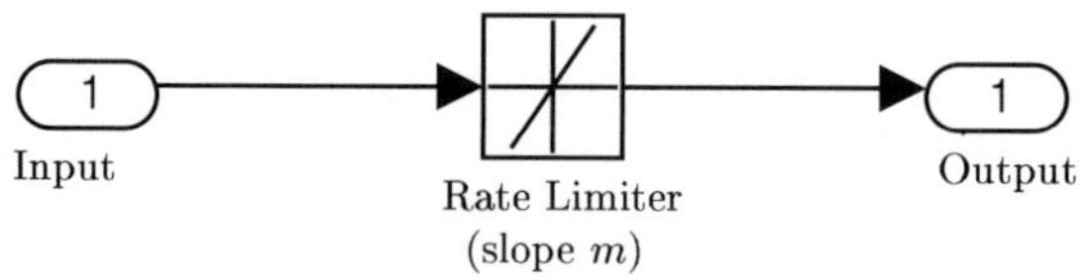

Fig. 1. Simulink block for a rate limiter.

In this way a saturation is applied not to the input of the nonlinearity but to the derivative of this input. Furthermore, if the output, ψ, is less than the input, u, the output will grow at the maximum speed, $\dot{\psi} = m$, and if $\psi > u$ then $\dot{\psi} = -m$.

A sinusoidal input is considered in the describing function method. In this case,

$$y(t) = a \sin \omega t \Longrightarrow \frac{dy(t)}{dt} = a\,\omega \cos \omega t \Longrightarrow -a\omega \leq \frac{dy(t)}{dt} \leq a\omega.$$

For small values of a and ω, the maximum slope of the input does not violate the bound $\pm m$ and, therefore, the output of the rate limiter $\psi(y(t))$ follows its input $y(t)$. This behaviour continues while $a\omega \leq m$. When this bound is violated, the output of the nonlinearity $\psi(y(t))$ is not able to follow exactly the input $y(t)$ during some periods in which the output evolves at the maximum speed ($\pm m$). Depending on the values of a and ω three qualitatively different behaviour modes of the output may exist:

mode (a) The slope of the input $y(t)$ never violates the constraint $\pm m$. The input $y(t)$ and the output $\psi(y(t))$ are overlapped during the whole period of the input. This behaviour mode is not longer valid when

$$\max\left\{\frac{dy(t)}{dt}\right\} = a\omega = m \quad \Longleftrightarrow \quad \frac{m}{\omega a} = 1. \tag{1}$$

mode (b) The output in this mode has time periods in which it follows the sinusoidal input and periods where it follows a straight line of slope $\pm m$ (see Fig. 2).

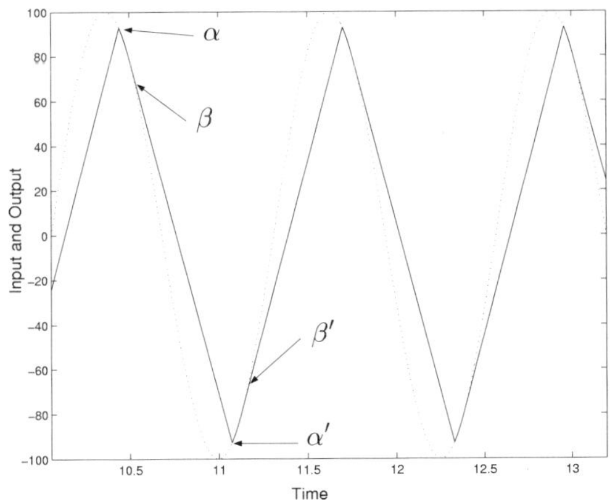

Fig. 2. Response of the rate limiter (solid line) to sinusoidal inputs (dashed) in **mode (b)**.

Assume that the steady state has been reached. Let α and α' be the values of ωt when $\psi(y(t))$ reaches $y(t)$ (in Fig. 2). Notice that $\alpha' = \alpha + \pi$. Analogously, let β and β' be the values of ωt when $\psi(y(t))$ leaves $y(t)$, at these points.

$$\frac{d\psi(y(t))}{dt} = \frac{dy(t)}{dt} = \begin{cases} -m \text{ for } \omega \quad t = \beta \\ m \text{ for } \omega \quad t = \beta'. \end{cases} \tag{2}$$

Notice that $\beta' = \beta + \pi$. It is evident that $\psi(y(t)) = y(t)$ in $[\alpha, \beta] \cup [\alpha', \beta']$. During $[\beta, \alpha']$ $\psi(y(t))$ is a straight line slope equal to $-m$ and during $[\beta', \alpha + 2\pi]$ $\psi(y(t))$ is a straight line with slope equal to m.

mode (c) In this case the output $\psi(y(t))$ is too slow and is not able to follow the input $y(t)$. Therefore , $\psi(y(t))$ evolves at the maximum rate in both directions $\pm m$. Figure 3 shows how, each time that the output reaches the input, the slope of the output changes. In the frontier between modes **(b)** and **(c)** $\alpha = \beta$ and $\alpha' = \beta'$. Furthermore, into this critical case, the slope of the input is equal to $-m$ in $\omega t = \alpha$ and equal to m in $\omega t = \alpha'$. Thus,

$$\begin{cases} \sin\beta = -\sin\alpha + \frac{m}{\omega a}(\alpha + \pi - \beta) \\ a\omega\cos\beta = -m \end{cases}$$

$$\alpha = \arctan\left(\frac{-\pi}{2}\right) \quad \Longrightarrow \quad \frac{m}{\omega a} = -\cos\alpha = \frac{2}{\sqrt{4+\pi^2}}.$$

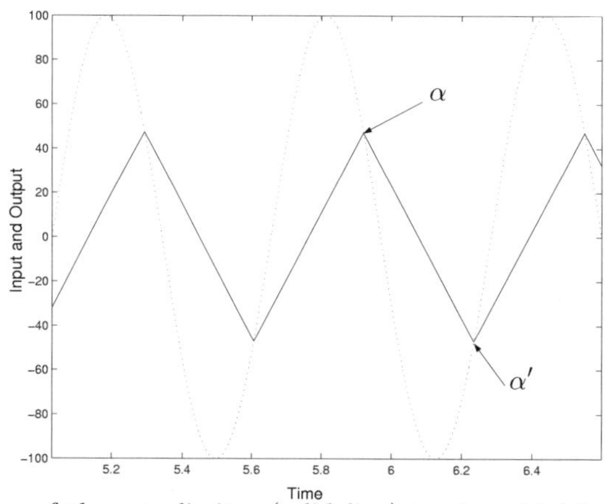

Fig. 3. Response of the rate limiter (solid line) to sinusoidal inputs (dashed) in **mode (c)**.

Notice the delay of $\psi(y(t))$ with respect to $y(t)$ in modes **(b)** and **(c)**. This delay has an important effect in the global behaviour of the system. As it can be seen, the value of the adimensional parameter $m/(\omega a)$ defines the qualitative behaviour mode of the nonlinearity, as appears in Table 1.

Modes **(a)** and **(c)** are particular cases of **mode (b)**. Indeed, in **mode (a)** $\alpha = 0$ and $\beta = \pi$ while in **mode (c)** $\alpha = \beta$. This fact will be used in the next section to obtain the describing function of rate limiters.

Modes	Range of $\frac{m}{\omega a}$
Mode (a)	$1 \leq \frac{m}{\omega a} < \infty$
Mode (b)	$\frac{2}{\sqrt{4+\pi^2}} < \frac{m}{\omega a} < 1$
Mode (c)	$0 < \frac{m}{\omega a} \leq \frac{2}{\sqrt{4+\pi^2}}$

Table 1. Qualitative behaviour modes of a rate limiter as a function of $\dfrac{m}{\omega a}$.

Remark 1. The analysis of systems with a rate limiter is very difficult. Notice that, at least, the nonlinearity is not C^1 since it contains a saturation. Furthermore, during some periods of time it is static (its output is equal to the input) while, during some others it is dynamical with one state $\dot{\psi} = \pm m$.

3 Describing function of rate limiters

The describing function of a rate limiter can be obtained computing

$$\begin{aligned}\Psi(a,\omega) &= \tfrac{1}{\pi a}\int_{-\pi}^{\pi} \psi(a\sin(\omega t))\sin(\omega t)d\omega t \\ &+ j\,\tfrac{1}{\pi a}\int_{-\pi}^{\pi} \psi(a\sin(\omega t))\cos(\omega t)d\omega t. \end{aligned} \tag{3}$$

In order to obtain both integrals of this expressions, angles α and β need to be computed:

- Since the slope of the input is $-m$ at $\omega t = \beta$, $\beta = \arccos(\frac{-m}{\omega a})$
- Angle α corresponds to the intersection of the line with slope m with the sinusoidal input $y(t)$. This condition can be expressed as

$$\sin\beta = -\sin\alpha + \frac{m}{\omega a}(\alpha + \pi - \beta). \tag{4}$$

This expression is an implicit definition of α which can be obtained numerically.

Performing some manipulations the describing function results:

$$\text{Re}\{\Psi(a,\omega)\} = \frac{1}{\pi}\Bigg[\Big(\sin\beta\cos\beta - \sin\alpha\cos\alpha\Big) + (\beta-\alpha) + 2\left(\frac{m}{\omega a}\right)(\sin\alpha + \sin\beta)\Bigg], \tag{5}$$

$$\text{Im}\{\Psi(a,\omega)\} = \frac{1}{\pi}\Bigg[(\sin^2\alpha - \sin^2\beta) + 2\left(\frac{m}{\omega a}\right)(\cos\alpha + \cos\beta)\Bigg]. \tag{6}$$

Remark 2. Notice that the describing function $\Psi(a,\omega)$ depends only on the parameter $m/(\omega a)$.

In particular, we obtain in modes **(a)** and **(c)**:

mode (a) $\alpha = 0$ y $\beta = \pi$: $\Psi(a,\omega) = 1$
mode (c) $\alpha = \beta$. Here α can be obtained analytically $\alpha = \sin^{-1}(\frac{\pi}{2}\frac{m}{\omega a})$.

Thus, the describing function has an analytical expression in **mode (c)**:

$$\Psi(a,\omega) = 4\left(\frac{m}{\omega\pi a}\right)\sin\alpha + j\,4\left(\frac{m}{\omega\pi a}\right)\cos\alpha.$$

4 Limit cycle analysis of systems with rate limiters

In order to apply the describing function method, it is interesting to compute $-1/\Psi(m/(\omega a))$. It is possible to obtain an analytical expression when the system works in **mode (c)**, that is, when $m/(\omega a) \leq 0.537$:

$$-\frac{1}{\Psi(a,\omega)} = \frac{-\frac{4m}{\omega\pi a}\sin\alpha}{(\frac{4m}{\omega\pi a})^2} + j\frac{\frac{4m}{\omega\pi a}\cos\alpha}{(\frac{4m}{\omega\pi a})^2} = -\frac{\omega\pi a}{4m}\sin\alpha + j\frac{\omega\pi a}{4m}\cos\alpha.$$

And taking into account that, in **mode (c)**, $\alpha = \sin^{-1}(\frac{\pi}{2}\frac{m}{\omega a})$:

$$-\frac{1}{\Psi(a,\omega)} = -\frac{\pi^2}{8} + j\frac{\omega\pi a}{4m}\cos\alpha. \tag{7}$$

Notice that the real part of $-1/\Psi(a,\omega)$ is constant in this mode. For the rest of the cases equations (5) and (6) must be used to compute $-1/\Psi(a,\omega)$. The full result is shown in Fig. 4. Notice that $\beta = \beta(m/(\omega a))$ and $\alpha = \alpha(m/(\omega a))$, this means that $\Psi(a,\omega) = \Psi\big(m/(\omega a)\big)$ and only one curve needs to be represented instead of a family of curves.

The intersections of the transfer function of the plant plus the controller and the curve $-1/\Psi(m/(\omega a))$ are limit cycle candidates. As the family $\Psi(m/(\omega a))$ lies in a single curve, any intersection corresponds to a prediction of a limit cycle, with the frequency of $G(j\omega)$ in the intersection point and amplitude $a = (\omega/m)/(m/(\omega a))_0$, where $(m/(\omega a))_0$ is the value of the adimensional parameter at the intersection point.

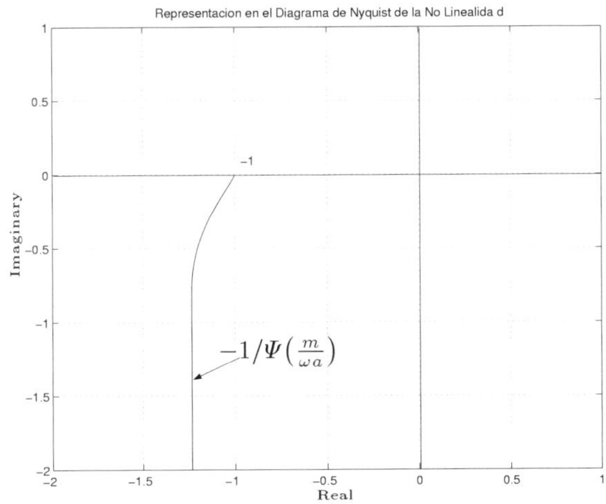

Fig. 4. Representation in the complex plane of $-1/\Psi(m/(\omega a))$ of a rate limiter.

5 Bifurcations in systems with a rate limiter

In this section, an approximate bifurcation analysis of three control systems is performed by means of the describing function method. The systems are linear except for the existence of a rate-limiting actuator.

The analysis is approximate, since the describing function is not a rigorous method, but it is difficult to perform an accurate analysis due to the complexity of the nonlinearity considered. On the other hand, the frequency domain in which the describing function stands, allows us to gain a graphical insight into the problem. Thus, some bifurcations related to the emergence of limit cycles can be easily detected in the examples.

5.1 Saddle-node bifurcation of periodic orbits

Consider a plant given by:

$$\begin{bmatrix} \dot{x}_1 \\ \dot{x}_2 \end{bmatrix} = \begin{bmatrix} -1.5 & -0.5 \\ 1 & 0 \end{bmatrix} \begin{bmatrix} x_1 \\ x_2 \end{bmatrix} + \begin{bmatrix} 1 \\ 0 \end{bmatrix} u,$$

$$y = \begin{bmatrix} 0 & 1 \end{bmatrix} \begin{bmatrix} x_1 \\ x_2 \end{bmatrix},$$

which corresponds to the transfer function:

$$G(s) = \frac{1}{(s+1)(s+2)},$$

with a proportional controller of value K_p, and with the reference equal to zero. Assume that the actuator introduces a rate limiter at the input of the

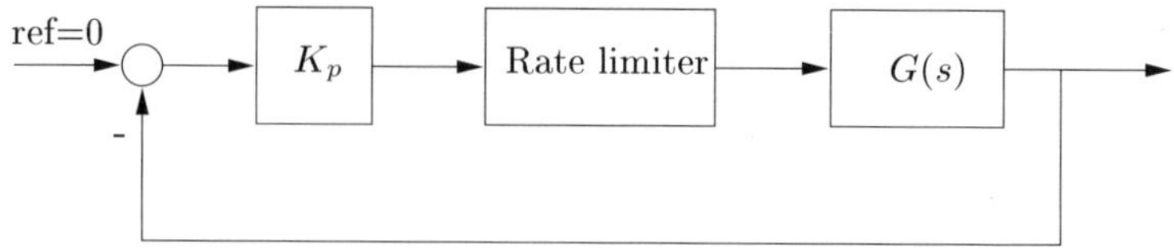

Fig. 5. Block diagram of the system.

plant. The resultant block diagram appears in Fig. 5. Figure 6 shows the Nyquist plot of $K_pG(s)$ for $K_p = 4$ and $K_p = 7$ and the plot of $-1/\Psi(m/(\omega a))$. For $K_p = 7$ two intersection points exist and, therefore, the describing function method predicts the existence of two limit cycles. Applying, as usual, the Loeb stability criterion [6], the limit cycle with smaller amplitude is unstable while the larger one is stable.

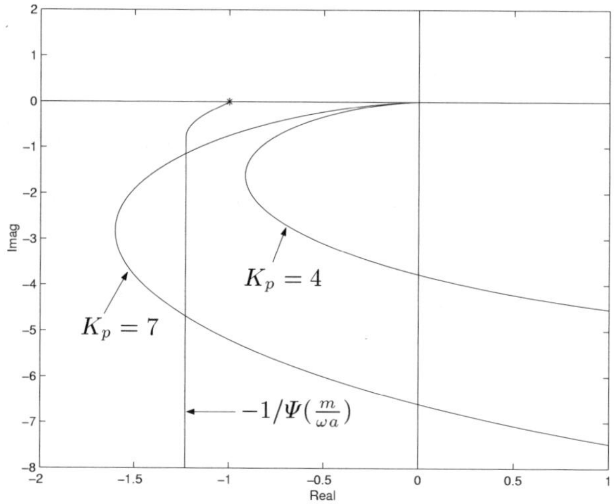

Fig. 6. Nyquist diagram of $G(j\omega)$ for $K_p = 4$ and $K_p = 7$ and $-1/\Psi(m/(\omega a))$.

Simulations corroborate this prediction as can be seen in Fig. 7. In the left graph, the system has two attractors: the equilibrium point at the origin; and a stable limit cycle. Depending on the initial condition the nonlinear system evolves towards the desired operating point (the origin) or towards the stable limit cycle. The reader may be surprised that the orbits intersect for a two dimensional $G(s)$. The reason is that the dimension if the system is higher than 2 since the rate limiter is dynamical (remember remark 1).

Notice that in spite of the fact that $G_c(j\omega)$ intersects $-1/\Psi(m/(\omega a))$ the desired operating point is stable. The effect of these intersections is that the stability is only local with a bounded attraction basin due to the existence

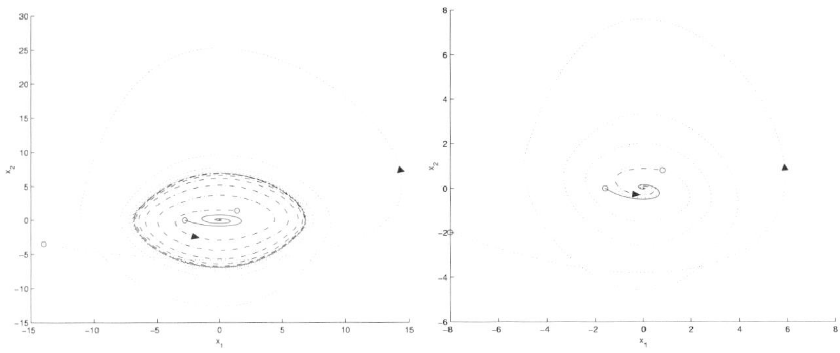

$K_p = 7$ $K_p = 4$

Fig. 7. State portrait for different values of K_p.

of an unstable limit cycle. In this way, the describing function method allows to study the global stability of the system.

For $K_p = 4$ the qualitative behaviour is completely different. The describing function method does not predict the existence of any limit cycle, see Fig. 6. The right graph of Fig. 7 corroborates this prediction. In this graph, simulations corresponding to $K_p = 4$ and the same initial conditions are presented. With $K_p = 4$, the origin is the only attractor.

It is evident that an intermediate value of K_p will exist such that both curves $G_c(j\omega)$ and $-1/\Psi(m/(\omega a))$ are tangent. At this point two limit cycles with the same amplitude, one of them stable and the other unstable, are predicted. This corresponds to the well-known saddle-node bifurcation of periodic orbits studied in the Qualitative Theory of Dynamical Systems [7,12] (see Fig. 8). The bifurcation would correspond with the tangent point of $G_c(j\omega)$ and $-1/\Psi(m/(\omega a))$, but the approximate character of the method, above all when the intersections are not transverse, should be taken into account. This approximate nature does not affect the qualitative classification of the behaviour modes of the system. As a matter of fact, the saddle-node bifurcation of limit cycles supplies an archetype that organizes in the simple scheme of Fig. 8 all the behaviour modes the system can display. This bifurcation is difficult to detect in any nonlinear system but, specially in systems with this complex nonlinearity. The use of an approximate method, the describing function, greatly simplifies the analysis. It should be emphasized the huge amount of information about the system behaviour summarized in Fig. 8 and that is inherent in Fig. 6. In this way it supplies a global perspective on the behaviour of the system.

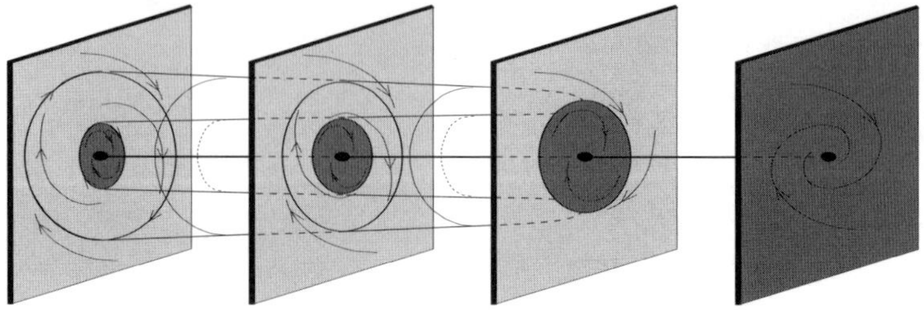

Fig. 8. Schematic representation of a saddle-node bifurcation of periodic orbits.

5.2 Subcritical Hopf bifurcation at infinity

Consider the same structure than in Fig. 5 but, now, the plant is given by:

$$\begin{bmatrix} \dot{x}_1 \\ \dot{x}_2 \end{bmatrix} = \begin{bmatrix} -1 & 0 \\ 1 & 0 \end{bmatrix} \begin{bmatrix} x_1 \\ x_2 \end{bmatrix} + \begin{bmatrix} 1 \\ 0 \end{bmatrix} u,$$
$$y = \begin{bmatrix} 0 & 1 \end{bmatrix} \begin{bmatrix} x_1 \\ x_2 \end{bmatrix},$$

which corresponds to the transfer function:

$$G(s) = \frac{1}{s(s+1)}.$$

The same analysis as before yields Fig. 9 where $K_pG(j\omega)$ has been plotted for K_p equal to 0.9 and 1.5. It can be seen that, in the first case, no limit cycles are predicted, while for $K_p = 1.5$ the method predicts an unstable limit cycle. Furthermore, the desired equilibrium is always stable and the amplitude of the predicted limit cycle grows as K_p approaches the bifurcation point (the value of K_p that is the boundary between the cases of prediction and no prediction of limit cycles). This behaviour corresponds to the case of the subcritical Hopf bifurcation at infinity whose bifurcation diagram is shown in Fig. 10 [13].

The predictions are corroborated by the simulations that appear in Fig. 11.

5.3 Supercritical Hopf-like bifurcation

Consider the same structure than in Fig. 5 but, now, the plant is given by:

$$\begin{bmatrix} \dot{x}_1 \\ \dot{x}_2 \end{bmatrix} = \begin{bmatrix} -5 & -6 \\ 1 & 0 \end{bmatrix} \begin{bmatrix} x_1 \\ x_2 \end{bmatrix} + \begin{bmatrix} 1 \\ 0 \end{bmatrix} u,$$
$$y = \begin{bmatrix} -1 & 1 \end{bmatrix} \begin{bmatrix} x_1 \\ x_2 \end{bmatrix},$$

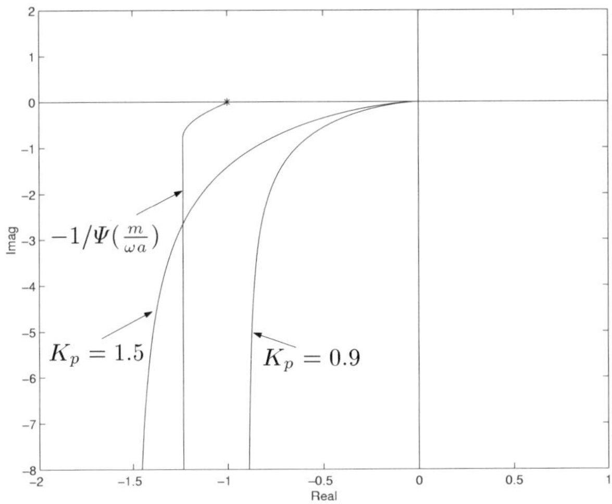

Fig. 9. Nyquist diagram of $G(j\omega)$ for $K_p = 0.9$ and $K_p = 1.5$ and $-1/\Psi(\frac{m}{\omega a})$.

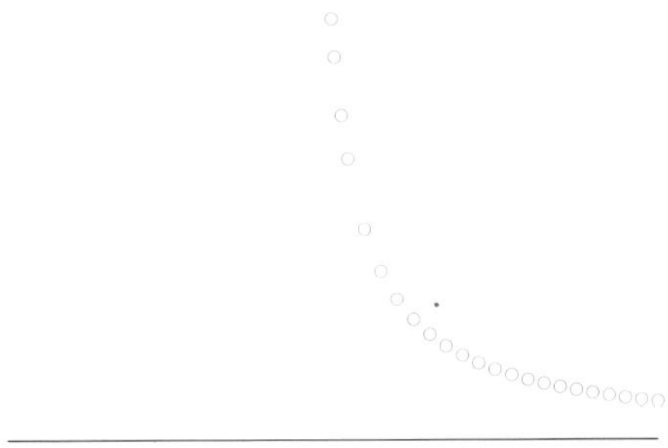

Fig. 10. Bifurcation diagram of a subcritical Hopf bifurcation at infinity.

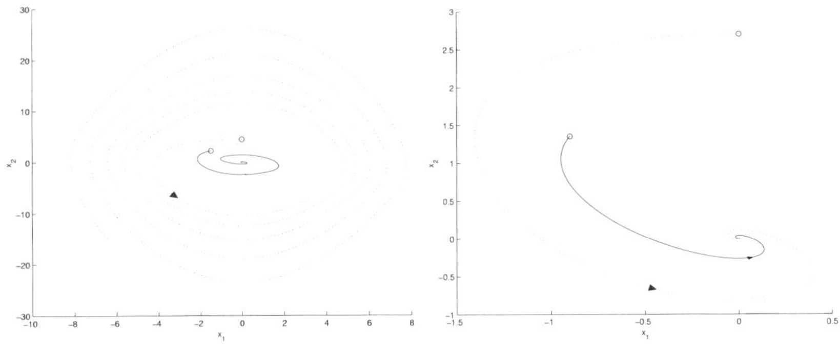

$K_p = 1.5$ $\qquad$ $K_p = 0.9$

Fig. 11. State portrait for different values of K_p.

which corresponds to the transfer function:

$$G(s) = \frac{-s+1}{(s+2)(s+3)}.$$

The graphs of $K_pG(j\omega)$ for $K_p = 4$ and $K_p = 6$ as well as the graph of $-1/\Psi(a,\omega)$ appear in Fig. 12. It can be seen that, for small values of K_p no limit cycles are predicted and the origin is stable while for large values of K_p a stable limit cycle is predicted and the equilibrium results unstable. This behaviour agrees with the Hopf bifurcation. Nevertheless, the predicted amplitude of the limit cycle for the bifurcation point (value of K_p that makes $K_pG(j\omega)$ to cross point -1) is greater than zero. Indeed, at this point

$$\frac{m}{\omega a} = 1 \Longrightarrow a = \frac{m}{\omega}$$

with $\omega > 0$. Furthermore, with a simple reasoning the following conclusion can be drawn: a system with a rate limiter as the only nonlinearity cannot present limit cycles with an arbitrarily small amplitude. The reason is that if $a\omega$ is very small the rate limiter is not active and, consequently, the whole system is linear. In any case, it can be seen that the predicted limit cycle will

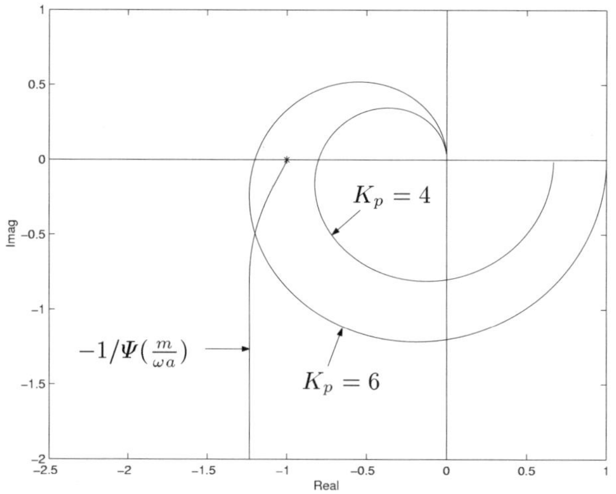

Fig. 12. Nyquist diagram of $G(j\omega)$ for $K_p = 4$ and $K_p = 6$ and $-1/\Psi(\frac{m}{\omega a})$.

have a larger amplitude for larger values of K_p. The resultant bifurcation diagram appears in Fig. 13. At the bifurcation point the origin is a center.

The predictions are corroborated by the simulations that appear in Fig. 14.

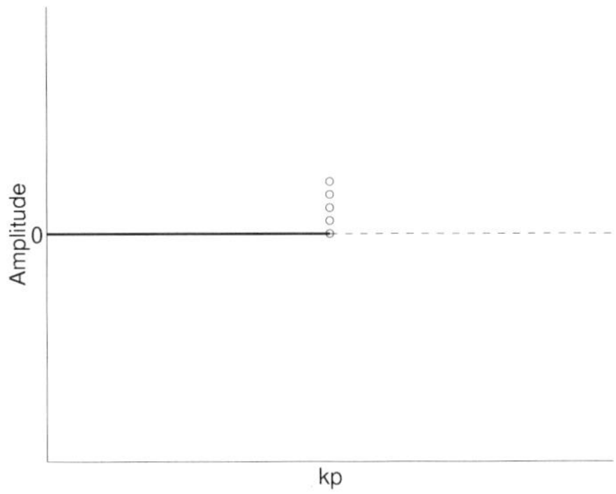

Fig. 13. Bifurcation diagram of the system.

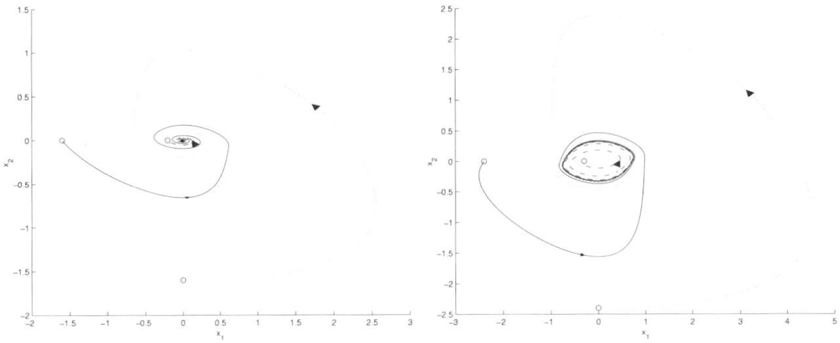

$K_p = 4$ $K_p = 6$

Fig. 14. State portrait for different values of K_p.

6 Conclusions

In this paper, it has been shown how the describing function method helps to analyze the bifurcations of control systems with a rate limiter. The complexity of this nonlinearity make the problem very involved but, it has been shown that with the describing function method interesting and intuitive conclusions can be drawn. In fact, this intuition has been very useful in the election of the three examples, which correspond to a saddle-node bifurcation of limit cycles, a subcritical Hopf bifurcation at infinity and a supercritical Hopf-like bifurcation. In practical applications, we can make use of the gained intuition in order to design compensators that avoid pathological behaviour.

Acknowledgments

This work has been supported by the Spanish Ministry of Science and Technology under grant DPI2000-1218-C04-01.

References

1. J. Aracil, F. Gordillo, and T. Álamo. Global stability analysis of second-order fuzzy control systems. In R. Palm, D. Driankov, and H. Hellendorn, editors, *Advances in Fuzzy Control*, pages 11–31. Physica-Verlag, 1997.
2. J. Aracil, E. Ponce, and T. Álamo. A frequency-domain approach to bifurcations in control systems with saturation. *International Journal of Systems Science*, 31(10):1261–1271, 2000.
3. D.P. Atherton. *Nonlinear Control Engineering. Describing Function Analysis and Design.* Van Nostrand Reinhold, 1975.
4. F. Bonani and M. Gilli. Analysis of stability and bifurcations of limit cycles in chua's circuit through the harmonic-balance approach. *IEEE Trans. on Circuits and Systems I- Fundamental Theory and Applications*, 46(8):881–890, 1999.
5. P.A. Cook. *Nonlinear Dynamical Systems.* Prentice Hall, 1994.
6. A. Gelb and W.E. Vander Velde. *Multiple-input Describing Functions and Nonlinear Systems Design.* Prentice-Hall, 1968.
7. J.K. Hale and H. Koçak. *Dynamics and Bifurcations.* Springer-Verlag, 1991.
8. H.K. Khalil. *Nonlinear Systems.* Prentice Hall, 1996.
9. Y.A. Kuznetsov. *Elements of Applied Bifurcation Theory.* Springer-Verlag, 1995.
10. I. Mareels, S. Van Gils, J.W. Polderman, and A. Ilchmann. Asymptotic dynamics in adaptive gain control. In *Advances in Control (Highlights of ECC'99)*, pages 29–63. Springer, 1999.
11. J. Moiola and G. Chen. *Hopf Bifurcation Analysis: a Frequency Domain Approach.* World Scientific, 1996.
12. M.G. Ortega, J. Aracil, F. Gordillo, and F.R. Rubio. Bifurcation analysis of a feedback system with dead zone and saturation. *IEEE Control System Magazine*, 20(4):91–101, 2000.
13. D.J. Pagano, E. Ponce, and J. Aracil. Bifurcation analysis of time-delay control systems with saturation. *International Journal of Bifurcations and Chaos*, 9(6):1089–1109, 1999.
14. C. Piccardi. Harmonic balance analysis of codimension-2 bifurcations in periodic systems. *IEEE Trans. on Circuits and Systems I- Fundamental Theory and Applications*, 43(12):1015–1018, 1996.
15. E. Ponce, J. Aracil, and D. Pagano. Control systems with actuator saturation and bifurcations at infinity. In *Proc. of the 7th Mediterranean Conference on Control and Automation (MED99)*, pages 1598–1608, 1999.
16. L. Rundqwist. Rate limiters with phase compensation. In *Proceedings of 20th Congress of ICAS*, pages 2634–2642, 1996.
17. L. Rundqwist, Karin S-G., and J. Enhagen. Rate limiters with phase compensation in Jas 39 Gripen. In *Proc. of the European Control Conference (ECC97)*, 1997.

Monitoring and Control of Bifurcations Using Probe Signals

Munther A. Hassouneh, Hassan Yaghoobi, and Eyad H. Abed

Department of Electrical and Computer Engineering, and the Institute for Systems Research, University of Maryland, College Park, MD 20742, USA;
E-mail: abed@eng.umd.edu

Abstract. Systems undergoing Hopf bifurcation are known to amplify nearly resonant perturbation signals. A lesser known fact is that such probe signals tend to also produce a shift in the parameter value where bifurcation occurs. In this paper, these rarely used phenomena are used as a basis for stability monitoring of systems that are susceptible to loss of stability through a Hopf bifurcation. The fact that the perturbation signals delay supercritical bifurcations and advance subcritical bifurcations is noted, and the amount of this shift is quantified analytically. This analysis is based on work by Gross that employs second order averaging. A monitoring system is developed that provides an early warning signal for subcritical Hopf bifurcation. Since subcritical bifurcations lead to large departures from normal operation, detection of an impending subcritical bifurcation is a valuable goal. The results are tested numerically on a second order system of van der Pol type. The monitoring system is further used to trigger a preventive control action moving the system away from the stability boundary and catastrophic bifurcation. The results are also applied to a power system model where the search for impending subcritical bifurcation is performed in a two dimensional bifurcation parameter space.

1 Introduction

Recently, Kim and Abed [1] considered design of monitoring systems for detection of impending loss of stability and bifurcation in uncertain systems. They employed results on so-called noisy precursors of bifurcations due to Wiesenfeld and co-workers [2].

In this paper, another approach is considered that makes use of features in the response of a system undergoing generic Hopf bifurcation to near-resonant perturbation signals.

Although the main focus in this paper is detection of impending Hopf bifurcations, the authors believe that the approach can be appropriately generalized to other static and dynamic bifurcations.

Resonant and non-resonant perturbation of dynamical systems undergoing Hopf bifurcation are investigated in [3–7]. In [8,9] it is shown that near-resonant perturbation of dynamical systems undergoing period-doubling bifurcation introduces a shift in the bifurcation. In the present work the averaged model of Gross [4] for near-resonant perturbation of a general low order

F. Colonius, L. Grüne (Eds.): Dynamics, Bifurcations, and Control, LNCIS 273, pp. 51–65, 2002.

system undergoing Hopf bifurcation is used to derive a formula for the parameter shift at the bifurcation. It is shown that this quantity is proportional to the stability coefficient of the bifurcation when resonant forcing is applied to the system. The proportionality factor is a function of the amplitude of the forcing signal and characteristic system parameters, and is always negative. The analysis shows that the supercritical bifurcation is delayed, while a subcritical is advanced by the perturbation signal.

A monitoring system based on the observations above can be applied to a model of the plant running in parallel with the plant itself. The bifurcation parameter of interest may change as the system is steered from one operating regime to another. This alleviates the need for massive off-line stability calculations, which can never address all possible system operating conditions. Depending on the operating regime, the bifurcation parameter in the monitoring system can be swept in a look-ahead operation to detect potential bifurcations. The forward looking information can provide adequate stability margin to the system if appropriate preventive control action is employed. Coupling between the physical plant and the monitoring system can be incorporated into a design to minimize the search space and focus on a small set of parameters. Strong vibrations of the plant variables can be used to trigger on-line changes in the monitoring system depending on the physical problem.

The remainder of the paper proceeds as follows. Section 2 continues with a brief background on the theory of Hopf bifurcation. Section 3 covers the main contribution of the paper where the averaged model of Gross [4] is used to derive the equations quantifying the shift of an impending bifurcation. In Section 4, an equation of van der Pol type is employed as an example to numerically test the results of Section 3. In Section 5, the early occurrence of the subcritical Hopf bifurcation of a system with a periodic probe signal is used as a precursor for applying a switching control to the system. In Section 6, the proposed method is used for detection of impending subcritical Hopf bifurcation in a power system model by performing a search in a two dimensional bifurcation parameter space.

2 Hopf bifurcation

The statement of the Hopf bifurcation theorem in [10] is well suited to the methodology used in the rest of the paper.

Theorem 3.1 Suppose that the system $\dot{x} = F_\mu(x)$, $\mu \in R$, $x \in R^n$ has an equilibrium $(\mu_c,\, x_0)$ at which the following properties are satisfied:
(H1) $D_x F_{\mu_c}(x_0)$ has a simple pair of pure imaginary eigenvalues and no other eigenvalues with zero real parts;
Then there is a smooth curve of equilibria $(\mu,\, x_\mu)$ with $x(\mu_c) = x_0$. The eigenvalues $\lambda(\mu)$, $\overline{\lambda(\mu)} = \alpha(\mu) \pm j\omega(\mu)$ of $D_x F_{\mu_c}(x(\mu))$ which are imaginary

at $\mu = \mu_c$, vary smoothly with μ. If, moreover,

$$\textbf{(H2)} \qquad \frac{d}{d\mu}(Re\lambda(\mu_c)) := d_0 \neq 0, \tag{1}$$

then there is a unique 3-dimensional center manifold passing through (μ_c, x_0) in $R \times R^n$ and a smooth system of coordinates (preserving the planes $\mu = const.$) for which the Taylor expansion of degree 3 on the center manifold is given by the following equation:

$$\begin{aligned} \dot{x}_1 &= (d_0\mu + a_0(x_1^2 + x_2^2))x_1 - (\omega_0 + c_0\mu + b_0(x_1^2 + x_2^2))x_2 \\ \dot{x}_2 &= (\omega_0 + c_0\mu + b_0(x_1^2 + x_2^2))x_1 + (d_0\mu + a_0(x_1^2 + x_2^2))x_2 \end{aligned} \tag{2}$$

If $a_0 \neq 0$, there is a surface of periodic solutions in the center manifold which has quadratic tangency with the eigenspace of $\lambda(\mu)$, $\overline{\lambda(\mu)}$ agreeing to second order with the paraboloid $\mu = -(a_0/d_0)(x_1^2 + x_2^2))$. If $a_0 < 0$, then these periodic solutions are stable limit cycles, while if $a_0 > 0$, the periodic solutions are repelling. Eq. (2) can be expressed in polar coordinates as

$$\begin{aligned} \dot{r} &= (d_0\mu + a_0 r^2)r \\ \dot{\theta} &= (\omega_0 + c_0\mu + b_0 r^2) \end{aligned} \tag{3}$$

where $d_0 = \alpha'(0)$ and $c_0 = \omega'(0)$ (derivatives are with respect to the bifurcation parameter μ), in the new coordinates.

The parameters a_0 and b_0 determine the stability (and local amplitude growth rate) of the bifurcated periodic solution and the amplitude dependent modification to its period. Parameters a_0 and b_0 are sometimes called the first Liapunov coefficient and the first frequency modifier, respectively. For simplicity, also assume that $d_0 = \alpha'(0) > 0$; that is, the critical mode crosses the $j\omega$ axis from left to right.

Next, consider the following model with linear part in Jordan form, representing a general 2-dimensional system at bifurcation:

$$\begin{bmatrix} \dot{x}_1 \\ \dot{x}_2 \end{bmatrix} = \begin{bmatrix} 0 & -\omega \\ \omega & 0 \end{bmatrix} \begin{bmatrix} x_1 \\ x_2 \end{bmatrix} + \begin{bmatrix} g(x_1, x_2) \\ h(x_1, x_2) \end{bmatrix} \tag{4}$$

The pair of pure imaginary eigenvalues occurs at $\pm j\omega$. Using normal form calculations [10], coefficients a_0 and b_0 can be obtained as follows:

$$\begin{aligned} a_0 = &\frac{1}{16}[g_{111} + g_{122} + h_{112} + h_{222}] + \frac{1}{16\omega}[g_{12}(g_{11} + g_{22}) \\ &-h_{12}(h_{11} + h_{22}) - g_{11}h_{11} + g_{22}h_{22}] \end{aligned} \tag{5}$$

$$\begin{aligned} b_0 = &\frac{1}{16}[h_{111} + h_{122} - g_{112} - g_{222}] - \frac{1}{24\omega}[g_{11}^2 + g_{12}^2 + h_{12}^2 + h_{22}^2] \\ &+\frac{5}{48\omega}[g_{22}^2 + h_{11}^2 + g_{11}g_{22} - g_{11}h_{12} - g_{12}h_{22}] \\ &-\frac{1}{48\omega}[g_{12}h_{11} + g_{22}h_{12}] \end{aligned} \tag{6}$$

Integer subscripts indicate partial derivatives with respect to the corresponding state variables. Derivatives are calculated at the equilibrium at criticality.

3 Analysis of the effects of near-resonant forcing

3.1 Averaged model

We begin by summarizing the results of Gross [4] on the averaged model for periodically forced nonlinear systems undergoing Hopf bifurcation. Consider the following 2-dimensional forced nonlinear system

$$\begin{bmatrix} \dot{x}_1 \\ \dot{x}_2 \end{bmatrix} = \begin{bmatrix} G(x_1, x_2, \mu) \\ H(x_1, x_2, \mu) \end{bmatrix} + k \begin{bmatrix} A_1 & B_1 \\ A_2 & B_2 \end{bmatrix} \begin{bmatrix} \cos\ \omega_F t \\ \sin\ \omega_F t \end{bmatrix} \tag{7}$$

where μ is the bifurcation parameter and k is the forcing amplitude. Assume that the unforced system undergoes a Hopf bifurcation. Without loss of generality we can assume that $\mu = 0$ at the bifurcation and $(0, 0)$ is an equilibrium of Eq. (7) for all μ close to 0 when $k = 0$. It is also assumed that an appropriate scaling of the time variable is used so that the resonant frequency of the system ω_0 at the Hopf bifurcation is 1.

The method of integral averaging [11] can be applied to the system (7) to yield a planar autonomous system. This method was used in [4] to derive the averaged equations corresponding to Eq. (7) in polar coordinates. Next, consider the system (7) in polar coordinates, where $x_1 = r\ \cos\ \theta$, $x_2 = r\ \sin\ \theta$, $\theta = \phi + \Omega t$ and $\Omega = \omega_0 + \nu = 1 + \nu$. Gross used the following variable and parameter scalings $r = \varepsilon\bar{r}$, $\mu = \varepsilon^2\bar{\mu}$, $k = \varepsilon^3\bar{k}$, $\nu = \varepsilon^2\bar{\nu}$, $\phi = \bar{\phi} + \cos^{-1}(\frac{(A_2 - B_1)}{2V})$ along with the time scaling $\bar{t} = \varepsilon^2 t$. Here $V = \frac{1}{2}[(A_1 + B_2)^2 + (A_2 - B_1)^2]^{1/2}$. It was shown that the averaged system in polar coordinates takes the form

$$\begin{aligned} \dot{\bar{r}} &= (d_0\bar{\mu} + a_0\bar{r}^2)\bar{r} + \bar{k}V\ \sin\ \bar{\phi} \\ \dot{\bar{\phi}} &= (c_0\bar{\mu} - \bar{\nu}) + b_0\bar{r}^2 + (\bar{k}/\bar{r})V\ \cos\ \bar{\phi}. \end{aligned} \tag{8}$$

The scaling parameter ε was assumed to be a small parameter, and ω_F was taken equal to Ω, with both assumed close to the resonant frequency $\omega_0 = 1$. The latter assumption implies that the small parameter $\bar{\nu}$ can be viewed as a detuning factor. Eq. (8) can be considered as a perturbation of the system of the form (3). For $k = 0$ the averaged model is exactly of the form (3) and the vector field corresponding to the amplitude variable $\bar{r}$ is decoupled from the phase variable $\bar{\phi}$. Note that the averaged equivalent model (8) is only valid locally and for small values of k for which the trajectories of system (7) stay within the basin of equilibria of interest.

Next, it will be shown, through analysis of the Jacobian matrix of system (8), that the mentioned perturbation implies a shift of the bifurcation. The amount of the shift is calculated and is valid at least for small values of $\bar{k}$ and $\bar{\nu}$. For the sake of convenience, we will omit the overbars in the system model in the sequel.

3.2 Calculation of shift in the critical value of the bifurcation parameter

The equilibrium points of the system (8) satisfy the following system of algebraic equations.

$$\begin{aligned} -(d_0\mu + a_0 r^2)r &= kV \sin\ \phi \\ -(c_0\mu - \nu)r - b_0 r^3 &= kV \cos\ \phi \end{aligned} \tag{9}$$

Equivalently, the following equation may be solved for the $r > 0$ component of the equilibria. By summing the squares of the respective sides of Eq. (9), r and ϕ are determined uniquely (ϕ is unique modulo 2π). Solutions to Eq.

$$E(r^2) = (kV)^2 \tag{10}$$

where

$$\begin{aligned} E(z) = {} & (a_0^2 + b_0^2)z^3 + 2(a_0 d_0\mu + b_0(c_0\mu - \nu))z^2 \\ & +((d_0\mu)^2 + (c_0\mu - \nu)^2)z. \end{aligned} \tag{11}$$

determine the r component of the equilibria and then Eq. (9) can be used to find the corresponding ϕ component. Stability of the equilibrium points is determined by analyzing the Jacobian matrix of the system (8). The Jacobian matrix at the equilibria can be shown to take the form

$$J(r) = \begin{bmatrix} d_0\mu + 3a_0 r^2 & -(c_0\mu - \nu)r - b_0 r^3 \\ 3b_0 r + \frac{(c_0\mu-\nu)}{r} & d_0\mu + a_0 r^2 \end{bmatrix}. \tag{12}$$

The trace and determinant of the Jacobian matrix are

$$tr(J(r)) = 2d_0\mu + 4a_0 r^2 \tag{13}$$

$$det(J(r)) = E'(r^2) \tag{14}$$

where E' is the derivative of E with respect to z. Since the determinant of the matrix J represents the product of the modes, the equilibrium cannot be a saddle if the determinant is positive and it is either a node or a focus depending on the sign of the trace (which equals the sum of the two modes). Therefore, when the sign of $tr(J(r))$ changes, the system undergoes a Hopf bifurcation if $det(J(r)) > 0$. It is clear from Eq. (13) that the sign of $tr(J(r))$ changes at $r^2 = (d_0\mu)/(-2a_0)$. Next, we proceed with an analysis to show that for sufficiently small values of $|\mu|$, $E'((d_0\mu)/(-2a_0)) > 0$ for all values of $a_0 \neq 0$, b_0, c_0, d_0, and $\nu \neq 0$. For this, consider $E'(r^2)$ when $r^2 = (d_0\mu)/(-2a_0)$.

$$\begin{aligned} E'(\frac{d_0\mu}{-2a_0}) = {} & 3(a_0^2 + b_0^2)(\frac{d_0\mu}{-2a_0})^2 + 4(a_0 d_0\mu + b_0(c_0\mu - \nu))(\frac{d_0\mu}{-2a_0}) \\ & +((d_0\mu)^2 + (c_0\mu - \nu)^2) \end{aligned} \tag{15}$$

Eq. (16) can be simplified further to the following form:

$$\begin{aligned} E'(\frac{(d_0\mu)}{(-2a_0)}) = {} & \frac{3}{4}(1 + (\frac{b_0}{a_0})^2)(d_0\mu)^2 - 2(d_0\mu + (\frac{b_0}{a_0})(c_0\mu - \nu))(d_0\mu) \\ & +((d_0\mu)^2 + (c_0\mu - \nu)^2) \end{aligned} \tag{16}$$

Consider the following two definitions:

$$\gamma = b_0/a_0 \tag{17}$$

$$S(\mu) = (c_0\mu - \nu)/d_0\mu \tag{18}$$

For $\mu \neq 0$, $E'(d_0\mu/(-2a_0)) > 0$ iff $E'(d_0\mu/(-2a_0))/(d_0\mu)^2 > 0$.
Next, we use Eqs. (17) and (18) to rewrite the latter condition as follows:

$$S(\mu)^2 - 2\gamma S(\mu) + \frac{3}{4}\gamma^2 - \frac{1}{4} > 0 \tag{19}$$

Inequality (19) is valid if and only if

$$S(\mu) < \gamma - \frac{1}{2}\sqrt{\gamma^2+1} \quad or \quad S(\mu) > \gamma + \frac{1}{2}\sqrt{\gamma^2+1} \tag{20}$$

Definition (18) implies that by choosing small enough values for μ, $S(\mu)$ can become arbitrary large in magnitude if $\nu \neq 0$; that is, satisfying one of the two inequalities (20). Consequently the important observation would be preservation of the bifurcation structure (i.e., Hopf bifurcation) for small values of μ and $\nu \neq 0$. Also note that r^2 has a meaningful value (i.e., positive) only when the numerator and the denominator of $(d_0\mu)/(-2a_0)$ agree in sign. This implies that, since $d_0 > 0$, the Hopf bifurcation experiences a delay if it was supercritical in the original unforced system and it experiences an advance, if it was subcritical in the unforced system. This observation will be re-confirmed later in this section when we calculate the critical value of μ_c in the periodically forced system.

Next, Eq. (10) is used to quantify the shift in the bifurcation. Substituting $r^2 = (d_0\mu)/(-2a_0)$ into Eq. (10) and solving for μ, it is found after some manipulation, that the corresponding shift in the critical value of the bifurcation parameter is the real solution of the following cubic equation where $\beta = c_0/d_0$.

$$(kV)^2(\frac{-2a_0}{d_0})^3 = [(a_0^2+b_0^2) - 4a_0b_0\beta + 4a_0^2\beta^2]\mu^3 \tag{21}$$
$$+[(4a_0b_0 - 8a_0^2\beta)\frac{\nu}{d_0}]\mu^2 + [4a_0^2\frac{\nu^2}{d_0^2}]\mu$$

Note that $\mu_c = 0$ for $k = 0$, therefore, by choosing k small, μ_c can be made small as well, some thing that was needed to preserve the local bifurcation structure.

Although we could proceed with applying the classical closed form solution method for cubic equations, the solution is intricate and in an algorithmic, conditional form. Therefore, a special but realistic case is considered next. Assume $\nu \to 0$, i.e., the detuning factor is very small or there is a good estimate of the resonant frequency of the Hopf bifurcation. In this case, the above cubic equation will be simplified to the following simple form at the limit:

$$[(a_0^2+b_0^2) - 4a_0b_0\beta + 4a_0^2\beta^2]\mu^3 = (kV)^2(\frac{-2a_0}{d_0})^3 \tag{22}$$

One can still solve numerically for μ_c in the general case where $\nu \neq 0$ as shown in the next section. The unique solution to the cubic equation above is

$$\mu_c = (\frac{(kV)^2}{\Lambda})^{1/3} \frac{-2a_0}{d_0} \tag{23}$$

where $\Lambda = (a_0^2 + b_0^2) - 4a_0 b_0 \beta + 4a_0^2 \beta^2$. The following observations are in order with respect to Eq. (23):

(i) First, Λ can be shown to be nonnegative regardless of the values of a_0, b_0 and β, and positive if either one of the two parameters a_0 or b_0 is nonzero, which is a realistic assumption. This claim can be proved by simply looking at the discriminant of Λ as a quadratic equation in β. The discriminant is $-16a_0^4$ and nonnegative. Therefore, there is no real root for Λ as a function β if $a_0 \neq 0$. Since for $\beta = 0$, $\Lambda = a_0^2 + b_0^2$ is nonzero by assumption, Λ is always positive.

(ii) Second, since $d_0 > 0$ by assumption, the sign of the bifurcation shift expressed by the right hand side of Eq. (23) is opposite that of a_0. *This means that the perturbation signal delays supercritical bifurcation, while it advances subcritical bifurcation.*

(iii) Third, for zero detuning, the amount of shift in the bifurcation as a function of forcing amplitude obeys a 2/3-power law.

(iv) Fourth, for zero detuning, the amount of shift in the bifurcation as a function of d_0 obeys a $-1/3$-power law, i.e., the shift is large for small values of $\alpha'(0)$.

Note that the results of this section are only valid when the perturbation signal does not drive the system outside the basin of the equilibrium of interest.

4 Numerical example

Consider the following second order system of van der Pol type where a nonlinearity in the damping allows the unforced autonomous system to exhibit a limit cycle:

$$\begin{aligned} \dot{y}_1 &= -y_2 \\ \dot{y}_2 &= 2\mu y_2 + (y_1^2 + y_2^2) y_2 + y_1 + k \sin t \end{aligned} \tag{24}$$

The unforced autonomous system corresponding to Eq. (24) undergoes a subcritical Hopf bifurcation at $\mu = 0$. At the bifurcation, $\lambda = \pm j$, $d_0 = 1$, $c_0 = 0$, $a_0 = 1/2$, $b_0 = 0$, $\beta = 0$, $V = 1/2$ and $\Lambda = 1/4$. The forcing frequency ω_F is 1. Substituting the system parameters into Eq. (22) results in the following simple cubic equation for the critical value of μ:

$$\mu^3 + 4\nu^2 \mu + k^2 = 0 \tag{25}$$

The classical closed form for the real solution to the special cubic Eq. (25) can be easily simplified to derive the following expression for the critical value of parameter μ where bifurcation occurs in system (24):

$$\mu_c = (-\frac{k^2}{2} + (\frac{64}{27}\nu^6 + \frac{k^4}{4})^{1/2})^{1/3} + (-\frac{k^2}{2} - (\frac{64}{27}\nu^6 + \frac{k^4}{4})^{1/2})^{1/3} \quad (26)$$

Note that, as expected, $\mu_c \to 0$ as $k \to 0$. Figure 1 illustrates the theoretical as well as the numerically observed values of the negative shift in the critical parameter value μ_c where bifurcation occurs in the forced system. Also note that the bifurcation shift is symmetric with respect to the detuning values and deviates from its value in the resonant case equally for positive or negative ν. The maximum absolute value of the shift corresponds to the zero detuning factor. It is interesting to see that all graphs corresponding to different detuning values converge as the forcing amplitude k increases.

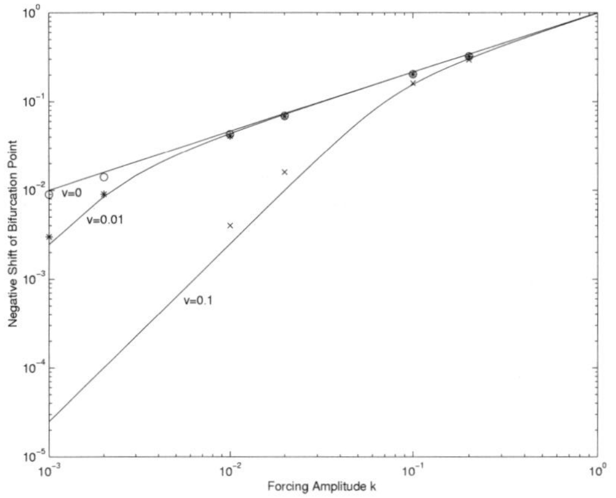

Fig. 1. *Negative shift of impending bifurcation as a function of forcing amplitude and detuning factor. Circle, asterisk and cross marks represent the numerical calculations corresponding to* $\nu = 0$, 0.01 *and* 0.1 *respectively. Solid lines represent the theoretically calculated values*

5 Combined Stability Monitoring and Control

Early occurrence of subcritical Hopf bifurcation in a periodically forced model of an actual physical system can be used as a warning signal for loss of stability and the associated bifurcation. Here we assume that a good system model is available. The idea considered next is to run a monitoring model of the system in parallel with the operation of the (physical) plant. The monitoring system is steered with identical control inputs as the plant. In addition, the monitoring system receives a periodic, near-resonant perturbation signal

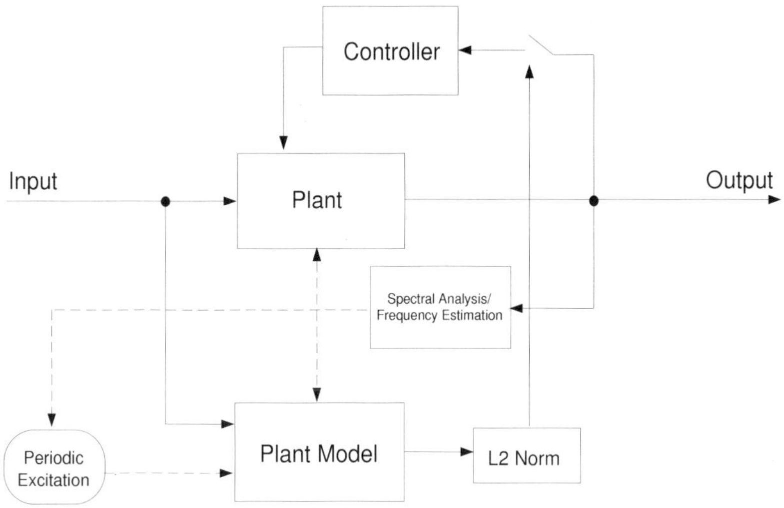

Fig. 2. *Preventive control of subcritical bifurcation*

(probe). An estimate of the Hopf bifurcation resonant frequency is used as the frequency of the periodic signal and can be found by performing spectral analysis at the output of the plant. The early occurrence of subcritical Hopf bifurcation in the monitoring system can be used as a precursor to trigger a preventive control action in the plant. This is preferable to other possible means for detecting nearness to instability in the model, such as reducing the damping in separate state equations until instability arises. The reason is that the present scheme also provides amplification of the probe signal even before instability. Moreover, this amplification occurs for systems with many parameters. There is no need to know which parameter would bring the system closer to instability should it be changed.

Next, the control strategy of Fig. 2 is applied to detect the impending bifurcation in system (25). Two copies of the system are simulated in SIMULINK; one is the plant (with states x_1, x_2), and one is the model (with states y_1, y_2). A small amplitude white Gaussian noise is used as the input to the plant to facilitate the estimation of the Hopf bifurcation frequency and a small amplitude sinusoidal signal is used as the input to the plant model. The forcing amplitude k is fixed at 0.05. The bifurcation parameter μ is slewed (increased linearly and slowly) in both systems starting from $\mu = -0.35$. As a control action, bifurcation parameter is simply pushed back to the starting level when bifurcation is detected in the model system. Note that availability of the bifurcation parameter for control purposes is not always an unrealistic assumption. For instance, in the power systems context, system load (a common bifurcation parameter) can be shed to prevent the system from losing stability.

Figure 3 shows the power spectral density of the plant output x_2 when $\mu = -0.1$. The peak value of the power spectrum provides the monitoring system with a good estimate of the resonant frequency. Here the estimated resonant frequency is 0.98, i.e., $\nu = -0.02$. Note that from Fig. 1, the level of deviation of the shift in bifurcation from its maximum for $\nu = 0$ is negligible for $k = 0.05$. Figure 4 shows the time response of the plant and the monitoring system

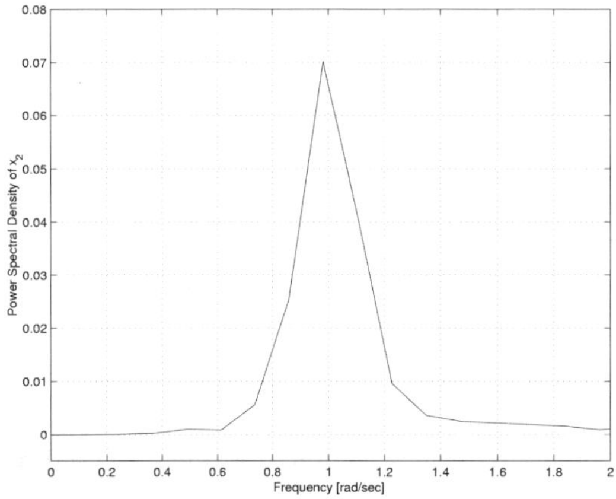

Fig. 3. *Estimation of the resonant frequency using the power spectral density of the output of the plant*

when the bifurcation parameter is slewed through its critical value and no control is applied. Here the monitoring system bifurcates subcritically when $\mu = -0.1191$. The theoretically estimated shift in the bifurcation is -0.1357. The plant itself bifurcates around $\mu = 0$ as expected. The monitoring system has a periodic time response (with an amplitude that is a function of μ) before the bifurcation. The time response of the plant is a filtered noise before bifurcation.

Figure 5 shows the time responses of the plant and the monitoring system when the bifurcation parameter is slewed and as a preventive control action, the bifurcation parameter is pushed back to the -0.35 level when the instability is detected in the monitoring system. Here the control action prevents the plant from experiencing the nearby subcritical bifurcation.

6 Detection of Impending Bifurcation in a Power System Model

Consider the simplified power system voltage dynamic model of Zaborszky and co-workers [12] with generator, voltage control, transmission line and

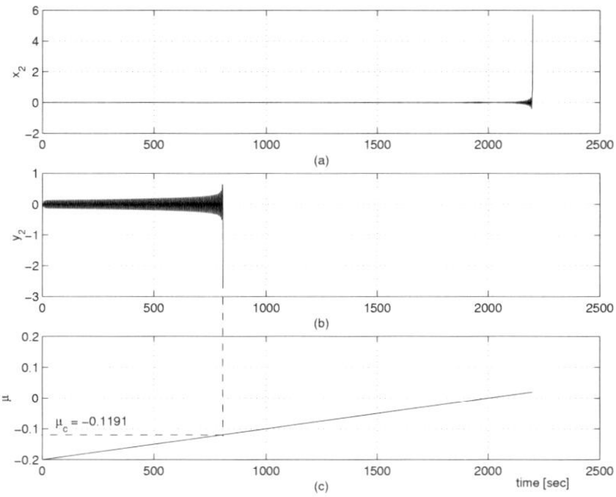

Fig. 4. *Time response of the plant and the monitoring system when no control is applied: (a) x_2 of the plant, (b) y_2 of the monitoring system, and (c) bifurcation parameter*

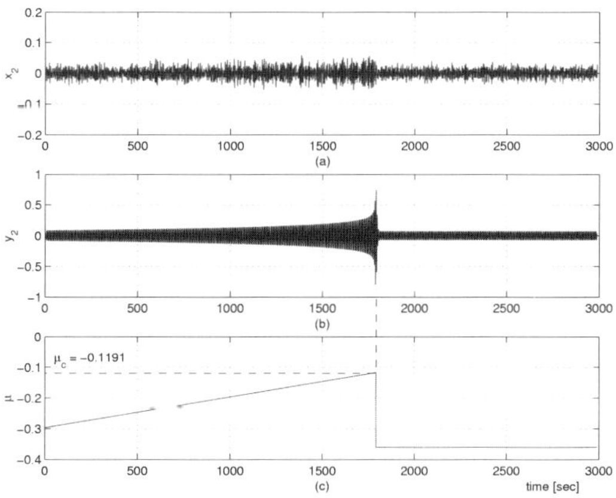

Fig. 5. *Time response of the plant and the monitoring system when preventive control action is applied: (a) x_2 of the plant, (b) y_2 of the monitoring system, and (c) bifurcation parameter*

matched load. The following second order parameter dependent differential-algebraic equations describe the dynamics:

$$
\begin{aligned}
T'_{d0}\dot{E}' &= -\frac{x_d}{x'_d}E' + \frac{x_d - x'_d}{x'_d}E_G\cos(\delta_G - \delta') + E_{fd} \\
T\dot{E}_{fd} &= -(E_{fd} - E^0_{fd}) - L(E_G - E_r)
\end{aligned}
\tag{27}
$$

$$
\begin{aligned}
0 &= \frac{E' E_G}{x'_d} \sin(\delta_G - \delta') + \frac{E_G E}{x} \sin(\delta_G - \delta) \\
0 &= \frac{1}{x'_d}(E_G^2 - E_G E' \cos(\delta_G - \delta') \\
&\quad + \frac{1}{x}(E_G^2 - E_G E \cos(\delta_G - \delta)) \\
0 &= \frac{E_G E}{x} \sin(\delta - \delta_G) + P \\
0 &= \frac{1}{x}(E^2 - E_G E \cos(\delta - \delta_G)) + Q \\
Q &= Q_0 + HE + BE^2
\end{aligned} \tag{28}
$$

Here, Eqs. (27) represent the dynamical portion of the model and Eqs.

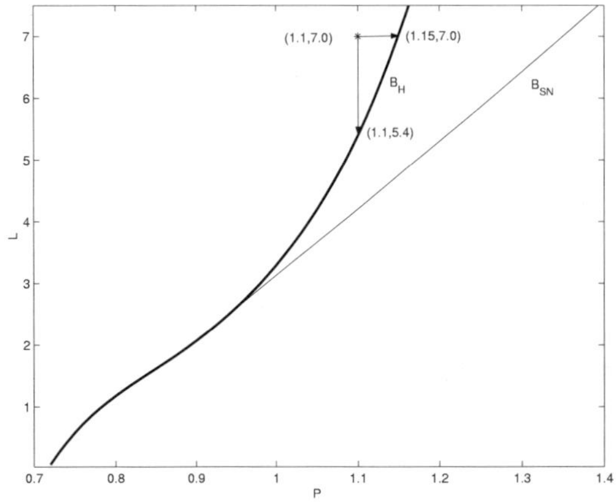

Fig. 6. *Bifurcation parameter space of the system (27)-(28) with* ($T'_{d0} = 5.0,\ T = 1.5,\ E^0_{fd} = 1.6,\ E_r = 1.0,\ x_d = 1.2,\ x'_d = 0.2,\ x = 0.1,\ x' = x - x'_d,\ Q_0 = 0.5P,\ H = 0,\ B = 0$)

(28) represent the algebraic constraints. This simplified model includes a one axis generator model and a first-degree simplification of the IEEE Type 1 excitation dynamics. Here the two bifurcation parameters of interest are L, the excitation amplifier gain and P, the load. The monitoring system framework discussed above will be applied to this model. The importance of having more than a single bifurcation parameter has to do with the fact that a probe signal will reveal an impending bifurcation regardless of which parameter's change will bring the system close to bifurcation.

Zaborszky and co-workers performed extensive bifurcation analysis on the model and identified the boundaries of several local and global bifurcations in the two dimensional bifurcation parameter space. Figure 6 shows one of their several bifurcation parameter spaces.

Considering P as the bifurcation parameter, it is clear from the bifurcation boundaries that the system typically undergoes a subcritical Hopf bifurcation followed by a saddle node bifurcation. The same scenario may happen if L is chosen as the bifurcation parameter. In Fig. 7, two sample bifurcation

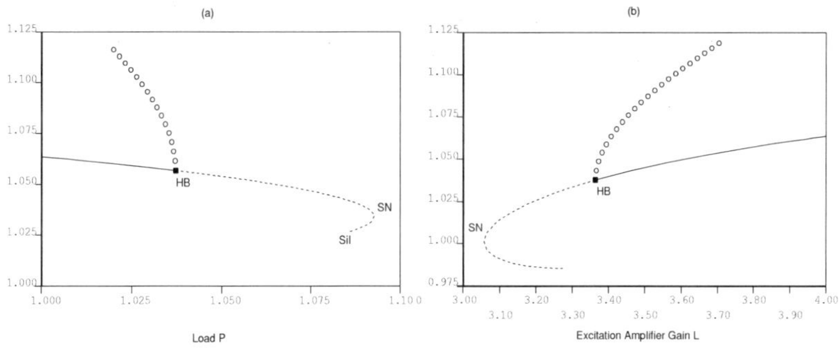

Fig. 7. *Power system bifuraction diagrams: (a) Bifurcation parameter P and (b) bifurcation parameter L*

diagrams are generated using the software package AUTO[13]. Next, a two dimensional search is performed in the bifurcation parameter space using the method of near-resonant perturbation to detect the impending subcritical bifurcations. Note that the bifurcation shifts in the direction of L is one

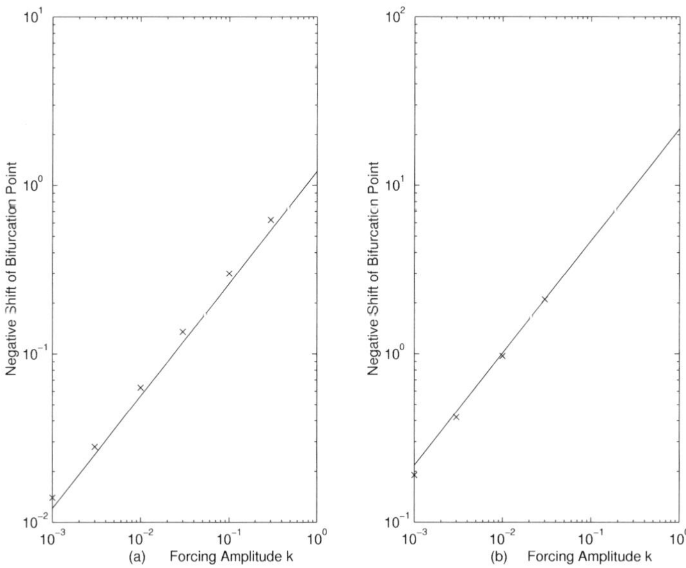

Fig. 8. *Negative shift of impending bifurcation in power system model as a function of forcing amplitude: (a) search direction P ($a_0 = 19.5296, b_0 = 31.2147, c_0 = -1.2760, d_0 = 1.10722, V = 1/2, \omega_0 = 0.544114$) and (b) search direction L ($a_0 = 18.4702, b_0 = 20.0079, c_0 = -0.11205, d_0 = 0.0462987, V = 1/2, \omega_0 = 0.450794$)*

order of magnitude larger than those in the direction of P due to small value of d_0. Note that the results of Eq. (23) predict large shifts in the bifurcation point for systems undergoing an almost degenerate bifurcation such as the one in power system model (27)-(28) where d_0 is close to zero. The bifurcation parameter P, i.e., the power system load, can be pushed back when the monitoring system warns of an impending instability.

7 Conclusions

Near-resonant perturbation of systems undergoing Hopf bifurcation results in a shift in the bifurcation. The shift in the bifurcation point is quantified for a general low order system undergoing Hopf bifurcation and it is shown to be a function of both the perturbation amplitude and the detuning factor. It is also shown that the impending bifurcation is advanced if it is subcritical, and is delayed if it is supercritical. These observations motivate a monitoring system that detects subcritical bifurcation prior to its occurrence, thus facilitating preventive control action. The results are applied to a second order system of van der Pol type through simulations, where numerical experiments agree well with the theoretical predictions. In an example van der Pol type system where subcritical Hopf bifurcation occurs, early detection of impending instability is employed to control the system. For this a near-resonant perturbation is applied to a replica of the system, and the early occurrence of bifurcation is used to trigger preventive control action in the plant. The results are also applied to a basic power system model where the search for impending subcritical bifurcation is performed in a two dimensional bifurcation parameter space.

The authors are considering extensions of this work in several directions. For example, if the frequency of the impending instability is not known with any accuracy, it may be possible to use a chaotic probe signal. The broadband nature of the chaotic signal can ensure that the critical frequency is excited. Another direction is to consider enducing other warning signals less severe than instability of a model system. For instance, if intermittent behavior [14] can be guaranteed in the vicinity of bifurcation, this would provide a relatively benign warning signal.

Acknowledgments. This research has been supported in part by the Office of Naval Research under Multidisciplinary University Research Initiative (MURI) Grant N00014-96-1-1123.

References

1. Kim, T., Abed, E.H. (2000) Closed-Loop Monitoring Systems for Detecting Impending Instability. IEEE Trans. Circuits and Systems–I: Fundamental Theory and Applications, *47*, 1479 -1493
2. Wiesenfeld K. (1985) Noisy Precursors of Nonlinear Instabilities. Journal of Statistical Physics, *38*, no. 5-6, pp

3. Gross P. (1993) On Harmonic Resonance in Forced Nonlinear Oscillators Exhibiting a Hopf Bifurcation. IMA Journal of Applied Mathematics, *50*, 1-12
4. Gross P. (1994) On Oscillation Types in Forced Nonlinear Oscillators Close to Harmonic Resonance. IMA Journal of Applied Mathematics, *53*, 27-43
5. Namachchivaya, N.S., Ariaratnam, S.T. (1987) Periodically Perturbed Hopf Bifurcation. SIAM Journal of Applied Mathematics, *47*, 15-39
6. Rosenblat, S., Cohen, D.S. (1981) Periodically Perturbed Bifurcation. II. Hopf Bifurcation. Studies in Applied Mathematics, *64*, 143-175
7. Smith H.L. (1981) Nonresonant Periodic Perturbation of the Hopf Bifurcation. Applicable Analysis, *12*, 173-195
8. Bryant P. (1986) Suppression of Period-Doubling and Nonlinear Parametric Effects in Periodically Perturbed Systems. Physical Review A, *33*, 2525-2543
9. Vohra, S.T., Fabiny, L. and Wiesenfeld, K. (1985) Observation of Induced Subcritical Bifurcation by Near-Resonant Perturbation. Physical Review Letters,*72*, 1333–1336
10. Guckenheimer, J., Holmes, P. (1983) Nonlinear Oscillations, Dynamical Systems, and Bifurcations of Vector Fields, *42*, Springer
11. Perko L. M. (1968) Higher Order Averaging and Related Methods for Perturbed Periodic and Quasi-Periodic Systems. SIAM Journal of Applied Mathematics, *17*, 698–724
12. Venkatasubramanian, V., Schattler, H. and Zaborszky, J. (1992) Voltage Dynamics: Study of a Generator with Voltage Control, Transmission, and Matched MW Load. IEEE Transactions on Automatic Control, *37*, 1717–1733
13. Doedel, E.J. (1981) AUTO: A Program for the Automatic Bifurcation Analysis of Autonomous Systems, Congressus Numerantium, *30*, 265–284
14. Pomeau, Y., Manneville, P. (1980) Intermittent Transition to Turbulence in Dissipative Dynamical Systems. Commun. Math. Phys., *74*, 189–197

Normal Form, Invariants, and Bifurcations of Nonlinear Control Systems in the Particle Deflection Plane

Wei Kang

Department of Mathematics, Naval Postgraduate School, Monterey, CA 93943, USA

Abstract. This paper addresses the problem of bifurcation analysis for nonlinear control systems. It is a review of some results published in recent years on bifurcations of control systems based on the normal form approach. We address three problems for systems with a single uncontrollable mode, namely the problem of normal forms, bifurcations of control systems, and bifurcation control by state feedback.

1 Introduction

Nonlinear dynamical systems exhibit complicated performance around bifurcation points. As the parameter of a system is varied, changes may occur in the qualitative structure of its solutions around a point of bifurcation. This paper addresses the problem of bifurcation analysis for nonlinear control systems. This is a review paper that summarizes the main results published in recent years on bifurcations of control systems based on normal forms and invariants. More specifically, we focus on the results from [15], [17], [18], [19] and [20].

The theoretical research on the bifurcation of control systems is a relatively young subject. Nevertheless, it has attracted increasing attention. For example, in [1], [7], [27], [33], and other related publications, bifurcation control based on the projection method was developed with some interesting applications. Many other theoretical approaches or engineering applications of bifurcation control can be found in the literature (e.g., [6], [10], [12], and [27]). A review of bifurcation and chaos in control systems can be found in [6]. More references on related topics can be found in [5], a bibliography of publications on bifurcation and chaos in control systems. Given the large volume of new developments in this field, it is impossible to review all the methods used for bifurcation control. Instead, this paper focuses on the approach of normal forms and invariants only.

The theory of bifurcation control based on normal forms was developed in mainly three parts, the normal form and invariants, the bifurcation of equilibrium sets, and bifurcation control by feedback. These results were published separately in several papers for the past ten years ([15], [17], [18], [19] and [20]). This review is to summarize these integral parts in one article so that

F. Colonius, L. Grüne (Eds.): Dynamics, Bifurcations, and Control, LNCIS 273, pp. 67–87, 2002.

the connections between the separately published papers and the philosophy of the overall approach can be illustrated efficiently. The paper is organized as follows. Several bifurcation problems are formulated in §2 for nonlinear control systems. Normal forms and invariants, the foundation of the framework, are introduced in §3. The bifurcation and topology of equilibrium points for systems with a single uncontrollable mode are summarized in §4. Bifurcation control by state feedback is addressed in §5, such as the stability of stationary bifurcations. In §6 and §7, we review some related results on the feedback control with perturbation, and the bifurcation control using nonsmooth feedback.

2 Problem formulation

The first question we have to answer is *what do we mean by the terminology "bifurcation of control systems?"* Let us review the bifurcation of a classical dynamic system $\dot{\xi} = f(\xi, \mu)$. The classical bifurcation theory studies the change of *qualitative properties* of the dynamical system as the parameter is varied. What are the qualitative properties? Some important properties include the topology of the equilibrium set, the stability, and the existence of periodic solutions. Therefore, if any of these properties is changed in a dynamical system as a parameter is varied, we say that the system exhibits a bifurcation. Typical bifurcations include saddle node, transcritical, pitchfork (both the topology of the equilibrium set and the stability are changed), and the Hopf bifurcation (a periodic solution is generated).

Let us follow the same way of thinking for control systems. The bifurcation theory of control systems should study the change of qualitative properties. What are the important qualitative properties for control systems? Some important properties of a control system include: the controllability, the stabilizability, and the topology of the equilibrium set. Therefore, the following problem of bifurcation for control systems is studied in this paper.

Problem 1. Bifurcation of control systems. It focuses on the change of qualitative properties of control systems such as controllability, stabilizability, and the topology of the equilibrium set.

So bifurcation of control systems may have a meaning that is different from bifurcation of dynamical systems without control. One may ask what is the relationship between the bifurcation of control systems and the classical bifurcation of dynamical systems? This question leads to another interesting topic that is addressed in this paper. We know that if a feedback $u = \alpha(\xi)$ is applied to a control system, it becomes a dynamical system without control (if feedforward is applied, it is a system with a parameter). It is shown in this paper that the bifurcation defined in Problem 1 is converted to a classical bifurcation in a closed-loop system. However, different feedbacks applied to the same system may result in completely different classical bifurcations. This phenomenon leads to the second problem of research.

Problem 2. Bifurcation control using state feedback. The problem focuses on the feedback design to achieve the stability around a critical point, or to achieve the desired performance by qualitatively changing a classical bifurcation.

We know that qualitative properties such as the controllability and stabilizability of a control system are invariant under changes of coordinates and the regular feedback $u = \alpha(\xi) + \beta(\xi)v$ if $\beta(\xi) \neq 0$. It is also known that systems may become simpler if suitable changes of coordinates and feedbacks are applied. Although two systems may look different, it is possible that they actually are equivalent to each other under a change of coordinates and feedback. So, the two systems have the same qualitative properties and the same bifurcation. Therefore, if we can find simple control systems that are equivalent to a family of complicated control systems, the bifurcation found for these simple systems represents the bifurcation of all systems in the family. This approach significantly simplifies the problem. This viewpoint leads to the following question: what are the simplest systems under changes of coordinates and feedbacks for a given family of control systems? This is equivalent to the following problem of normal forms.

Problem 3. Given a family of nonlinear control systems, find a set of normal forms equivalent to systems in the family under changes of coordinates and feedbacks.

Although Problem 3 is raised here for the purpose of bifurcation analysis, it is an interesting question by itself. Actually when this problem was addressed in [15], it was introduced as an extension of feedback linearization and Brunovsky form. In [32], normal forms were also used in the study of feedforward systems. Problem 1 was addressed in [17] and [18] for systems with a single uncontrollable mode. Problem 2 was addressed in [19] and [20].

Consider the following control system with a parameter

$$\dot{\xi} = f(\xi, \mu) + g(\xi, \mu)u, \; f(0,0) = 0, \tag{1}$$

where $\xi \in \Re^n$ is the state variable, $u \in \Re^m$ is the control input and μ is the parameter. We assume that the rank of $g(\xi, \mu)$ is m at the point of interest. Unless it is otherwise specified, all vector fields and state feedbacks in this paper are C^k for some $k > 0$ sufficiently large. System (1) is said to be *linearly controllable* at $(\xi, \mu) = (0,0)$ if its linearization (A, B),

$$A = \frac{\partial f}{\partial \xi}(0,0), \; B = g(0,0)$$

is controllable. The origin $(\xi, \mu) = (0,0)$ is called an equilibrium or equilibrium point of (1) because $\xi(t) = 0$ is a constant solution if $\mu = 0$ and $u = 0$. A constant solution may exist for other values of (ξ, μ, u). The *equilibrium set* is defined by

$$E = \{(\xi, \mu) | \text{ there exists } u_0 \in \Re \text{ such that } f(\xi, \mu) + g(\xi, \mu)u_0 = 0\}.$$

A point in E is called an *equilibrium* or *equilibrium point*. Feedbacks are not involved in this definition. If the control input u is substituted by a feedback $u = u(\xi, \mu)$, a *closed-loop equilibrium*, (ξ_0, μ_0), is defined by $f(\xi_0, \mu_0) + g(\xi_0, \mu_0)u(\xi_0, \mu_0) = 0$. The set of all closed-loop equilibria is

$$E_c = \{(\xi, \mu) | f(\xi, \mu) + g(\xi, \mu)u(\xi, \mu) = 0\}.$$

The concept of the equilibrium set plays an important role in this paper. It is known that the closed-loop equilibrium set E_c, in general, is changed if the feedback is varied. However, the set E_c under any state feedback must be a subset of E. Therefore, the set E consists of all possible closed-loop equilibria.

It is assumed throughout this paper that there exists a single uncontrollable mode (denoted by λ) in the linearization. The dimension of the state space is at least two ($n \geq 2$). The dimension of the control input is one. If $\lambda \neq 0$, the sign of λ determines the stabilizability of the uncontrollable dynamics. Therefore, a small variation of μ does not change the stability, i.e. there is no stationary bifurcation at $\mu = 0$. If $\lambda = 0$, the stability of the system depends on the value of the parameter. Several kinds of bifurcations occur in the performance. Thus, we focus on systems with $\lambda = 0$ in the following sections.

3 Normal form and invariants

In this section, Problem 3 is addressed. While we focus on the results from [15], [16], [17], [18], and [22], we would like to point out that some concepts from these papers were generalized to a larger family of systems such as discrete-time systems ([2]), systems with multiple uncontrollable modes ([9]) and normal forms of higher degree terms ([21] and [24]).

In the dissertation of Poincaré, normal forms were derived for dynamical systems without a control input. Poincaré's idea is to simplify the linear part of a system first, using a linear change of coordinates. Then, the quadratic terms in the system are simplified, using a quadratic change of coordinates, then cubic terms, and so on. For control systems, we will use a similar idea. However, the normal form is different. Someone may ask: is it necessary to derive a different normal form for control systems, rather than using the Poincaré normal form? The answer is yes. In fact, even for a linear control system $\dot{\xi} = A\xi + Bu$, the controller normal form is more useful than the diagonal form of A in the feedback design. For a nonlinear affine control system $\dot{\xi} = f(\xi) + g(\xi)u$, it has two vector fields $f(\xi)$ and $g(\xi)$. Therefore, the normal form of a control system requires the simplification of both f and g simultaneously. The simplification of f does not necessarily result in a simple form for g. Furthermore, the transformation group of control systems consists of changes of coordinates and feedbacks. This is different from the Poincaré normal form of dynamical systems where feedbacks are not considered.

3.1 Normal form

Consider system (1). If the system is linearly controllable at the origin, the system is linearly controllable for all equilibrium points in E around $(\xi, \mu) = (0, 0)$. Therefore, Problem 1 in §2 is interesting if (1) is not linearly controllable at the origin.

Assumption 1: System (1) has one uncontrollable mode, $\lambda = 0$, i.e.

$$rank(\begin{bmatrix} B & AB & A^2B & \cdots & A^{n-1}B \end{bmatrix}) = n - 1, \lambda = 0. \tag{2}$$

The first step of finding normal forms is to simplify the linear part in its Taylor expansion. A linear transformation is the following linear change of coordinates and linear feedback

$$\begin{aligned} \bar{\xi} &= T_0\mu + T_1\xi \\ \bar{u} &= \alpha_0\mu + \alpha_1\xi + \beta u \end{aligned} \tag{3}$$

where T_1 is invertible and $\beta \neq 0$. Once the linear part of the system is transformed into its normal form, a quadratic transformation in the form of (4) is used to simplify the quadratic part

$$\begin{aligned} \bar{\xi} &= \xi + \phi^{[2]}(\mu, \xi) \\ \bar{u} &= u + \alpha^{[2]}(\mu, \xi) + \beta^{[1]}(\mu, \xi)u. \end{aligned} \tag{4}$$

The superscripts of $\phi^{[2]}(\mu, \xi)$, $\alpha^{[2]}(\mu, \xi)$, and $\beta^{[1]}(\mu, \xi)$ denote that the functions are homogeneous polynomials of second and first degree in (μ, ξ). Similar superscript is also applied to other vector fields and functions. In (4), the new and old variables agree at the linear level. This is important because such a transformation leaves the linear part of a system unchanged. So, it does not change the linear normal form while simplifying the quadratic part. The linear and quadratic normal forms are summarized in Theorem 1. In the normal form, the new variable has two components $\bar{\xi} = \begin{bmatrix} z & x \end{bmatrix}^T$.

Theorem 1. *([18]) Consider system (1) satisfying Assumption 1. Then, there exists a linear and quadratic transformation that transforms the system into one of the following normal forms.*

(1) Double-zero uncontrollable mode.

$$\begin{aligned} \dot{z} &= \mu + \sum_{i=1}^{n-1} \gamma_{x_i x_i} x_i^2 + \gamma_{z x_1} z x_1 + \gamma_{x_1 \mu} x_1 \mu + \gamma_{zz} z^2 + O(z, x, \mu, u)^3 \\ \dot{x} &= A_2 x + B_2 u + \tilde{f}^{[2]}(x) + O(z, x, \mu, u)^3. \end{aligned} \tag{5}$$

(ii) Simple-zero uncontrollable modes.

$$\begin{aligned} \dot{z} &= \sum_{i=1}^{n-1} \gamma_{x_i x_i} x_i^2 + \\ &\quad + \gamma_{\mu\mu}\mu^2 + \gamma_{zx_1} z x_1 + \gamma_{x_1\mu} x_1 \mu + \gamma_{z\mu} z\mu + \gamma_{zz} z^2 + O(z, x, \mu, u)^3 \\ \dot{x} &= A_2 x + B_2 u + \tilde{f}^{[2]}(x) + O(z, x, \mu, u)^3. \end{aligned} \tag{6}$$

In (5) and (6), $\tilde{f}^{[2]}(x)$ is from [15], and (A_2, B_2) is in Brunovsky form.

$$\begin{aligned} \tilde{f}_i^{[2]}(x) &= \sum_{j=i+2}^{n-1} a_{ij} x_j^2, & 1 \le i \le n-3, \\ \tilde{f}_i^{[2]}(x) &= 0, & i = n-2, n-1. \end{aligned}$$

$$A_2 = \begin{bmatrix} 0\,1\,0\,0 \cdots 0 \\ 0\,0\,1\,0 \cdots 0 \\ \cdot\;\cdot\;\cdot\;\cdot\;\;\cdot\;\;\cdot \\ 0\,0\,0\,0 \cdots 1 \\ 0\,0\,0\,0 \cdots 0 \end{bmatrix}_{(n-1)\times(n-1)} \qquad B_2 = \begin{bmatrix} 0 \\ 0 \\ \vdots \\ 0 \\ 1 \end{bmatrix}.$$

3.2 Invariants

In the matrix theory, the normal form of a matrix is the Jordan canonical form. Its invariants are eigenvalues. The eigenvalues are invariant under matrix transformations. In dynamical systems, eigenvalues are used to characterize the qualitative properties (properties that are invariant under change of coordinates) such as stability. As for nonlinear systems without control, the linear invariants are eigenvalues and the nonlinear invariants are the homogeneous resonant terms in the Poincaré normal form. It was proved by Poincaré that the resonant terms in his nonlinear normal form are invariant under homogenous changes of coordinates.

We know that qualitative properties of control systems, such as the topology of the equilibrium set and the controllability, are invariant under a change of coordinates and feedback. It is expected to characterize these properties by invariant numbers of control systems. In the linearization of a control system, the uncontrollable mode and the controllability index are invariant. What are the invariants in the nonlinear terms of a control system? The following result from [18] answers this question. For the quadratic invariants, we assume that a system is already transformed into its linear normal form (7) or (8).

$$\begin{aligned} \dot{z} &= \mu + f_1(z, x, \mu) + g_1(z, x, \mu)u \\ \dot{x} &= A_2 x + B_2 u + f_2(z, x, \mu) + g_2(z, x, \mu)u, \end{aligned} \tag{7}$$

$$\begin{aligned} \dot{z} &= f_1(z, x, \mu) + g_1(z, x, \mu)u \\ \dot{x} &= A_2 x + B_2 u + f_2(z, x, \mu) + g_2(z, x, \mu)u. \end{aligned} \tag{8}$$

The nonlinear terms f_i and g_i, for $i = 1, 2$, are not necessarily in the quadratic normal form. If a quadratic transformation is applied to (7) or (8), the nonlinear vector fields are changed. However, the following numbers obtained from the nonlinear system are left invariant.

Invariants	In Normal Form	Remark
$\frac{1}{2} C_x A_e^{t-1} [ad_{f_e}^r(g_e), ad_{f_e}^{r-1}(g_e)]\|_{(0,0,0)}$	$a_{t\ n-r}$	$1 \leq r \leq n-3$ $1 \leq t \leq n-r-2$
$\frac{1}{2} C_z [ad_{f_e}^r(g_e), ad_{f_e}^{r-1}(g_e)]_{(0,0,0)}$	$\gamma_{x_{n-r} x_{n-r}}$	$1 \leq r \leq n-1$
$(-1)^{n-1} C_z [X_\mu, ad_{f_e}^{n-1}(g_e)]\|_{(0,0,0)}$	$\gamma_{\mu x_1}$	
$(-1)^{n-1} C_z [X_z, ad_{f_e}^{n-1}(g_e)]\|_{(0,0,0)}$	$\gamma_{z x_1}$	
$\frac{1}{2} C_z ad_{X_\mu}^2 (f)\|_{(0,0,0)}$	$\gamma_{\mu\mu}$	(8)
$C_z ad_{X_z} ad_{X_\mu} (f)\|_{(0,0,0)}$	$\gamma_{\mu z}$	(8)
$\frac{1}{2} C_z ad_{X_z}^2 (f)\|_{(0,0,0)}$	γ_{zz}	(7) and (8)

(9)

In (9), f_e (or g_e) is the extended vector field including f (or g) in (7) or (8) and the parameter equation $\dot{\mu} = 0$. The notations C_z, C_x, X_z and X_μ are the following row and column vectors in $\Re^{n+1}$,

$$\begin{aligned} C_z &= \begin{bmatrix} 1 \ 0 \ 0 \cdots 0 \end{bmatrix}, \quad C_x = \begin{bmatrix} 0 \ 1 \ 0 \cdots 0 \end{bmatrix}, \\ X_z &= \begin{bmatrix} 1 \ 0 \ 0 \cdots 0 \end{bmatrix}^T, X_\mu = \begin{bmatrix} 0 \ 0 \cdots 0 \ 1 \end{bmatrix}^T. \end{aligned} \tag{10}$$

The following theorem implies that the quadratic invariants characterize the quadratic part of a system.

Theorem 2. *([18]) Given a system satisfying Assumption 1. Suppose its linearization is (7) or (8).*

(i) Quadratic transformations defined by (4) do not change the value of the quadratic invariants.

(ii) The quadratic invariants of normal form (5) or (6) are the corresponding coefficients of the quadratic terms listed in (9).

(iii) Two systems of the form (7) (or (8)) are equivalent under quadratic transformations if and only if they have the same quadratic invariants.

Cubic invariants and cubic normal forms were derived in several publications such as [21] and [24]. However, we do not review these results here. In the following, a set of special invariants are identified, which are not only easy to compute but also critical for the bifurcation analysis performed in the sections that follow.

3.3 Resonant terms

The Poincaré normal form consists of resonant terms. The coefficients of resonant terms are left invariant under homogeneous transformations. For control systems, there is a set of homogeneous polynomials that are invariant under homogeneous transformations. The coefficients of resonant terms form a subset of the invariants in (9). Furthermore, the resonant terms are critical to the feedback control of bifurcations.

Given a system in the form of (7) or (8). The homogeneous parts of degree d in f_i and g_i of (7) and (8) are denoted by $f_i^{[d]}$ and $g_i^{[d]}$. A homogeneous transformation of degree $[d]$ is defined by a change of coordinates and state feedback in the following form

$$\begin{array}{ll} z = \bar{z} + \phi_1^{[d]}(\bar{z}, \bar{x}, \mu), & x = \bar{x} + \phi_2^{[d]}(\bar{z}, \bar{x}, \mu) \\ u = \bar{u} + \alpha^{[d]}(\bar{z}, \bar{x}, \mu) + \beta^{[d-1]}(\bar{z}, \bar{x}, \mu)\bar{u} & \end{array} \tag{11}$$

where $\bar{z}$ and $\bar{x}$ are the new coordinates, $\bar{u}$ is the new control input introduced by the regular feedback.

Definition 1. Consider (7) or (8). A homogeneous term in $f_i^{[d]}(z, x, \mu)$ or $g_i^{[d-1]}(z, x, \mu)u$ is called a resonant term if transformations defined by (11) leave the coefficient of the term invariant.

For instance, if (11) is applied to (7), it can be proved that the term $\bar{z}^2$ in $\bar{f}_1^{[2]}$ of the resulting system has the same coefficient as the term z^2 in $f_1^{[2]}$ of (7). So, z^2 in $f_1^{[2]}$ is resonant. Notice that the resonant terms of control systems are different from the resonant terms in the classical theory of dynamical systems. A resonant term in Definition 1 is invariant under both changes of coordinates and state feedback. This is different from the theory of classical dynamic systems that does not deal with feedback. Define

$$\begin{array}{l} R^{[d]}(\mu, z, x_1) = f_1^{[d]}(\mu, z, x)|_{x_2 = x_3 = \cdots = x_{n-1} = 0}, \\ R_1^{[d]}(z, x_1) = R(z, x_1, 0) \end{array} \tag{12}$$

where $f_1^{[d]}(\mu, z, x)$ is the homogeneous vector field of degree d from the Taylor expansion of $f_1(\mu, z, x)$ in (7) and (8).

Theorem 3. *([18]) In (7), all terms of $R_1^{[d]}(z, x_1)$ are resonant. In (8), all terms of $R^{[d]}(\mu, z, x_1)$ are resonant.*

The coefficients of quadratic resonant terms are part of the invariants in (9). For example, the coefficient of z^2 in $R^{[2]}$ equals the invariant γ_{zz} in (9). The quadratic functions of resonant terms $R^{[2]}(\mu, z, x_1)$ and $R_1^{[2]}(z, x_1)$ are also denoted by $Q(\mu, z, x_1)$ and $Q_1(z, x_1)$, i.e. $Q(\mu, z, x_1) = R^{[2]}(\mu, z, x_1)$ and $Q_1(z, x_1) = R_1^{[2]}(z, x_1)$. The coefficients in $R^{[2]}$ and $R_1^{[2]}$ form two matrices,

$$Q = \begin{bmatrix} \gamma_{\mu\mu} & \frac{\gamma_{\mu z}}{2} & \frac{\gamma_{\mu x_1}}{2} \\ \frac{\gamma_{\mu z}}{2} & \gamma_{zz} & \frac{\gamma_{z x_1}}{2} \\ \frac{\gamma_{\mu x_1}}{2} & \frac{\gamma_{z x_1}}{2} & \gamma_{x_1 x_1} \end{bmatrix}, \quad Q_1 = \begin{bmatrix} \gamma_{zz} & \frac{\gamma_{z x_1}}{2} \\ \frac{\gamma_{z x_1}}{2} & \gamma_{x_1 x_1} \end{bmatrix}. \tag{13}$$

The matrices satisfy

$$Q(\mu, z, x_1) = \left[\mu\ z\ x_1\right] Q \left[\mu\ z\ x_1\right]^T, Q_1(z, x_1) = \left[z\ x_1\right] Q_1 \left[z\ x_1\right]^T. \quad (14)$$

4 Bifurcation of control systems

In §3, normal forms of systems satisfying Assumption 1 are derived. The bifurcation of the normal form represents the bifurcation of all systems satisfying Assumption 1. In this section, we study the bifurcation of the equilibrium set E and the bifurcation of the controllability. Later in §5, we will prove that these bifurcations correspond to classical stationary bifurcations of closed-loop systems. The relationship between the bifurcation of control systems and the classical stationary bifurcations will also be discussed in §5.

4.1 Systems with a double-zero uncontrollable mode

Consider (7). It can be transformed into the normal form (5) by a suitable quadratic transformation. In [18], we focus on the resonant terms in the normal form (5) rather than studying arbitrary nonlinear terms in (7). It significantly simplifies the analysis.

Theorem 4. *([18]) Let a system in the form of (7) be given.*
(i) Its equilibrium set satisfies

$$\begin{aligned} \mu &= -Q_1(z, x_1) \\ &= -\gamma_{x_1x_1}x_1^2 - \gamma_{zx_1}x_1 z - \gamma_{zz}z^2 + O(z, x_1)^3 \\ x_i &= O(z, x_1)^2 \qquad\qquad 2 \le i \le n-1. \end{aligned} \quad (15)$$

(ii) There exists a function $c(z, x_1)$ in the following form

$$c(z, x_1) = \gamma_{zx_1}z + 2\gamma_{x_1x_1}x_1 + O(z, x_1)^2 \quad (16)$$

such that the system is linearly controllable at a point $(z, x, \mu) \in E$ if and only if $c(z, x_1) \neq 0$.

Equation (15) is a quadratic approximation of the equilibrium set E (projected to $\mu z x_1$-space). If Q_1 is sign definite (or equivalently $det(Q_1) > 0$), E is approximately a paraboloid. If Q_1 is not sign definite but it has full rank (or equivalently $det(Q_1) < 0$), then E is approximately a saddle. If we define

$$E_{\mu_0} = \{(x, \mu) \in E | \mu = \mu_0\},$$

the topology of E_μ changes as μ passes through zero. If E is approximately a paraboloid, E_μ is empty for the values of μ on one side of zero and it is a closed curve if μ is on the other side of $\mu = 0$. If E is approximately a saddle, then E_0 is approximately a set with two lines which meet at the origin. It is a connected set. However, E_μ is approximately a hyperbola for $\mu \neq 0$, which is not a connected set. Examples of a paraboloid and saddle are shown in Figure 1. The system is uncontrollable at the intersection of $c(z, x_1) = 0$ and E. The uncontrollable equilibrium points are shown in the figure by the curve on the surface E.

4.2 Systems with simple-zero uncontrollable modes

For system (8), the set E is approximated by the resonant terms $Q(\mu, z, x_1)$ as proved in the following theorem.

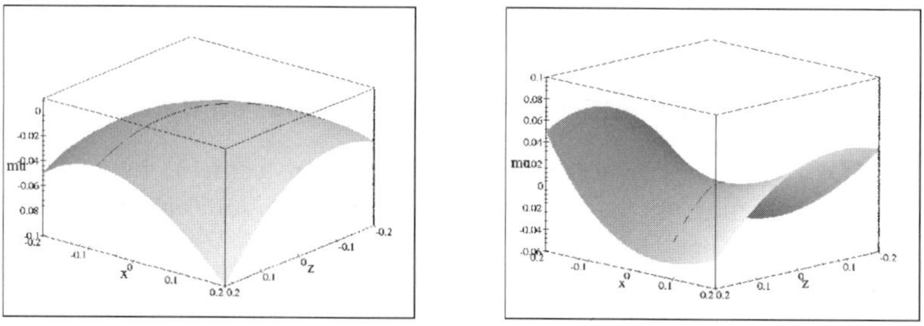

Fig. 1. Equilibrium set

Theorem 5. *([18]) Given a system (8).*

(i) If Q is sign definite, the origin is an isolated equilibrium point, i.e. $E = \{(0,0,0)\}$.

(ii) If Q is indefinite and if Q has full rank, its equilibrium set satisfies

$$\begin{aligned} &\left[\mu \; z \; x_1\right] Q \left[\mu \; z \; x_1\right] + O(\mu, z, x_1)^3 = 0. \\ &x_i = O(\mu, z, x_1)^2 \qquad\qquad 2 \leq i \leq n-1. \end{aligned} \tag{17}$$

(iii) In the case of (ii), there exists a function $c(\mu, z, x_1)$ in the following form

$$c(\mu, z, x_1) = 2 \left[0\;0\;1\right] Q \left[\mu \; z \; x_1\right] + O(\mu, z, x_1)^2 \tag{18}$$

such that the system is linearly controllable at a point $(\mu, z, x) \in E$ if and only if $c(\mu, z, x_1) \neq 0$.

In (i), the system has an equilibrium point only if $\mu = 0$. When $\mu \neq 0$, the system has no equilibrium point in a neighborhood of the origin. In (ii) and (iii), the equilibrium set E is approximately a cone. The uncontrollable equilibrium points are approximately the intersection of the cone with a plane.

The intersection has two generic cases, it may consist of two lines (as shown in Figure 2), or it consists of a single point that is the origin.

To summarize, the topology of the equilibrium set E and the set of uncontrollable points are completely determined by the resonant terms Q or Q_1 of the system. The relationship between Q, Q_1 and E is summarized in the following table.

Table 1. The classification of equilibrium sets

Uncontrollable Mode	Condition	equilibrium set
Double-zero	$det(Q_1) > 0$	paraboloid
	$det(Q_1) < 0$	saddle
Simple-zero	Q is sign definite	single point
	Q is indefinite, $det(Q) \neq 0$	cone

5 Bifurcation control using state feedback

It has been observed in engineering and scientific applications that a feedback is able to change not only the stability of a bifurcation but also its type. The same system may exhibit more than one type of bifurcations, depending on the selection of the feedback. In [19], all possible bifurcations with quadratic and cubic degeneracies that can be generated by (7) and (8) are found. It is proved that the bifurcations are completely determined by the feedback and the resonant terms. Actually, the bifurcation and its stability is closely related to the equilibrium set E, which is summarized in §4.

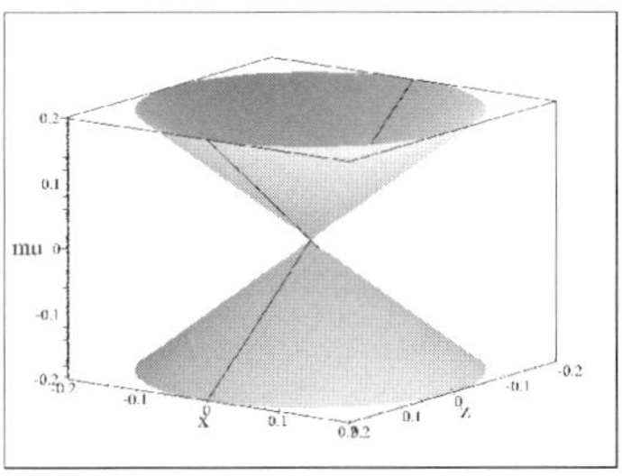

Fig. 2. Equilibrium set

In this section, we use system (8) and its normal form (6) to illustrate the relationship between the feedback and the bifurcation of the closed-loop system. The results for both (7) and (8) are summarized in Table 5.2. A state feedback used for bifurcation control is defined by

$$u = a_\mu \mu + a_z z + \sum_{i=1}^{n-1} a_i x_i + \alpha^{[2]}(\mu, z, x) + O(z, x, \mu)^3. \tag{19}$$

To guarantee that the system is operated around the origin, we assume,
Assumption 2: State feedback (19) places the controllable poles in the left half plane, i.e. the eigenvalues of the matrix $A_2 + B_2 \begin{bmatrix} a_1 & a_2 & \cdots & a_{n-1} \end{bmatrix}$ are all in the left half plane.

5.1 Bifurcations with quadratic degeneracy

For a closed-loop system under feedback (19), the input u is replaced by (19). For the closed-loop system, the equilibrium set is denoted by E_c, which consists of all (μ, z, x) so that the vector field with feedback equals zero. A definition of E_c is introduced in §2. Given (8), it is easy to check that E_c, the closed-loop equilibrium set, is the intersection of E and the set $u(\mu, z, x) = 0$. In a neighborhood of the origin, the intersection is approximately the intersection of the plane

$$a_\mu \mu + a_z z + \sum_{i=1}^{n-1} a_i x_i = 0 \tag{20}$$

with E. We know that E is either a single point or a cone. So, the intersection between (20) and E is either a single point or two lines passing through the origin. The later case obviously implies a transcritical bifurcation. It is proved to be true and the result is summarized in the following theorem. The following 2×2 matrix is critical, which is determined by the invariants Q and the linear coefficients of the feedback,

$$\bar{Q} = \begin{bmatrix} a_1 & 0 & -a_\mu \\ 0 & a_1 & -a_z \end{bmatrix} Q \begin{bmatrix} a_1 & 0 \\ 0 & a_1 \\ -a_\mu & -a_z \end{bmatrix}. \tag{21}$$

Theorem 6. *Consider a closed-loop system (8)-(19) satisfying Assumption 2. Suppose*

$$Q_1(a_1, -a_z) \neq 0. \tag{22}$$

(i) If $\bar{Q}$ is sign definite, then $(z, x, \mu) = (0, 0, 0)$ is an isolated equilibrium point of the closed-loop system. It is unstable.

(ii) If $\bar{Q}$ is indefinite with full rank, then the closed-loop system has a transcritical bifurcation around the origin.

(iii) Assume that the feedback satisfies the condition in (ii). Given any $(z, x, \mu) \in E_c$ in a neighborhood of the origin, it is locally asymptotically stable if

$$\left[0 \; a_1 \; -a_z\right] Q \left[\mu \; z \; x_1\right]^T > 0 \tag{23}$$

The system is unstable if

$$\left[0 \; a_1 \; -a_z\right] Q \left[\mu \; z \; x_1\right]^T < 0. \tag{24}$$

An example of engine compressors is given in [19]. The compressor model is based on the 3D Moore-Greitzer model. A transcritical bifurcation occurs in this model. To achieve the stability on a given branch of the equilibrium, the controllers are derived on the basis of Theorem 6.

For system (7), the set E is either a paraboloid or a saddle. The intersection between (20) and E is a parabola. Therefore, it indicates a saddle node bifurcation. This is proved to be true (Table 2). The stability of an equilibrium point is determined by the sign of $Q_1(z, x_1)z$. Details can be found in [19].

5.2 Bifurcations with cubic degeneracy

On the center manifold of (8)-(19), condition (22) implies that the quadratic terms of the reduced system dominate the local performance. However, if $Q_1(a_1, -a_z) = 0$, the performance of the reduced system on the center manifold is affected by cubic terms in the system. It is proved in [19] that the pitchfork bifurcation occurs and the stability is closely related to the quadratic feedback as well as cubic resonant terms in $R^{[3]}(\mu, z, x_1)$. In fact, only the cubic resonant terms of z and x_1 are important. So, we define

$$C(z, x_1) = R_1^{[3]}(0, z, x_1) = f_1^{[3]}(\mu, z, x)|_{x_2 = x_3 = \cdots = x_{n-1} = \mu = 0}, \tag{25}$$

where $f_1^{[3]}$ is the cubic part in the uncontrollable equation of (6). The coefficients in the quadratic feedback $\alpha^{[2]}(\mu, z, x)$ are denoted by a_{zz}, $a_{z\mu}$, $a_{x_1x_1}$, $a_{x_1\mu}$, etc. The following quadratic function from $\alpha^{[2]}$ in (19) is useful,

$$\alpha_{zx_1}^{[2]}(z, x_1) = a_{zz}z^2 + a_{zx_1}zx_1 + a_{x_1x_1}x_1^2, \tag{26}$$

i.e. $\alpha_{zx_1}^{[2]}$ is the restriction of $\alpha^{[2]}$ to the zx_1-plane. The following number, determined by invariants and feedback, is critical for bifurcation control,

$$D = a_1 C(a_1, -a_z) + (2a_z\gamma_{x_1x_1} - a_1\gamma_{zx_1})\alpha_{zx_1}^{[2]}(a_1, -a_z) \tag{27}$$

where $\gamma_{x_1x_1}$ and γ_{zx_1} are quadratic resonant coefficients in (6).

Theorem 7. *Consider a closed-loop system (6)-(19) satisfying Assumption 2. Suppose*

$$Q_1(a_1, -a_z) = 0. \tag{28}$$

(i) The closed-loop system has a pitchfork bifurcation at the origin if

$$\begin{aligned} &D \neq 0, \\ &\left[0 \; a_1 \; -a_z\right] Q \left[a_1 \; 0 \; -a_\mu\right]^T \neq 0. \end{aligned} \tag{29}$$

(ii) The pitchfork bifurcation is supercritical if $D < 0$, and subcritical if $D > 0$.

Suppose that a control system has a transcritical bifurcation. An interesting conclusion of Theorem 7 is that a feedback is capable of changing the transcritical bifurcation into a pitchfork bifurcation, a different bifurcation type. For the normal form (5), the change is even more dramatic. The condition $Q_1(a_1, -a_z) \neq 0$ implies a saddle node bifurcation. On one side of $\mu = 0$, the system has no equilibrium point, and therefore cannot be stabilized. However, it is proved in [19] that a feedback satisfying $Q_1(a_1, -a_z) = 0$ removes the saddle node bifurcation. The resulting system has no bifurcation if $D \neq 0$. Locally the system has a unique equilibrium point for every value of μ in a neighborhood of $\mu = 0$. Furthermore, if $D < 0$, all the equilibrium points around the origin are stable. Notice that a linear controller cannot remove the bifurcation because the system is not linearly controllable. The stabilization is done by the quadratic part of the feedback. The relationship between the feedback, the resonant terms and the bifurcation is summarized in the following table.

Table 2. The bifurcations under state feedback

System	Condition		Bifurcation	Stability
Simple-zero Uncontrollable Mode	$Q_1(a_1, -a_z) \neq 0$	$\det(\tilde{Q}) > 0$	The origin is an isolated equilibrium point	The system is unstable
		$\det(\tilde{Q}) < 0$	Transcritical bifurcation	The system is stable at (z, x, μ) if $\tilde{Q}(\mu, z, x_1) > 0$
				The system is unstable at (z, x, μ) if $\tilde{Q}(\mu, z, x_1) < 0$
	$Q_1(a_1, -a_z) = 0$	$D \neq 0$ $\tilde{Q}(a_1, 0, -a_\mu) \neq 0$	Pitchfork bifurcation	Supercritical if $D < 0$
				Subcritical if $D > 0$
Double-zero Uncontrollable Mode	$Q_1(a_1, -a_z) \neq 0$		Saddle-node bifurcation	The system is stable at (z, x, μ) if $Q_1(z, x_1)z < 0$
				The system is unstable at (z, x, μ) if $Q_1(z, x_1)z > 0$
	$Q_1(a_1, -a_z) = 0$	$D \neq 0$	No bifurcation, there exists a unique equilibrium point for every fixed value of μ	All equilibrium points are stable if $D < 0$
				All equilibrium points are unstable if $D > 0$

In the table, Q_1 is defined by (14), $\bar{Q}$ is defined by (21), D is defined by (27), and $\tilde{Q}$ is a function with three variables defined by

$$\tilde{Q}(a,b,c) = \left[0 \; a_1 \; -a_z\right] Q \left[a \; b \; c\right]^T.$$

6 The cusp bifurcation and hysteresis

From Table 2, system (5) has no bifurcation if $Q_1(a_1, -a_z) = 0$ and $D \neq 0$. However, this statement is based on the assumption that μ is the only parameter in the system. If the system has other parameters, it may exhibit rich bifurcation phenomena. For example, it was proved in [19] that the bifurcation has qualitative change if the feedback is perturbed by a constant. In other words, if the feedback is not zero at the origin, it might result in a fundamental change in the bifurcations of the system. In this section, we use (5) as an example to illustrate the phenomenon.

Given a system in the form of (5), suppose that

$$Q_1(a_1, -a_z) = 0, \; D \neq 0. \tag{30}$$

Suppose that the feedback equals ν at the origin, i.e.

$$u = \nu + a_\mu \mu + a_z z + \sum_{i=1}^{n-1} a_i x_i + \alpha^{[2]}(\mu, z, x) + O(z, x, \mu)^3. \tag{31}$$

It was proved in [19] that the center manifold of (5)-(31) and the reduced system satisfy

$$\begin{aligned} x_1 &= -\frac{a_z}{a_1} z - \frac{a_1 a_\mu - a_2 a_z}{a_1^2} \mu - \frac{1}{a_1} \nu + O(z,\mu)^2, \\ x_2 &= -\frac{a_z}{a_1} \mu + O(z,\mu)^2, \\ x_i &= O(z,\mu)^2 \qquad for \quad i \geq 3, \end{aligned} \tag{32}$$

$$\dot{z} = f_c(z, \mu, \nu) = \nu_1 + \nu_2 z + D_2(\mu, \nu) z^2 + D z^3 + \cdots, \tag{33}$$

where D is defined by (27), and

$$\begin{aligned} \nu_1 &= \mu + O(\mu,\nu)^2, \\ \nu_2 &= \left(\frac{2\gamma_{x_1 x_1} a_z - \gamma_{z x_1} a_1}{a_1^2} (a_\mu - \frac{a_2 a_z}{a_1}) - \frac{\gamma_{x_1 \mu} a_z}{a_1} \right) \mu + \\ &\quad + \frac{2 a_z \gamma_{x_1 x_1} - a_1 \gamma_{z x_1}}{a_1^2} \nu + \cdots, \end{aligned} \tag{34}$$

On the center manifold, the reduced system is (33). If ν_1 and ν_2 in (33) are free variables, the system exhibits a cusp bifurcation ([13]). The system has cubic degeneracy. The set of degenerate equilibrium points is a cusp. The figure of f_c for fixed values of (ν_1, ν_2) and the stability at the equilibrium are illustrated in the following figure.

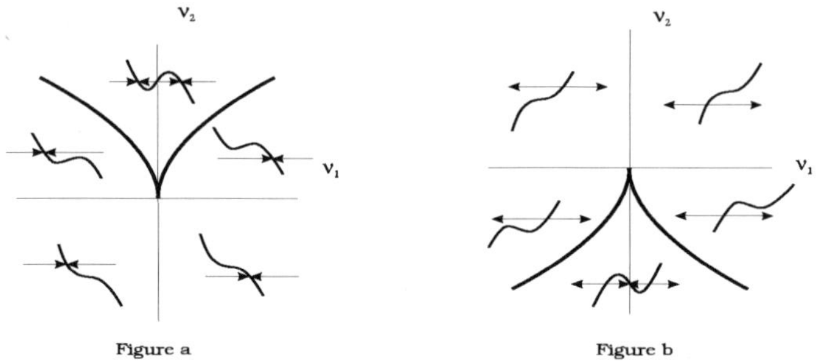

Fig. 3. A cusp and phase portrait of f_c. $D_3(0) < 0$ in (a). $D_3(0) > 0$ in (b).

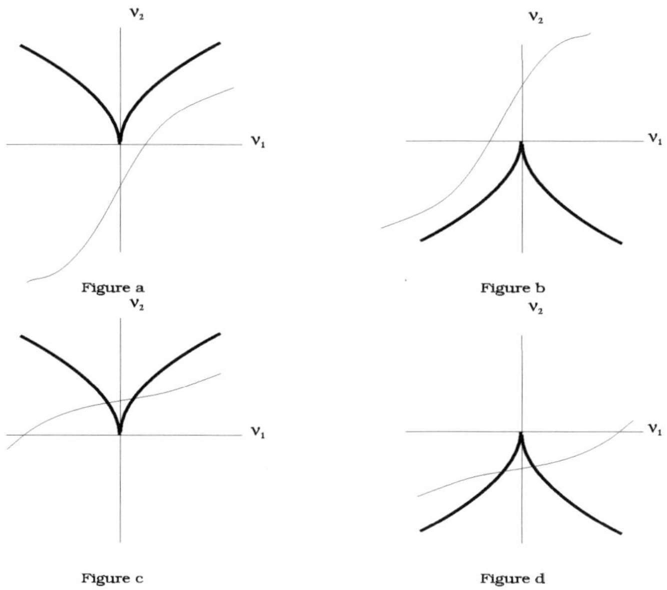

Fig. 4.

From (34), ν_i, $i = 1, 2$, depends on the value of (ν, μ). For any fixed value of ν, (34) represents a curve in the $\nu_1\nu_2$-plane. The bifurcation of (5)-(31) is determined by this curve and the cusp in Figure 3. If the curve has no intersection with the cusp, the cusp bifurcation diagram implies that there is no bifurcation around the origin (Figure 4a-b). If the curve meets with the cusp (Figure 4c-d), the bifurcation diagram is shown in Figure 5, which is called a hysteresis ([34]). We summarize the results in the following theorem.

Theorem 8. *Consider (5)-(31) satisfying Assumption 2 and (30). Fix any value of ν around $\nu = 0$, (5)-(31) has the following properties.*

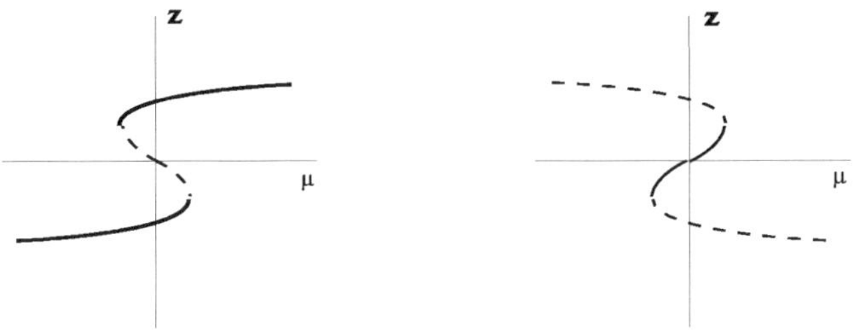

Figure a Figure b

Fig. 5. Hysteresis. $D < 0$ in (a), $D > 0$ in (b).

(i) If $(2a_z\gamma_{x_1x_1} - a_1\gamma_{zx_1})D\nu > 0$, *(5)-(31) has no bifurcation around the origin and* $\mu = 0$. *There exists a neighborhood of* $(z, x, \mu) = (0, 0, 0)$ *in which the system has a unique closed-loop equilibrium point for every value of* μ. *Furthermore, the system is stable if* $D < 0$, *and the system is unstable if* $D > 0$.

(ii) If $(2a_z\gamma_{x_1x_1} - a_1\gamma_{zx_1})D\nu < 0$, *the system exhibits hysteresis (Figure 5). Its stability is determined by the value of* D *as shown in Figure 5.*

In [19], perturbed feedback is applied to (6) also. If E is a cone, it is proved that the transcritical bifurcation splits to two saddle node bifurcations.

7 Other related issues

If a continuous but nonsmooth feedback is applied to the system, it is possible to stabilize a bifurcation that cannot be stabilized by smooth feedbacks. In [25], a transcritical bifurcation is converted to a "birdfoot" bifurcation, which does not exist for smooth systems. The birdfoot bifurcation looks similar to the pitchfork bifurcation. The equilibrium set has a nonsmooth corner at the origin. The system can be made supercritical. It is more stable than a transcritical bifurcation.

In [19], nonsmooth feedback is applied to a system with saddle-node bifurcation. Consider a normal form with double-zero uncontrollable mode. The nonsmooth feedback is

$$a_z|z| + a_1x_1 + \cdots + a_{n-1}x_{n-1} + a_\mu x_\mu + O(z, x, \mu)^2.$$

Suppose a_z satisfies

$$Q_1(a_1, -a_z) < 0,\ Q_1(a_1, a_z) > 0.$$

Then the closed-loop system has a unique equilibrium point for every value of μ around the origin. The equilibrium point is stable if $\mu \neq 0$. So, the system is stable for both $\mu > 0$ and $\mu < 0$. It compares sharply against the saddle-node bifurcation, where only one side of $\mu = 0$ has equilibrium points.

In [20], the Hopf bifurcation is studied. The paper assumes that a control system has two uncontrollable modes at $\mu = 0$, both on the imaginary axis and conjugate to each other. Since the imaginary eigenvalues are uncontrollable, the Hopf bifurcation cannot be removed by pole placement. For a nonzero parameter μ, the system may exhibit periodic solutions. Since the periodic solution lies on the center manifold, the orientation of the center manifold at the origin indicates the orientation of the periodic solution around the origin. It is proved in [20] that the orientation of the center manifold at the origin is determined by the linear feedback. Necessary and sufficient conditions are derived for all orientations that can be achieved by feedbacks. The stability of the periodic solution in the Hopf bifurcation is determined by its quadratic and cubic invariants, and the quadratic feedback. Quadratic normal forms are derived in [20] for systems with two imaginary uncontrollable modes. Normal forms of higher degrees are derived in [24]. In [20], a family of feedbacks is derived to stabilize the periodic solution of the Hopf bifurcation. Since the stability depends on the quadratic center manifold, the control of a quadratic center manifold by nonlinear feedback is addressed in [20]. An explicit formula that bridges the quadratic center manifold and the feedback is derived. Similar work about center manifold control was first studied in [9].

All results introduced so far assume that the system has a parameter μ. However, a control system without a parameter could have a bifurcation too. This is different from the classical bifurcation theory. In [17], it was proved that the topology of the equilibrium set is changed even without a parameter. The control input u changes the equilibrium set. Furthermore, *control bifurcation* was introduce in [24]. In this work, a family of smooth feedbacks on the equilibrium set is applied to the system. As the equilibrium point is changed, the stability of the feedback changes, thus the control exhibits a bifurcation. There is no explicit parameter in the system. However, the change of input value or the change of equilibrium points results in the qualitative change of the performance of the given feedback.

Normal forms and bifurcation control of discrete-time systems have been studied in several recent publications (see, for instance, [2] and [14]). In addition to bifurcations, normal forms and invariants have other applications. For example, in [32] and references therein, the authors addressed the problems of control system symmetry, and feedforward forms based on invariants and normal forms.

8 Conclusions

We believe that the normal form and its invariants represent the essential nonlinearity in a control system. Therefore, the improvement in understanding normal forms and invariants will lead to a better understanding of nonlinear performance of control systems. We know that bifurcations are purely nonlinear performance because the linear part of a system at a bifurcation point does not dominate even the local performance due to the imaginary or zero eigenvalues. Therefore, the stability of bifurcations has to be closely related to its core nonlinear terms in the system. The results in this paper prove that invariants play a key role in the characterization of the qualitative performance. The proofs of these results are made possible by the normal form.

The research on normal forms, invariants, bifurcations and their applications has been fruitful for the last few years. However, the research on this topic is still far from mature. Normal forms of multiple uncontrollable modes are one of our next research target. Recent work on this issue is reported in [24]. Normal forms of higher degree terms are also under investigation. The convergence of normal forms is still largely open. While the distribution of eigenvalues of a system without control determines the convergence of the Poincaré normal form, the eigenvalues in a controllable system can be changed by feedback. So, what is the key that determines the convergence of a control system normal form is a tough challenge that we face. Bifurcations and their control is still a young field that has a great potential in the future. In this review, the bifurcation of the topology of the equilibrium set is thoroughly studied. However, there are certainly many other qualitative properties that are important to control systems, such as stabilizability, observability, and zero dynamics. The bifurcations of these qualitative properties are interesting problems for future research. Bifurcation and chaos of discrete control systems is another interesting area for future research.

References

1. E. H. Abed and J.-H. Fu, "Local feedback stabilization and bifurcation control,I-II. Stationary bifurcation", *Systems & Control Letters*, 8, 1987, 467-473.
2. J.P. Barbot, S. Monaco and D. Normand-Cyrot, "Quadratic forms and approximative feedback linearization in discrete time, *International Journal of Control*, 67, pp. 567-586, 1997.
3. S. Behtash and S. Sastry, "Stabilization of nonlinear systems with uncontrollable linearization", *IEEE Trans. Automat. Contr.*, vol. AC-33, pp. 585-591.
4. J. Carr, Applications of Center Manifold Theory, Springer-Verlag, New York, 1981.
5. G. Chen, "Control and synchronization of chaotic systems (a bibliography)", ECE Dept., Univ. of Houston, TX - available from ftp: "ftp.egr.uh.edu/pub/TeX/chaos.tex".
6. G. Chen and J. L. Moiola, "An overview of bifurcation, chaos and nonlinear dynamics in control systems", *J. of Franklin Instit.* 331B, 1994, 819-858.

7. F. Colonius, and W. Kliemann (1995). Controllability and stabilization of one-dimensional systems near bifurcation points. *Systems & Control Letters*, **24**, 87-95.
8. K. M. Eveker, C. N. Nett, "Control of compression system surge and rotating stall: a laboratory-based 'hands-on' introduction", Proceedings of the American Control Conference, Baltimore, MD, 1994, 1307-1311.
9. O. E. Fitch, *The control of bifurcations with engineering applications*, Ph.D. Dissertation, Naval Postgraduate School, Monterey, CA 93943, 1997.
10. R. Genesio, A. Tesi, H. O. Wang and E. H. Abed, *Control of period doubling bifurcations using harmonic balance*, Proc. IEEE Conf. Decision and Control, San Antonio, Texas, 1993, 492-497.
11. P. Glendinning, Stability, Instability and Chaos: an introduction to the theory of nonlinear differential equations, Cambridge University Press, 1994.
12. G. Gu, X. Chen, A. Sparks and S. Banda, "Bifurcation stabilization with local output feedback," *SIAM J. Contr. Optimiz.*, Vol.37, 934-956, 1999.
13. J. Hale and H. Koçak, Dynamics and Bifurcations, Springer-Verlag, New York, 1991.
14. B. Hamzi, J.P. Barbot, S. Monaco, and D. Normand-Cyrot, "Normal forms versus Naimark-Sacker bifurcation control, NOLCOS'01, to appear.
15. W. Kang and A. J. Krener, "Extended quadratic controller normal form and dynamic feedback linearization of nonlinear systems", *SIAM J. Control and Optimization*, vol. 30, No. 6, pp 1319-1337, 1992.
16. W. Kang, *Extended controller normal form, invariants and dynamic feedback linearization of nonlinear control systems*, Ph.D. Dissertation, University of California at Davis, 1991.
17. W. Kang, Bifurcation and normal form of nonlinear control systems – part I, *SIAM J. Control and Optimization*, **36**, (1998), 193-212.
18. W. Kang, Bifurcation and normal form of nonlinear control systems – part II, *SIAM J. Control and Optimization*, **36**, (1998), 213-232.
19. W. Kang, Bifurcation control via state feedback for systems with a single uncontrollable mode, *SIAM J. Control and Optimization*, **38**, (2000), 1428-1452.
20. B. Hamzi, W. Kang, and J-P. Barbot, On the Control of Hopf Bifurcations, *Proc. IEEE Conference on Decision and Control*, Sydney, Australia, December 12-15, 2000.
21. W. Kang, Extended controller form and invariants of nonlinear control systems with a single input, *J. of Mathematical Systems, Estimation and Control*, **6** (1996), 27-51.
22. W. Kang, *Bifurcation and topology of equilibrium sets*, IEEE Conf. on Decision and Control, Kobe, Japan, 1996, 2151-2155.
23. W. Kang, F. Papoulias, *Bifurcation and normal forms of dive plane reversal of submersible vehicles*, Proc. International Offshore and Polar Engineering Conference, Honolulu, Hawaii, May 25-30, 1997, 62-69.
24. A. J. Krener, W. Kang and D. E. Chang, *Control Bifurcations*, preprint.
25. A. J. Krener, The Feedbacks which Soften the Primary Bifurcation of MG 3, PRET Working Paper D95-9-11, 1995.
26. M. Krstic, J. M. Protz, J. D. Paduano, P. V. Kokotovic, *Backstepping designs for jet engine stall and surge control*, Proc. IEEE Conf. Decision and Control, New Orleans, LA, 1995, 3049-3055.

27. D.-C. Liaw and E. H. Abed, *Stability analysis and control of rotating stall*, Proc. IFAC Nonlinear Control Systems Design Symposium, Bordeaux, France, June, 1992.
28. F. E. McCaughan, *Bifurcation analysis of axial flow compressor stability*, SIAM J Applied Mathematics, vol. 20, 1990, 1232-1253.
29. F. K. Moore and E. M. Greitzer, *A theory of post-stall transients in axial compression systems -- Part I: Development of equations*, ASME J. Engineering for Gas Turbines and Power, vol. 108, 1986, 68-76.
30. M. A. Pinsky and B. Essary, *Normal forms, averaging and resonance control of flexible structures*, J. of Dynamic Systems, Measurement and Control, v. 116, N3, 1994, pp. 357-366.
31. M. A. Pinsky and I. Shumyatsky, *Feedback stabilization of bifurcation phenomena and its application to the control of voltage instabilities and collapse*, Proc. NOLCOS'95, Lake Tahoe, California, 1995, pp. 58-63.
32. I. A. Tall and W. Respondek, "Normal forms, canonical forms, and invariants of single input nonlinear systems under feedback," Proc. IEEE Conf. on Decision and Control, Sydney, Australia, 2000.
33. H. O. Wang, E. H. Abed, *Bifurcation control of a chaotic system*, Automatica, vol. 31, No. 9, 1995, 1213-1226.
34. S. Wiggins, Introduction to Applied Nonlinear Dynamical Systems and Chaos, Springer-Verlag, 1990.

Bifurcations of Reachable Sets Near an Abnormal Direction and Consequences

Emmanuel Trélat

Université de Bourgogne
Laboratoire de Topologie, UFR Sciences et Techniques
BP47870, 21078 Dijon Cedex, France
e-mail : trelat@topolog.u-bourgogne.fr

Abstract. We describe precisely, under generic conditions, the contact and the bifurcations of the reachable set at time T along an abnormal direction, first for a single-input affine control system with constraint on the control, and then as an application for a sub-Riemannian system of rank 2. As a consequence we obtain in sub-Riemannian geometry a new splitting-up of the sphere near an abnormal minimizer γ into two sectors, bordered by the first Pontryagin's cone along γ, called the L^∞-sector and the L^2-sector. Moreover we find again necessary and sufficient conditions of optimality of an abnormal trajectory for such systems, for any optimization problem.

1 Setup and dcfinitions

1.1 Abnormal trajectories

Consider a control system on $\mathbb{R}^n$:

$$\dot{x}(t) = f(x(t), u(t)),\ x(0) = x_0 \tag{1}$$

where $f : \mathbb{R}^n \times \mathbb{R}^m \longrightarrow \mathbb{R}^n$ is smooth, $x_0 \in \mathbb{R}^n$, and the set of admissible controls $\mathcal{U}$ is made of measurable bounded functions $u \ : [0, T(u)] \longrightarrow \Omega \subset \mathbb{R}^m$.

Definition 1. Let $T > 0$. The *end-point mapping* at time T of system (1) is the mapping

$$E_T \ : \begin{array}{l} \mathcal{U} \longrightarrow \mathbb{R}^n \\ u \longmapsto x_u(T) \end{array}$$

where x_u is the trajectory associated to u.

It is a very classical fact that E_T is smooth in the L^∞ topology, if $\mathcal{U} \subset L^\infty([0,T])$.

Definition 2. The *time-extended end-point mapping* (see [3]) of system (1) is the mapping

$$F \ : \begin{array}{r} \mathbb{R}^+ \times \mathcal{U} \longrightarrow \mathbb{R}^n \\ (T, u) \longmapsto E_T(u) \end{array}$$

F. Colonius, L. Grüne (Eds.): Dynamics, Bifurcations, and Control, LNCIS 273, pp. 89–99, 2002.

Definition 3. A control u (or the corresponding trajectory x_u) is said to be *abnormal* on $[0,T]$ if (T,u) is a singular point of the mapping F.

An equivalent definition may be given using the well-known *Pontryagin Maximum Principle* (see [9]), which parametrizes the previous condition :

Definition 4. A control u on $[0,T]$ (or the corresponding trajectory x_u) is said to be *abnormal* if there exists a non trivial absolutely continuous function $p(.) : [0,T] \longrightarrow \mathbb{R}^n \times \mathbb{R}$ called adjoint vector, such that the trajectory x satisfies almost everywhere the system :

$$\dot{x} = \frac{\partial H}{\partial p}, \ \dot{p} = -\frac{\partial H}{\partial x}, \ \frac{\partial H}{\partial u} = 0 \tag{2}$$

where $H(x,p,u) =< p, f(x,u) >$ is the Hamiltonian of the system, and moreover :

$$H(x,p,u) = 0 \quad \text{a.e. on } [0,T] \tag{3}$$

Remark 1. If a control u is abnormal on $[0,T]$ then it is abnormal on $[0,t]$ for any $t \in [0,T]$.

Definition 5. Let u be an abnormal control on $[0,T]$, and x_u its associated trajectory. The subspace Im $dE_t(u)$ is called the *first Pontryagin's cone* at $x_u(t)$.

Remark 2. An abnormal control on $[0,T]$ is in particular a singular point of the end-point mapping E_T. Conversely if u is a singularity of the end-point mapping and if moreover the trajectory x_u associated to u is almost everywhere tangent to its first Pontryagin's cone then u is abnormal.

1.2 Accessibility sets

Definition 6. Consider the control system (1), and let $T > 0$. The *accessibility set at time* T, denoted by $Acc(T)$, is the set of points that can be reached from x_0 in time T by solutions of system (1), i.e. this is the image of the end-point mapping E_T.

Let γ be a reference trajectory on $[0,T]$, solution of (1), associated to a control u. Our aim is to describe $Acc(T)$ near $\gamma(T)$.

1.3 Single-input affine systems

Consider a smooth *single-input affine system in* $\mathbb{R}^n$ *with constraint on the control* :

$$\begin{aligned} &\dot{x}(t) = X(x(t)) + u(t)Y(x(t)), \quad |u(t)| \leqslant \eta \\ &x(0) = x_0 \end{aligned} \tag{4}$$

where X, Y are smooth vector fields in $\mathbb{R}^n$, and u is a scalar measurable function. Suppose that the trajectory γ associated to the control $u = 0$ is abnormal on $[0, T]$. Let $p(.)$ be an adjoint vector associated to γ. Then the condition $H = 0$ is equivalent to :

$$< p(t), X(\gamma(t)) >= 0 \quad \text{a.e.}$$

i.e. $X \in \text{Im } dE_t(u)$ along γ.

Remark 3. In this case the first Pontryagin's cone along γ is obtained obviously by derivating the previous condition, and we get :

$$\text{Im } dE(u) = \text{Vect } \{ad^k X.Y(\gamma(T)) \ / \ k \in \mathbb{N}\}$$

The aim of this article is to describe precisely the accessibility set $Acc(T)$ at time T near $\gamma(T)$.

The essential tool needed for the proofs is the so-called *intrinsic second-order derivative* of the end-point mapping (see [3]). Using the formalism and normal forms of [5], the intrinsic second-order derivative may be represented by an explicit differential operator along the abnormal trajectory γ ; then a spectral analysis of this operator leads to a precise description of the boundary of $Acc(T)$ near $\gamma(T)$. Applying this result to sub-Riemannian systems of rank 2 then leads to a new splitting-up of the sub-Riemannian sphere near an abnormal minimizer γ into two sectors, in which the behaviours of minimizing trajectories near γ are topologically different. On the other part we can find again and improve slightly some well-known results on optimality of abnormal trajectories.

2 Asymptotics of the reachable sets

In this section we describe precisely the boundary of reachable sets for a single-input affine system with constraint on the input near a reference abnormal trajectory.

Consider a smooth *single-input affine control system* in $\mathbb{R}^n$, $n \geqslant 3$:

$$\dot{x}(t) = X(x(t)) + u(t)Y(x(t)), \ x(0) = 0 \tag{5}$$

with the *constraint* on the control

$$|u(t)| \leqslant \eta \tag{6}$$

Let $Acc^\eta(T)$ denote the *reachable set* at time T for this affine system with constraint η on the control. Let γ be a reference trajectory defined on $[0, T]$. In the sequel we make the following assumptions along γ :

(H_0) γ is injective, corresponding to the control $u = 0$ on $[0, T]$.

(H_1) $\forall t \in [0,T]$ $K(t) = \text{Vect}\ \{ad^k X.Y(\gamma(t))\ /\ k \in \mathbb{N}\}$ (first Pontryagin's cone along γ) has codimension 1, and is spanned by the first $n-1$ vectors, i.e. :

$$K(t) = \text{Vect}\ \{ad^k X.Y(\gamma(t))\ /\ k = 0 \ldots n-2\}$$

(H_2) $\forall t \in [0,T]$ $ad^2 Y.X(\gamma(t)) \notin K(t)$.
(H_3) $\forall t \in [0,T]$ $X(\gamma(t)) \notin \text{Vect}\ \{ad^k X.Y(\gamma(t))\ /\ k = 0 \ldots n-3\}$.
(H_4) $\forall t \in [0,T]$ $X(\gamma(t)) \in K(t)$.

In these conditions γ is *abnormal* and its first Pontryagin's cone $K(t)$ is an hyperplane in $\mathbb{R}^n$. Actually assumptions $(H_1 - H_3)$ are generic along γ, see [6].

The following theorem is founded on a very precise spectral analysis of the intrinsic second-order derivative of the end-point mapping along the abnormal direction γ (started in [5]), which actually leads to a contact theory of reachable sets (see [12]).

Theorem 1. *Consider the affine system (5) with the constraint (6), and suppose that assumptions $(H_0 - H_4)$ are fulfilled along the reference abnormal trajectory γ on $[0,T]$. Then there exist coordinates $(x_1, \ldots, x_n)$ locally along γ such that in these coordinates :*

1. *$\gamma(t) = (t, 0, \ldots, 0)$, and the first Pontryagin's cone along γ is : $K(t) = \text{Vect}\ \{\frac{\partial}{\partial x_1}, \ldots, \frac{\partial}{\partial x_{n-1}}\}_{|\gamma}$.*
2. *If T is small enough then for any point $(x_1, \ldots, x_n)$ of $Acc^\eta(T) \backslash \{\gamma(T)\}$ close to $\gamma(T)$ we have : $x_n > 0$ (see Fig. 1).*

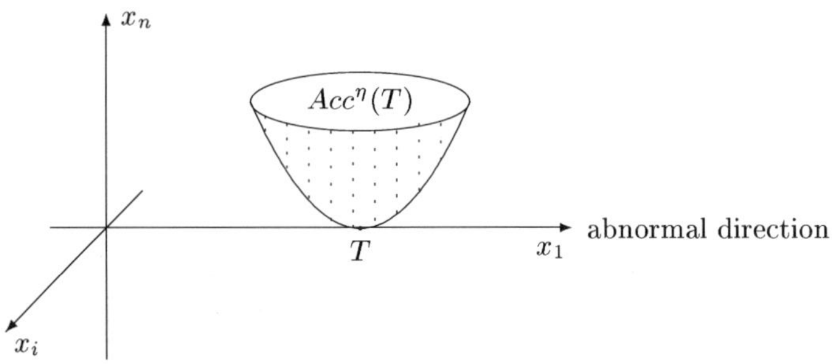

Fig. 1. Shape of $Acc^\eta(T)$, T small

3. *There exist two times t_{cc}, t_c such that $0 < t_{cc} < t_c$, called conjugate times or bifurcation times along γ, such that the following holds.*
   If $T < t_c$, then in the plane (x_1, x_n), near the point $(T, 0)$, the boundary of $Acc^\eta(T)$ does not depend on η, is a curve of class C^2 tangent to the abnormal direction, and its first term is

$$x_n = A_T(x_1 - T)^2 + o((x_1 - T)^2)$$

The function $T \mapsto A_T$ is continuous and strictly decreasing on $[0, t_c[$. It is positive on $[0, t_{cc}[$ and negative on $]t_{cc}, t_c[$.

4. *If $T > t_c$ then $Acc^\eta(T)$ is open near $\gamma(T)$.*

The evolution as a function of T of the intersection of $Acc^\eta(T)$ with the plane (x_1, x_n) is represented on Fig. 2. The contact with the abnormal direction is of order 2 ; the coefficient A_T describes the *concavity* of the curve. Beyond t_c the reachable set is open.

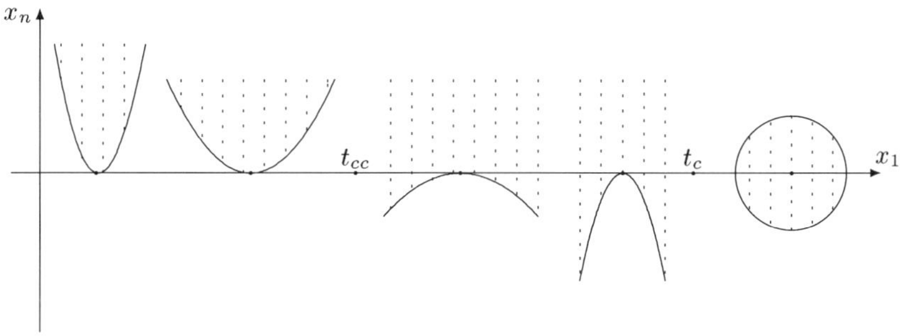

Fig. 2. Evolution of $Acc^\eta(T)$ in function of T

Remark 4. The proof of this result consists in representing the intrinsic second-order derivative of the end-point mapping along the abnormal direction γ by an *explicit* differential operator. In [5] the authors then studied the sign of the quadratic form by making a spectral analysis of this operator, and this leads to optimality results for an abnormal trajectory, see next section. The contribution of [12] is a refinement of this spectral theory, studying precisely this quadratic form we obtain the contact of the accessibility sets with the abnormal direction. Indeed under assumptions $(H_1 - H_4)$ we may construct a normal form for system (5), in which the first $n-1$ coordinates represent the *controllable part* of the system, i.e. the part that can be linearized, and the nth coordinate represents the intrinsic second order derivative.

The coefficient A_T can be explicitly computed. It is an invariant of the system, see [12]. Moreover the bifurcation times t_{cc} and t_c can be computed using an algorithm, see [5], and actually they correspond to times when the intrinsic second-order derivative is degenerate.

3 Applications

3.1 Application to the optimality status of an abnormal trajectory

In this section we apply our previous theory on reachable sets to studying optimality of abnormal trajectories, and we find again and slightly improve some well-known results. Indeed, in the notations of Theorem 1, this Theorem implies in particular that if $T < t_{cc}$ then γ is *isolated in C^0-topology* in the space of all trajectories which connect given end-points and thus it is *optimal for any cost* in this topology. This well-known property, called *rigidity*, was intensively studied. The main results concerning this analysis in a generic context were given first in [10] and [5] for single-input affine systems, then in [1,4,8,13] for sub-Riemannian systems and in [3] in general. Moreover, these authors developed a Morse theory in order to characterize conjugate points, that is, points beyond which the abnormal trajectory is no more optimal.

Hence the results given in this section are not really new. However, they are slightly different from the results cited above. Indeed, on the one hand in [5] were obtained necessary and sufficient conditions for C^0-time-optimality of abnormal trajectories of single-input affine systems without any constraint on the control. Here we improve their statement by adding a *constraint on the control* and studying the problem of minimizing *any cost*. On the other hand, in [3] was made a general theory (i.e., for nonlinear systems) on optimality of abnormal trajectories in L^∞ topology on the controls. The results given here are valid in the C^0-topology on the trajectories (but only for single-input affine systems). Moreover, we study the equivalence between the time-optimality problem and the problem of minimizing any cost, the final time being fixed or not. Finally, Theorem 6, which concerns optimality of abnormal trajectories for sub-Riemannian systems of rank 2, makes a link between the works of [5] and [1,2].

Consider the single-input affine system (5) with constraint (6) and suppose assumptions $(H_0 - H_4)$ are fulfilled along a reference abnormal trajectory γ. We first investigate the time-optimal problem and then the problem of minimizing any cost.

Time optimality The trajectory γ is called *C^0-time-minimal* on $[0,T]$ if there exists a C^0-neighborhood of γ such that T is the minimal time to steer $\gamma(0)$ to $\gamma(T)$ among the solutions of the system (5) with the constraint (6) that are entirely contained in this neighborhood.

We have the following result (compare with [2,3,5]) :

Theorem 2. *Under the assumptions of Theorem 1, the trajectory γ is C^0-time-minimal if and only if $T < t_{cc}$.*

Remark 5. Inspecting Fig. 2 obviously leads to the conclusion in the L^∞ topology on the controls. But note that the result is valid in a stronger topology. This fact is due to the construction of a normal form along the abnormal trajectory, see [5].

Optimization of any cost Let us now consider the problem of minimizing some cost $C(T, u)$, also denoted by $C_T(u)$, where C is a smooth function satisfying the following additional assumption along the reference abnormal trajectory γ :

$$(H_5)\ \forall T \quad \text{rank}\ (dE_T(0), dC_T(0)) = n$$

We distinguish between two optimization problems.

1. Final time not fixed The trajectory γ is said to be C^0-*cost-minimal* on $[0, T]$ if there exists a C^0-neighborhood of γ such that for any trajectory q contained in this neighborhood, with $q(0) = \gamma(0)$ and $q(t) = \gamma(T)$, we have : $C(t, v) \geqslant C(T, 0)$, where v is the control associated to q.

We have the following (compare with [3]) :

Theorem 3. *Under assumptions* $(H_0 - H_5)$, *the trajectory* γ *is* C^0-*cost-minimal if and only if it is* C^0-*time-minimal, i.e. if and only if* $T < t_{cc}$.

2. Final time fixed The trajectory γ is said to be C^0-*cost-minimal* with fixed final time on $[0, T]$ if there exists a C^0-neighborhood of γ such that for any trajectory q contained in this neighborhood, with $q(0) = \gamma(0)$ and $q(T) = \gamma(T)$, we have : $C_T(v) \geqslant C_T(0)$, where v is the control associated to q.

Theorem 4. *The trajectory* γ *is* C^0-*cost-minimal with fixed final time if and only if* $T < t_c$.

Hence in this case the time-optimal problem is not equivalent to the problem of minimizing some cost. The trajectory γ ceases to be C^0-time-optimal *before* it ceases to be C^0-cost-optimal (since $t_{cc} < t_c$).

Remark 6. The proof of both last results consists in considering the cost as a coordinate of the system, and to verify the assumptions of Theorem 1 using assumption (H_5). The importance of fixing the final time or not can again be understood by inspecting Fig. 2.

Remark 7. The role of each bifurcation time t_{cc} or t_c is now clear. The first bifurcation time t_{cc} is related to the time-optimality status of the reference abnormal trajectory γ, whereas the second bifurcation time t_c is related to the controllability of the system along γ (indeed the system becomes controllable along γ beyond t_c).

3.2 Application to the sub-Riemannian case

Consider a smooth sub-Riemannian structure (M, Δ, g) where M is a Riemannian n-dimensional manifold, $n \geqslant 3$, Δ is a rank 2 distribution on M, and g is a metric on Δ. Let $x_0 \in M$; our point of view is local and we can assume that $M = \mathbb{R}^n$ and $x_0 = 0$. Suppose there exists a smooth injective abnormal trajectory γ passing through 0. Up to changing coordinates and reparametrizing we can assume that :

- $\gamma(t) = (t, 0, \dots, 0)$,
- $\Delta = \text{Span}\,\{X, Y\}$ where X, Y are g-orthonormal,
- γ is the integral curve of X passing through 0.

Under these assumptions, the sub-Riemannian problem is equivalent to the *time-optimal problem* for the system :

$$\dot{x} = vX(x) + uY(x), \;\; x(0) = 0 \tag{7}$$

where the controls v, u satisfy the *constraint* :

$$v^2 + u^2 \leqslant 1 \tag{8}$$

The reference abnormal trajectory γ corresponds to the control : $v = 1, u = 0$.

Definition 7. We call *affine system associated to the sub-Riemannian system* (7) the following system :

$$\dot{x} = X(x) + wY(w) \tag{9}$$

where the control w satisfies a constraint of the form : $|w| \leqslant \eta$.

Our aim is to compare the system (7) with its associated affine system (9). We use the following reparametrizing :

$$\frac{ds}{dt} = v$$

which holds only if v does not vanish. This condition is satisfied when the control (v, u) is in a α-neighborhood in L^∞ metric of the abnormal reference control $(1, 0)$, for in this case v is close to 1 in L^∞-topology. Using this method it is possible to describe the accessibility set of system (7) in a α-neighborhood in L^∞ metric of the reference abnormal control. This reasoning, combined with a general statement of [11], leads to the following result.

Theorem 5. *Consider the sub-Riemannian problem for the system given by* $\dot{x} = vX(x) + uY(x)$. *Let* γ *be an abnormal reference trajectory. Suppose assumptions* $(H_0 - H_4)$ *hold along* γ. *Then there exist coordinates* $(x_1, \dots, x_n)$ *locally along* γ *in which, if* T *is small enough:*

- *$\gamma(t) = (t, 0, \dots, 0)$,*
- *the first Pontryagin's cone along γ is $K_\gamma(t) = \{x \in \mathbb{R}^n \ / \ x_n = 0\}$,*
- *The sub-Riemannian sphere $S(0,T)$ splits into two sectors near $\gamma(T)$:*
  1. *the L^∞-sector : $(x_n > 0) \cap S(0,T)$, made of end-points of minimizing trajectories associated to controls which are close to the abnormal reference control in L^∞-topology. Hence minimizing trajectories steering 0 to these points are close to γ in C^1-topology. Moreover, in the plane (x_1, x_n) its graph is*

     $$x_1 \geqslant T, \ x_n \sim A_T.(x_1 - T)^2$$

     where $T \mapsto A_T$ is continuous, positive and decreasing.
  2. *the L^2-sector : $(x_n < 0) \cap S(0,T)$, made of end-points associated to minimizing controls which are close to the abnormal reference control in L^2-topology, but not in L^∞-topology. Hence trajectories steering 0 to these points are close to γ in C^0-topology, but not in C^1-topology. This sector is tangent to the abnormal direction.*

  These two sectors are separated by the first Pontryagin's cone $\{x \in \mathbb{R}^n \ / \ x_n = 0\}$ along γ (see Fig. 3).

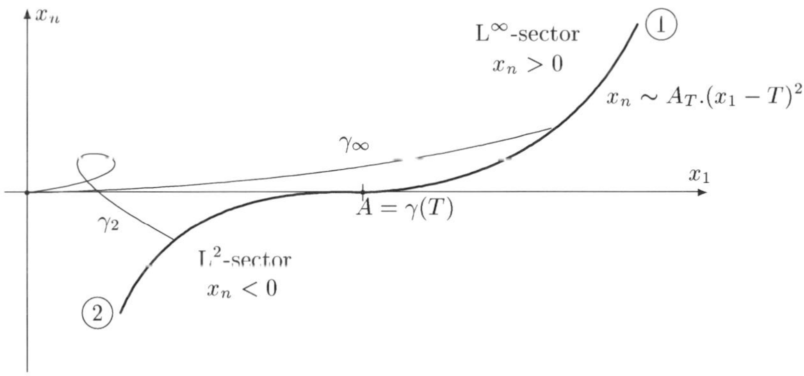

Fig. 3.

Remark 8. The abnormal trajectory γ is optimal for the sub-Riemannian problem if and only if $T < t_{cc}$, where $T < t_{cc}$ is the first bifurcation time for the associated affine system (9).

Typical example : the Martinet case.

Consider the two following vector fields in $\mathbb{R}^3$:

$$X = \frac{\partial}{\partial x} + \frac{y^2}{2}\frac{\partial}{\partial z}, \ Y = \frac{\partial}{\partial y}$$

and endow the distribution spanned by these vector fields with an analytic metric g of the type :

$$g = adx^2 + cdy^2$$

where $a = (1+\alpha y)^2$ and $c = (1+\beta x+\gamma y)^2$. The abnormal reference control for the sub-Riemannian system $\dot{x} = vX(x)+uY(x)$ with constraint $v^2+u^2 \leqslant 1$ is $v = 1, u = 0$, and corresponds to the trajectory $\gamma : x(t) = t, y(t) = z(t) = 0$. We have, see [7] :

Lemma 1. *In the Martinet case, the assumptions ($H_0 - H_4$) are fulfilled along γ if and only if $\alpha \neq 0$. In this case branches* 1 *and* 2 *(see Fig. 3 with $x_1 = x, x_n = z$) have the following contacts with the abnormal direction :*

- *branch* 1 : $x \geqslant T,\ z = \frac{1}{2T\alpha^2}(x-T)^2 + o((x-T)^2)$
- *branch* 2 : $x \leqslant T,\ z \sim \frac{1}{6}(1+O(T))(x-T)^3$

Finally we have the following result, consequence, as previously, of Theorem 5 and of comparison with the affine system (9) :

Theorem 6. *Under the assumptions of Theorem 5, the abnormal reference trajectory γ is C^0-optimal for the sub-Riemannian system (7) if and only if it is C^0-time-minimal for its associated affine system (9). Moreover γ is abnormal for this affine system.*

In particular, *bifurcation times are the same along γ for the sub-Riemannian system (7) and its associated affine system (9).* Therefore the whole formalism that was introduced for affine systems is still valid in sub-Riemannian geometry. Hence the bifurcation time of the sub-Riemannian problem can be computed using an algorithm. This result makes a link between works of [5] and [1], [2].

Remark 9. If T is small enough (depending on the choice of the Riemannian structure, and smaller than t_{cc}), then γ is moreover *globally* optimal among all sub-Riemannian trajectories steering 0 to $\gamma(T)$ (see for instance [1]).

Remark 10. It should be noted that the loss of optimality holds in L^∞. Hence *the L^2-sector plays no role in the optimality of the reference abnormal trajectory.* The loss of optimality holds in the L^∞-sector.

References

1. Agrachev A., Sarychev A. (1995), Strong minimality of abnormal geodesics for 2-distributions, J. of Dynamical and Control Systems, Vol. 1, No. 2, 139-176.
2. Agrachev A., Sarychev A. (1996), Abnormal sub-Riemannian geodesics : Morse index and rigidity, Annales de l'IHP, Vol. 13, 635-690.
3. Agrachev A., Sarychev A. (1998), On abnormal extremals for Lagrange variational problems, Journal of Mathematical Systems, Estimation, and Control, Vol. 8, No. 1, 87-118.

4. Bryant R. L., Hsu L. (1993), Rigidity of integral curves of rank 2 distributions, Invent. Math. 114, 435-461.
5. Bonnard B., Kupka I. (1993), Théorie des singularités de l'application entrée/sortie et optimalité des trajectoires singulières dans le problème du temps minimal, Forum Math. 5, 111-159.
6. Bonnard B., Kupka I. (1997), Generic properties of singular trajectories, Annales de l'IHP, Analyse non linéaire, Vol. 14, No. 2, 167-186.
7. Bonnard B., Trélat E. (1999), Role of abnormal minimizers in sub-Riemannian geometry, , to appear in Annales de la Faculté des Sciences de Toulouse.
8. Liu W. S., Sussmann H. J. (1995), Shortest paths for sub-Riemannian metrics of rank two distributions, Memoirs AMS, No. 564, Vol. 118.
9. Pontryagin L. and al., Théorie mathématique des processus optimaux, Eds Mir, Moscou (1974).
10. Sarychev A. (1982), The index of the second variation of a control system, Math. USSR Sbornik, Vol. 41, No. 3.
11. Trélat E. (2000), Some properties of the value function and its level sets for affine control systems with quadratic cost, Journal of Dynamical and Control Systems, Vol. 6, No. 4, 511-541.
12. Trélat E. (2001), Asymptotics of accessibility sets along an abnormal trajectory, ESAIM: COCV, June 2001, Vol. 6, 387-414.
13. Zhong Ge (1993), Horizontal path space and Carnot-Caratheodory metric, Pacific J. Math., Vol. 161, 255-286.

Part II

Stabilization and Robustness

Oscillation Control in Delayed Feedback Systems

Fatihcan M. Atay

Moda Cad. 214-A, Kadikoy, Istanbul 81300, Turkey

Abstract. The control of limit cycle oscillations is discussed for weakly nonlinear delayed feedback systems. Through a center manifold reduction the effects of the control parameters on the periodic behavior are determined. The results are applied to the perturbed harmonic oscillator under the action of delayed feedback of position, which is a prototype system commonly arising in diverse biological and industrial settings. It is shown that, using a cubic feedback function, unwanted periodic solutions can be annihilated, and an asymptotically stable limit cycle can be created oscillating at an arbitrary prescribed amplitude. The frequency of the oscillations can also be controlled to a limited extent. The presence of the delay in the control signal turns out to be crucial for achieving these goals using position feedback.

1 Introduction

Delays in the feedback action is an inevitable feature of many natural and man-made control mechanisms. The presence of the delay usually enriches the possible dynamics and increases the mathematical complexity of the system, even though the undelayed behavior may be deceptively simple. It is a challenging control task to compensate for the destabilizing effects of the delays, or otherwise deal with the added intricacies. On the other hand, it is equally interesting to use the delays to elicit a desired dynamical behavior which may not be possible in an undelayed system.

Positive uses of delays have been noted several decades ago [21], and more recent works have used delays to enhance the system performance in various ways [19,20,18,14]. Traditionally most results in this area seem to focus on linear systems and stability. In the present article, we consider a nonlinear phenomenon, namely limit cycle oscillations, for a class of systems under the action of delayed feedback. Our purpose is to show that the feedback parameters can be so chosen that the controlled system exhibits asymptotically stable periodic oscillations at an arbitrary prescribed amplitude and a range of possible frequencies. Interestingly, the existence of a delay turns out to be a necessary condition for this property if the feedback uses only position information.

We consider the classical harmonic oscillator and its perturbations controlled by delayed feedback, described by equations of the form

$$\ddot{x} + x + \varepsilon g(x, \dot{x}) = f(x(t - \tau)), \tag{1}$$

F. Colonius, L. Grüne (Eds.): Dynamics, Bifurcations, and Control, LNCIS 273, pp. 103–116, 2002.

where $x \in \mathbf{R}$. Here, g represents the nonlinearities in the system, ε is a nonnegative parameter, and f is the feedback function whose action is delayed by an amount $\tau \geq 0$. Systems of the form (1) arise in various biological and industrial settings, for example in the production of proteins [12,1], orientation control in the fly [17,16], neuromuscular regulation of movement and posture [1,5,7], acousto-optical bistability [22], metal cutting [4], vibration absorption [15], and control of an inverted pendulum [3]. Note that, although the uncontrolled system ($f \equiv 0$) is planar, the closed-loop system is infinite-dimensional if the delay τ is nonzero.

The problem considered in this article is to determine the feedback law f so that (1) has an asymptotically stable periodic solution with desired properties and which attracts all initial conditions in a reasonably large region. Our main result is that this objective can be achieved by using a simple cubic feedback function. Furthermore, the amplitude of the resulting oscillations can be set arbitrarily, while the frequency can be modified to some extent. The conclusion holds for general nonlinearities g and sufficiently small values of ε. Note that f uses only the position information x. Thus, when the derivative $\dot{x}$ is not available for feedback, which is the case in many biological applications, its absence can be compensated for by a suitable delay in the feedback action.

Our strategy can be summarized as follows. Suppose that the uncontrolled system has an equilibrium solution, which we take to be 0 without loss of generality, which is independent of ε. We consider (1) as a perturbation of the linear equation

$$\ddot{x} + x = bx(t - \tau), \tag{2}$$

with ε playing the role of a perturbation parameter. Equation (2) has periodic solutions if it has purely imaginary characteristic values. Such a situation is typical in the study of Hopf bifurcations, where ε is introduced as a scaling factor to blow up some neighborhood of the equilibrium point. In Sect. 3.1 we determine the values of $b, \tau \in \mathbf{R}$ so that (2) has a pair of purely imaginary characteristic values, while the other characteristic values have negative real parts. Thus the long term behavior is governed by the dynamics on the center manifold. In Sect. 3.2 the reduced dynamics on the center manifold is derived for the full nonlinear equation (1). Using the method of averaging, we then investigate the periodic solutions via the fixed points of a scalar equation. Finally, in Sect. 4 we determine a feedback function so that the periodic solutions have a prescribed amplitude and attract all initial conditions in a large domain. While the concept of center manifold reduction and averaging is familiar from the theory of ordinary differential equations, the theoretical and computational issues here are more involved due to the infinite-dimensionality of delayed systems. In the next section we give a brief review of the theory and introduce the notation for the rest of the article.

2 Perturbations of linear retarded equations

Consider the linear system with discrete delays

$$\dot{x}(t) = \sum_{j=0}^{m} A_j x(t - \tau_j), \tag{3}$$

where $x \in \mathbf{R}^n$, $A_j \in \mathbf{R}^{n \times n}$, $\tau_j \geq 0$. Letting $\tau = \max\{\tau_j\}$, an appropriate state-space for (3) is $\mathcal{C} = \mathcal{C}([-\tau, 0], \mathbf{R}^n)$, the space of continuous functions mapping the interval $[-\tau, 0]$ into $\mathbf{R}^n$, equipped with the usual sup norm. Points x_t in $\mathcal{C}$ correspond to "segments" of the solutions x of (3) through $x_t(s) = x(t + s)$ for $s \in [-\tau, 0]$. The characteristic values of (3) are the roots λ of the characteristic equation

$$\det\left(\lambda I - \sum_{j=0}^{m} A_j \exp(-\lambda \tau_j)\right) = 0\,. \tag{4}$$

For each characteristic value λ , there exists a solution of (3) in the form $e^{\lambda t} b$, where b is some constant vector and $t \in \mathbf{R}$. In general (4) has an infinite number of solutions; however, the number of solutions with nonnegative real parts is finite [11].

Associated with (3) is the *adjoint equation*

$$\dot{z}(t) = -\sum_{j=0}^{m} A_j^T z(t + \tau_j), \tag{5}$$

where the superscript T denotes the matrix transpose. Defining $z_t(s) = z(t+s)$ for $s \in [0, \tau]$, the points z_t form a trajectory in $\mathcal{C}^* = \mathcal{C}([0, \tau], \mathbf{R}^n)$ corresponding to the solution z of (5). The spaces $\mathcal{C}$ and $\mathcal{C}^*$ are related through the bilinear form

$$(z_t, x_t) = z_t^T(0) x_t(0) + \sum_{j=0}^{m} \int_0^{\tau_j} z_t^T(s) A_j x_t(s - \tau_j)\, ds\,, \tag{6}$$

so that whenever x and z are solutions of (3) and (5), respectively, which are defined on some interval J, then $(z_t, x_t) \equiv$ constant for all $t \in J$ [13]. The characteristic values of (5) are the roots of the equation

$$\det\left(\lambda I - \sum_{j=0}^{m} A_j^T \exp(-\lambda \tau_j)\right) = 0\,. \tag{7}$$

For each characteristic value λ, there exists a solution of (5) in the form $e^{-\lambda t}b$, where b is some constant vector and $t \in \mathbf{R}$. Note that the characteristic values of (3) and (5) coincide.

Suppose that (3) has d characteristic values (counting multiplicity) with nonnegative real parts, and let Φ be an $(n \times d)$ matrix whose columns form a basis for the generalized eigenspaces corresponding to these characteristic values. Similarly, let Ψ be a matrix whose columns span the generalized eigenspaces of those characteristic values of (5) with nonnegative real parts. It can be shown that the matrix (Ψ, Φ) is nonsingular [9]. Hence, replacing Ψ with $\Psi[(\Psi,\Phi)^{-1}]^T$ if necessary, it may be assumed that (Ψ,Φ) is the identity matrix. Furthermore, there exists a matrix $B \in \mathbf{R}^{d\times d}$, whose eigenvalues are precisely the characteristic values of (3) with nonnegative real parts, such that $\Phi(\theta) = \Phi(0)e^{B\theta}$ for $\theta \in \mathbf{R}$. The bases Φ and Ψ allow a decomposition of the space $\mathcal{C}$: For any $\varphi \in \mathcal{C}$, one has

$$\varphi = \Phi c + \bar{\varphi}, \quad c = (\Psi, \varphi).$$

This decomposition is unique for any given φ in $\mathcal{C}$ [10].

Now consider the following perturbation of the linear equation (3)

$$\dot{x}(t) = \sum_{j=0}^{m} A_j x(t-\tau_j) + \varepsilon h(t, x_t), \quad |\varepsilon| \ll 1. \tag{8}$$

where $h : \mathbf{R} \times \mathcal{C} \to \mathbf{R}^n$ is a continuous function. Assume for simplicity that for $\varepsilon = 0$ (8) has no characteristic values with positive real parts. Let x be a solution of (8) with initial value $x_0 \in \mathcal{C}$, and define $\bar{x}_t$ and $y(t) \in \mathbf{R}^d$ by

$$x_t = \Phi y(t) + \bar{x}_t, \quad y(t) = (\Psi, x_t). \tag{9}$$

where Φ and Ψ are as in the previous paragraph. Then y satisfies

$$\dot{y}(t) = By(t) + \varepsilon \Psi^T(0) h(t, \Phi y(t) + \bar{x}_t), \quad y(0) = (\Psi, x_0),$$

where the matrix B has purely imaginary eigenvalues. Furthermore, it can be shown that bounded solutions are such that $\bar{x}_t = O(\varepsilon)$ as $\varepsilon \to 0$ [10]. Hence, to first order in ε, the analysis of (8) is reduced to the investigation of the d-dimensional ordinary differential equation

$$\dot{y}(t) = By(t) + \varepsilon \Psi^T(0) h(t, \Phi y(t)). \tag{10}$$

Since the eigenvalues of B are purely imaginary, the qualitative dynamics are best displayed by changing to polar coordinates and averaging the slowly-varying terms.

3 The harmonic oscillator under delayed feedback

We now return to the harmonic oscillator under the action of delayed feedback

$$\ddot{x} + x = bx(t-\tau), \tag{11}$$

and its perturbations of the form

$$\ddot{x} + x + \varepsilon g(x(t), \dot{x}(t)) = bx(t-\tau) + \varepsilon \bar{f}(x(t-\tau)) . \tag{12}$$

Using the foregoing ideas, we shall determine the center manifolds of (11), and then obtain the dynamics on the center manifold for the perturbed system (12).

3.1 The linear equation

Letting $(x_1, x_2) = (x, \dot{x})$ puts (11) into the form (3)

$$\frac{d}{dt}\begin{bmatrix} x_1(t) \\ x_2(t) \end{bmatrix} = A_0 \begin{bmatrix} x_1(t) \\ x_2(t) \end{bmatrix} + A_1 \begin{bmatrix} x_1(t-\tau) \\ x_2(t-\tau) \end{bmatrix}, \tag{13}$$

with

$$A_0 = \begin{bmatrix} 0 & 1 \\ -1 & 0 \end{bmatrix}, \quad \text{and} \quad A_1 = \begin{bmatrix} 0 & 0 \\ b & 0 \end{bmatrix}.$$

The characteristic equation is

$$\chi(\lambda) = \lambda^2 + 1 - be^{-\lambda\tau} = 0 . \tag{14}$$

The roots $\lambda(b, \tau)$ vary smoothly with the parameters b and τ, and we shall investigate this dependence in order to determine the parameter values for which a two-dimensional center subspace exists. Related studies of (14) have also appeared in [3,6].

If $\lambda = \pm i\omega$ for some $\omega > 0$, it follows from (14) that

$$1 - \omega^2 - b\cos\omega\tau = 0 \tag{15}$$

$$b\sin\omega\tau = 0 . \tag{16}$$

The possible values of ω are of interest since these give the frequency of periodic solutions. The following lemma summarizes the properties of (14) relevant for the present study. Its proof will also establish the stability diagram given in Fig. 1.

Lemma 1. *Let $\omega \in (\sqrt{2/5}, \sqrt{2})$ and $b = \omega^2 - 1$. Let τ be an arbitrary nonnegative number if $\omega = 1$, otherwise let $\tau = \pi/\omega$. Then the characteristic equation (14) has precisely two roots $\lambda = \pm i\omega$ on the imaginary axis, and no roots with positive real parts.*

Proof. For $b = 0$ and $\tau \in \mathbf{R}$, the only solution to (15)–(16) is $\omega = 1$. When $b \neq 0$, (16) implies that $\omega = \omega_n = n\pi/\tau$ for some nonnegative integer n, and substituting into (15) gives

$$(-1)^n b = 1 - \omega_n^2 = 1 - n^2\left(\frac{\pi}{\tau}\right)^2 . \tag{17}$$

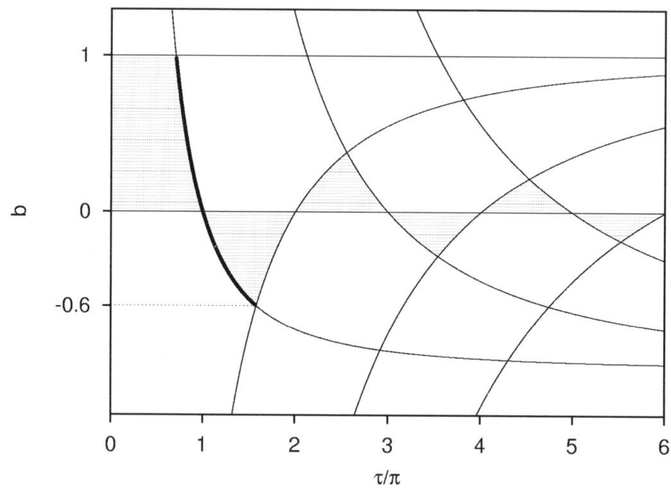

Fig. 1. Regions of stability in the parameter plane

For $n = 0, 1, 2, \ldots$ these equations define a curve γ_n on the b-τ parameter plane, as shown in Fig. 1. Together with the line $b = 0$, they form the set of parameter values for which (14) has a root on the imaginary axis. As these curves are crossed transversally, a pair of roots move across the imaginary axis into the left or the right half plane (except for γ_0, which is the line $b = 1$, where only a single root $\omega = 0$ crosses). The direction of movement is found by implicit differentiation of (14):

$$\frac{\partial \lambda}{\partial b}|_{\lambda=i\omega} = \frac{1}{2\lambda e^{\lambda} + b\tau}|_{\lambda=i\omega} = \frac{1}{(b\tau - 2\omega \sin \omega\tau) + 2i\omega \cos \omega\tau}.$$

Clearly, the sign of the real part of the above expression is determined that by the sign of $(b\tau - 2\omega \sin \omega\tau)$. Thus,

$$\text{sgn}\left(\frac{\partial(\text{Re}\,\lambda)}{\partial b}|_{\lambda=i\omega}\right) = \begin{cases} -\,\text{sgn}(\sin \tau) & \text{if } b = 0 \\ \quad \text{sgn}(b) & \text{if } b \neq 0 \end{cases}. \tag{18}$$

It follows that any region of stability must border the $b = 0$ line, since vertically moving away from this line cannot decrease the number of unstable roots. With the knowledge that on the line $b = 0$ there is one pair $\pm i$ on the imaginary axis and no roots on the right half plane, one obtains the regions of stability shown shaded in Fig. 1. On the boundaries of these regions there are purely imaginary roots of the form $\pm i\omega$ and no unstable roots. It is also easy to see that such roots on the imaginary axis are simple. Indeed, if $\chi(\lambda) = 0$, then by (14)

$$\chi'(\lambda) = 2\lambda + b\tau e^{-\lambda\tau} = 2\lambda + \tau(\lambda^2 + 1)$$

Hence $\chi'(i\omega) = 2i\omega + \tau(1-\omega^2) \neq 0$ whenever $\omega \neq 0$. Therefore, the boundaries of the stability regions, other than γ_0 and the intersections of the γ_i, are the loci of parameter values for which there are precisely two complex conjugate roots on the imaginary axis, and no roots with positive real parts.

To complete the proof, consider the particular boundary segment consisting of the part of the curve γ_1 for which $-3/5 < b < 1$, as shown with a heavy line in Fig. 1. By (17) $b = \omega^2 - 1$ and $\omega = \pi/\tau$ on γ_1; so ω assumes values in the interval $(\sqrt{2/5}, \sqrt{2})$ as the segment is traversed. Therefore, for each parameter pair (τ, b) on the segment there exist precisely two roots $\lambda = \pm i\omega$, except possibly at the intersection with the line $b = 0$. However, it is clear that no additional roots are introduced at the intersection since when $b = 0$ (14) has the unique and unrepeated pair of roots $\lambda = \pm i$.

Now suppose the values of b and τ are chosen as in Lemma 1. A basis for the center subspace of (13) corresponding to the pair of characteristic values $\pm i\omega$ is given by the columns of

$$\Phi = \begin{bmatrix} \cos\omega s & \sin\omega s \\ -\omega\sin\omega s & \omega\cos\omega s \end{bmatrix}, \quad s \in [-\tau, 0]. \tag{19}$$

It is easily checked that $\Phi(s) = \Phi(0)\exp(Bs)$, where

$$B = \begin{bmatrix} 0 & \omega \\ -\omega & 0 \end{bmatrix}. \tag{20}$$

Similarly, the columns of

$$\hat{\Psi} = \begin{bmatrix} \omega\sin\omega s & -\omega\cos\omega s \\ \cos\omega s & \sin\omega s \end{bmatrix}, \quad s \in [0, \tau]$$

span the eigenspace for the adjoint equation for $\lambda = \pm i\omega$. Calculating the bilinear form (6) for the basis vectors and using (15)–(16) gives

$$\begin{aligned}(\hat{\Psi}, \Phi) &= \hat{\Psi}^T(0)\Phi(0) + \int_0^\tau \hat{\Psi}^T(s)A_1\Phi(s-\tau)\,ds \\ &= \begin{bmatrix} \frac{1}{2}\tau(1-\omega^2) & \omega \\ -\omega & \frac{1}{2}\tau(1-\omega^2) \end{bmatrix}\end{aligned}$$

with inverse

$$\frac{4}{\tau^2(1-\omega^2)^2 + 4\omega^2}\begin{bmatrix} \frac{1}{2}\tau\left(1-\omega^2\right) & -\omega \\ \omega & \frac{1}{2}\tau\left(1-\omega^2\right) \end{bmatrix}. \tag{21}$$

Hence if $\hat{\Psi}$ is replaced by $\Psi = \hat{\Psi}[(\hat{\Psi}, \Phi)^{-1}]^T$, one has $(\Psi, \Phi) = I$. Then,

$$\Psi^T(0) = \frac{4}{\tau^2(1-\omega^2)^2 + 4\omega^2}\begin{bmatrix} \omega^2 & \frac{1}{2}\tau\left(1-\omega^2\right) \\ -\frac{1}{2}\omega\tau\left(1-\omega^2\right) & \omega \end{bmatrix}. \tag{22}$$

3.2 The reduced equation and averaging

Consider now the perturbed harmonic oscillator (12), which we rewrite in vector form

$$\frac{d}{dt}\begin{bmatrix} x_1(t) \\ x_2(t) \end{bmatrix} = A_0 \begin{bmatrix} x_1(t) \\ x_2(t) \end{bmatrix} + A_1 \begin{bmatrix} x_1(t-\tau) \\ x_2(t-\tau) \end{bmatrix} + \varepsilon \begin{bmatrix} 0 \\ \bar{f}\left(x_1(t-\tau)\right) - g\left(x_1(t), x_2(t)\right) \end{bmatrix}.$$

Using (19), (20), and (22), the reduced equation (10) in this case is calculated as

$$\begin{aligned} \dot{y}(t) &= \begin{bmatrix} 0 & \omega \\ -\omega & 0 \end{bmatrix} y(t) + \varepsilon \Psi^T(0) \begin{bmatrix} 0 \\ \bar{f}\left(\Phi_1(-\tau)y(t)\right) - g\left(\Phi(0)y(t)\right) \end{bmatrix} \\ &= \begin{bmatrix} 0 & \omega \\ -\omega & 0 \end{bmatrix} y(t) + 4\varepsilon \frac{\bar{f}\left(\Phi_1(-\tau)y(t)\right) - g\left(\Phi(0)y(t)\right)}{\tau^2(1-\omega^2)^2 + 4\omega^2} \begin{bmatrix} \frac{1}{2}\tau\left(1-\omega^2\right) \\ \omega \end{bmatrix}, \end{aligned} \tag{23}$$

where Φ_1 denotes the first row of Φ. We switch to polar coordinates

$$y = (r\cos\theta\,, -r\sin\theta)^T \tag{24}$$

in order to put (23) into a form suitable for averaging. Noting that

$$\Phi(-\tau)y = \begin{bmatrix} r\cos\left(\theta-\omega\tau\right) \\ -\omega r\sin\left(\theta-\omega\tau\right) \end{bmatrix}, \quad \text{and} \quad \Phi(0)y = \begin{bmatrix} r\cos\theta \\ -\omega r\sin\theta \end{bmatrix},$$

(23) is transformed into

$$\begin{aligned} \begin{bmatrix} \dot{r} \\ \dot{\theta} \end{bmatrix} &= \begin{bmatrix} 0 \\ \omega \end{bmatrix} + \frac{4\varepsilon}{\tau^2(1-\omega^2)^2 + 4\omega^2} \begin{bmatrix} \frac{1}{2}\tau\left(1-\omega^2\right)\cos\theta - \omega\sin\theta \\ -\frac{1}{2r}\tau\left(1-\omega^2\right)\sin\theta - \frac{1}{r}\omega\cos\theta \end{bmatrix} \\ &\quad \times \left(\bar{f}\left(r\cos\left(\theta-\omega\tau\right)\right) - g\left(r\cos\theta, -\omega r\sin\theta\right)\right). \end{aligned} \tag{25}$$

Note that the right hand side is 2π-periodic in θ, and r is slowly varying for small ε. Averaging (25) with respect to θ over one period leads to

$$\dot{r} = -\varepsilon(F(r) + G(r)) \tag{26}$$

$$\dot{\theta} = \omega + \mathcal{O}(\varepsilon)\,, \tag{27}$$

where

$$F(r) = -\frac{1}{2\pi}\int_0^{2\pi} \frac{4\left(\frac{\tau}{2}\left(1-\omega^2\right)\cos\theta - \omega\sin\theta\right)}{\tau^2(1-\omega^2)^2 + 4\omega^2} \bar{f}\left(r\cos\left(\theta-\omega\tau\right)\right)\, d\theta \tag{28}$$

$$G(r) = \frac{1}{2\pi}\int_0^{2\pi} \frac{4\left(\frac{\tau}{2}\left(1-\omega^2\right)\cos\theta - \omega\sin\theta\right)}{\tau^2(1-\omega^2)^2 + 4\omega^2} g\left(r\cos\theta, -\omega r\sin\theta\right)\, d\theta\,. \tag{29}$$

Making a change of variables $\theta \mapsto (\theta - \omega\tau)$ in (28), expanding the trigonometric functions, and noting that the integral of any term of the form $\sin\theta \cdot \bar{f}(r\cos\theta)$ is zero, $F(r)$ can be written as

$$F(r) = \gamma \frac{1}{2\pi} \int_0^{2\pi} \cos\theta \, \bar{f}(r\cos\theta) \, d\theta\,, \tag{30}$$

with

$$\gamma = \frac{4\omega \sin\omega\tau - 2\tau\left(1-\omega^2\right)}{\tau^2(1-\omega^2)^2 + 4\omega^2}. \tag{31}$$

Clearly, the quantity γ depends only on the parameters b and τ appearing in the unperturbed linear equation, while the rest of the expression in (30) depends only on the feedback perturbation $\bar{f}$. Hence, $\bar{f}$ can influence the dynamics of (26) only if $\gamma \neq 0$. The next lemma shows that γ is nonzero for almost all values of b, τ under consideration.

Lemma 2. *Suppose b and τ are chosen as in Lemma 1. Then $\gamma = 0$ if and only if $b = 0$ and $\sin\tau = 0$.*

Proof. If $b = \omega^2 - 1 \neq 0$ and $\tau = \pi/\omega$, then

$$\gamma = \frac{1}{2} \frac{\pi\omega(\omega^2-1)}{\pi^2(1-\omega^2)^2 + 4\omega^4} \neq 0,$$

whereas if $b = 0$, then $\omega = 1$ by (15), so that $\gamma = \sin\tau$.

On the other hand, it is obvious from (31) that $\gamma = 0$ if $\tau = 0$, so that the presence of a positive delay is a necessary condition for controlling the dynamics on the center manifold by position feedback.

4 Controlling the amplitude and frequency of oscillations

The averaged equations (26)–(27) contain important information about the periodic solutions of (1). Indeed, by the averaging theorem [8,10] and through (9) and (24), a hyperbolic equilibrium $R > 0$ of (26) corresponds to a nontrivial hyperbolic periodic solution of (1) whose amplitude is near R and frequency is near ω. The problem to address now is how to choose the control f so that the averaged equation (26) has desired dynamics. In particular, we would like to ensure that (26) has a unique equilibrium point at R in some bounded (but large) interval, which attracts all initial conditions there. Our main result shows that this can be achieved by a cubic feedback function.

Theorem 1. *Suppose $g : \mathbf{R}^2 \to \mathbf{R}$ is a C^2 function such that $g(0,0) = 0$. Let $R > 0$ and $\omega \in (\sqrt{2/5}, \sqrt{2})$. Then there exist $\tau > 0$, a feedback function of the form*

$$f(x) = bx + \varepsilon(b_1 x + b_3 x^3)\,, \tag{32}$$

and $\varepsilon_0 > 0$ such that for each $\varepsilon \in (0, \varepsilon_0)$, (1) has an attracting periodic solution $x(t) = R\cos(\omega t) + \mathcal{O}(\varepsilon)$.

Proof. Given $\omega \in (\sqrt{2/5}, \sqrt{2})$, choose b and τ as in Lemma 1. Using the notation of (12), we have $\bar{f}(x) = b_1 x + b_3 x$, and the corresponding averaged function F is calculated through (30) as

$$F(r) = q_1 r + q_3 r^3,$$

with

$$q_1 = \frac{1}{2}\gamma b_1 \quad \text{and} \quad q_3 = \frac{3}{8}\gamma b_3\,. \tag{33}$$

Let G be given by (29), and define

$$\bar{G}(r) = \begin{cases} G(r)/r & \text{if } r \neq 0, \\ G'(0) & \text{if } r = 0, \end{cases} \tag{34}$$

Consequently,

$$F(r) + G(r) = r(q_1 + q_3 r^2 + \bar{G}(r)), \tag{35}$$

Now let $R > 0$ and fix $c > R$. By the assumptions on g it is easy to see that $\bar{G}$ is continuously differentiable on $\mathbf{R}$; so, the following nonnegative numbers exist:

$$\mu_1 = \max\{|\bar{G}(r)| : r \in [0, c]\}, \tag{36}$$

$$\mu_2 = \max\{|\bar{G}'(r)| : r \in [0, c]\}. \tag{37}$$

Choose q_1 such that

$$q_1 < -\frac{5}{3}\mu_1 - R\mu_2, \tag{38}$$

and define q_3 by

$$q_3 = \frac{-q_1 - \bar{G}(R)}{R^2} > 0. \tag{39}$$

Note from (35) that positive roots of $F + G$ and of the function

$$H(r) := q_1 + q_3 r^2 + \bar{G}(r)$$

coincide. Using (39), one has

$$H(r) = q_1 \left(1 - \frac{r^2}{R^2}\right) + \bar{G}(r) - \frac{r^2}{R^2}\bar{G}(R), \tag{40}$$

so, $H(R) = 0$. We claim that H has no other positive roots in $(0, c)$. Indeed, for $r \in (0, R/2]$:

$$\begin{aligned} H(r) &\leq q_1 \left(1 - \frac{r^2}{R^2}\right) + \mu_1 \left(1 + \frac{r^2}{R^2}\right) \\ &\leq \frac{3}{4}q_1 + \frac{5}{4}\mu_1 < -\frac{3}{4}R\mu_2 \leq 0, \end{aligned}$$

where the last inequality follows by (38). Hence, $H(r)$ has no roots on the interval $(0, R/2]$. On the other hand, for $r \in [R/2, c)$,

$$\begin{aligned} H'(r) &= \frac{-2(q_1 + \bar{G}(R))}{R^2} r + \bar{G}'(r) \geq \frac{-2(q_1 + \mu_1)}{R^2}\frac{R}{2} - \mu_2 \\ &\geq \frac{1}{R}\left(-q_1 - \frac{5}{3}\mu_1 - R\mu_2\right) > 0, \end{aligned} \tag{41}$$

Thus H is strictly increasing for $r \in [R/2, c]$, so that any root in this interval is necessarily unique. Hence, R is the unique root of H (and consequently of $F + G$) in the interval $(0, c)$. It remains to show that R is attracting in $(0, c)$. Furthermore,

$$F'(R) + G'(R) = H(R) + RH'(R) = 0 + RH'(R) > 0$$

by (41), and

$$F'(0) + G'(0) = H(0) = q_1 + \bar{G}(0) \leq q_1 + \mu_1 < 0,$$

by (35) and (38). So, in the scalar equation (26) the origin is unstable and R is asymptotically stable. It then follows by the averaging theorem (e.g. Theorems 3.1 and 3.3 in [10], Theorem 4.1.1 in [8]) that for all sufficiently small and positive values of ε, (1) has an asymptotically stable periodic solution of the form $x(t) = R\cos\omega t + \mathcal{O}(\varepsilon)$. Since $\gamma \neq 0$ by Lemma 2, the coefficients of (32) are found from (33) as $b_1 = 2q_1/\gamma$ and $b_3 = 8q_3/(3\gamma)$.

Remark. It follows by Lemma 1 and the above proof that if $\omega = 1$, then for any τ satisfying $\sin\tau \neq 0$ a feedback function of the form (32) can be found with the same conclusion as in the statement of the theorem.

Theorem 1 shows that delayed position feedback offers complete control over the amplitude of the periodic solutions and only limited control over the frequency. The case when frequency need not be modified ($\omega = 1$) is particularly interesting since the cubic feedback scheme (32) then works for almost all values of the delay, as remarked above. In addition, the design of the feedback law can then be reduced to the tuning of a single parameter b_1, using the idea of the proof of the theorem. Indeed, by (33) and (39), the coefficient b_3 can be determined from the knowledge of b_1:

$$b_3 = -\frac{8}{3R^2}\left(\frac{1}{2}b_1 + \frac{G(R)}{\gamma R}\right), \tag{42}$$

with $\gamma = \sin\tau$. To obtain oscillations at an amplitude of R, by (38) it is sufficient (but not necessary) to choose b_1 such that γb_1 is large and negative. We present some examples.

Example. For the van der Pol's oscillator under linear feedback

$$\ddot{x} + \varepsilon(x^2 - 1)\dot{x} + x = \varepsilon b_1 x(t - \tau), \tag{43}$$

the function $G(r) = \frac{1}{8}r\left(r^2 - 4\right)$; so for small ε the uncontrolled system has the familiar limit cycle solution close to $2\cos t$. Putting $\omega = 1$ in (30), one has $F(r) = \frac{1}{2}b_1 \sin\tau$, and the averaged equation (26) takes the form

$$\dot{r} = -\frac{1}{8}\varepsilon r(r^2 - 4(1 - b_1 \sin\tau)).$$

Choosing $b_1 \sin\tau < 1$, the amplitude of the limit cycle is changed to $2\sqrt{1 - b_1 \sin\tau}$. On the other hand, choosing $b_1 \sin\tau > 1$ destroys the limit cycle and stabilizes the zero solution. Both these feats can be accomplished for any value of the delay satisfying $\sin\tau \neq 0$. For more details the reader is referred to [2].

Example. Consider the system

$$\ddot{x} - \varepsilon\sin(\dot{x}) + x = f(x(t - \tau)). \tag{44}$$

Here, the function $G(r)$ has infinitely many zeros, and thus the uncontrolled system has an infinite number of (stable and unstable) limit cycles, and the origin is unstable. Suppose it is desired to destroy these limit cycles and have an attracting stable limit cycle oscillating at an amplitude of $R = 2$ and frequency $\omega = 1$. A numerical calculation gives $G(2) = -0.577$. Assuming a delay value of $\tau = \pi/2$, we have $\gamma = 1$ by (31). Choosing $b = 0$ and $b_1 = -0.5$, and determining $b_3 = 0.36$ from (42), the cubic feedback (32) gives the limit cycle shown in Fig. 2. Other trajectories starting from a variety of initial conditions converge to the limit cycle, shown by dotted lines in the same figure. Hence, the domain of attraction of the limit cycle is reasonably large. The extent of the domain depends in general on the values of b_1, ε, and τ, and increases with decreasing ε when other parameters are fixed.

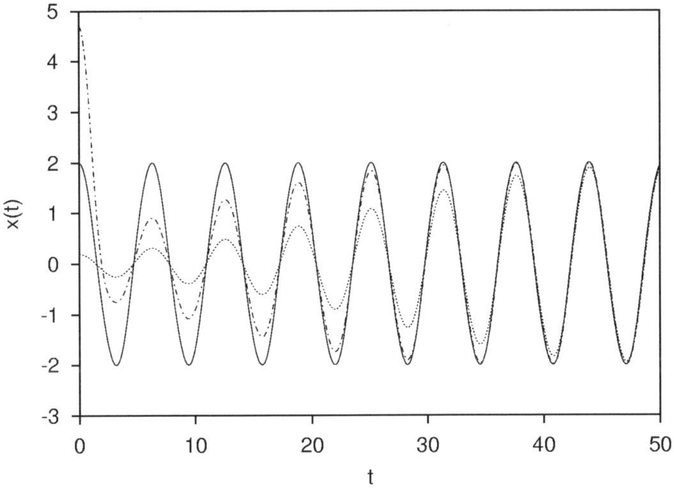

Fig. 2. The limit cycle with amplitude 2, shown with the solid line, obtained for the parameter values $\varepsilon = 0.1$ and $\tau = \pi/2$ with the cubic feedback (32). The dotted lines are trajectories starting from arbitrary initial conditions and converging to the limit cycle

5 Conclusion

It is seen that nonlinear delayed feedback of position can be effective in the control of oscillations. By using a simple cubic feedback function, it is possible to suppress unwanted oscillations or create a limit cycle with a desired amplitude. These observations may help explain the mechanism for the observed oscillatory behavior in biological systems, or design controllers in the absence of derivative information. It thus seems worthwhile to generalize the results presented here to more general classes of systems, which may be higher dimensional, or may contain more general (e.g. distributed) delays. Since our results are largely independent of the particular nonlinearities in the system, or in some cases the actual value of the delay, the proposed feedback scheme may also carry certain robustness properties, which need to be investigated further. The delayed feedback thus presents an interesting and challenging study in the area of nonlinear control.

References

1. an der Heiden U. (1979) Delays in physiological systems. J. Math. Biol. **8**:345–364
2. Atay F.M. (1998) Van der Pol's oscillator under delayed feedback. J. Sound Vib. **218**:333–339

3. Atay F.M. (1999) Balancing the inverted pendulum using position feedback. Appl. Math. Lett. **12**:51–56
4. Berger B.S., Rokni M., Minis, I. (1993) Complex dynamics in metal cutting. Q. Appl. Math. **51**:601–612
5. Beuter A., Bélair J., Labrie C. (1993) Feedback and delays in neurological diseases: A modeling study using dynamical systems. Bull. Math. Biol. **55**:525–541
6. Campbell S.A., Bélair J. (1999) Resonant codimension two bifurcation in the harmonic oscillator with delayed forcing. Can. Appl. Math. Q. **7**:218-238
7. Eurich C.W., Milton J.G. (1996) Noise-induced transitions in human postural sway. Phys. Rev. E **54**:6681–6684
8. Guckenheimer J., Holmes P. (1983) Nonlinear Oscillations, Dynamical Systems, and Bifurcations of Vector Fields. Springer-Verlag, New York
9. Hale J.K. (1963) Linear functional-differential equations with constant coefficients. Contrib. Diff. Eqs. **2**:291–319
10. Hale J.K. (1966) Averaging methods for differential equations with retarded arguments and a small parameter. J. Diff. Eqs. **2**:57–73
11. Hale J.K. (1977) Theory of Functional Differential Equations. Springer-Verlag, New York
12. Hastings S., Tyson J., Webster D. (1977) Existence of periodic solutions for negative feedback cellular control systems. J. Diff. Eqs. **25**:39–64
13. Kolmanovskii V., Myshkis A. (1992) Applied Theory of Functional Differential Equations. Kluwer Academic, Dordrecht
14. Kwon W.H, Lee G.W., Kim S. W. (1990) Performance improvement using time delays in multivariable controller design. Int. J. Control **52**:1455–1473
15. Olgac N., Holm-Hansen B.T. (1994) A novel active vibration absorption technique: Delayed resonator. J. Sound Vib. **176**:93–104
16. Poggio T., Reichardt W. (1976) Visual control of orientation behavior in the fly. Part II: Towards the underlying neural interactions. Q. Rev. Biophys. **9**:377–438
17. Reichardt W., Poggio T. (1976) Visual control of orientation behavior in the fly. Part I: A quantitative analysis. Q. Rev. Biophys. **3**:311–375
18. Shanmugathasan N., Johnston R.D. (1988) Exploitation of time delays for improved process control. Int. J. Control, **48**:1137–1152
19. Suh I.H., Bien Z. (1979) Proportional minus delay controller. IEEE Trans. Aut. Control **AC-24**:370–372
20. Suh I.H., Bien Z. (1980) Use of time-delay actions in the controller design. IEEE Trans. Aut. Control **AC-25**:600–603
21. Tallman G.H., Smith O. J. M. (1958) Analog study of dead-beat posicast control. IRE Trans. Aut. Control, **AC-3**:14–21
22. Vallée R., Dubois M., Coté M., Delisle C. (1987) Second-order differential-delay equation to describe a hybrid bistable device. Phys. Rev. A **36**:1327–1332

Nonlinear Problems in Friction Compensation

Antonio Barreiro[1], Alfonso Baños[2], Francisco Gordillo[3], and Javier Aracil[3]

[1] Dept. Ingeniería de Sistemas y Automática, University of Vigo, Vigo, Spain
[2] Dept. Informática y Sistemas, Universidad de Murcia, Murcia, Spain
[3] Dept. Ingeniería de Sistemas y Automática, University of Sevilla, Sevilla, Spain

Abstract. Friction is an elusive, difficult problem in control systems. Due to this, using several techniques is, in principle, a good idea which may help in understanding all the various, related effects, which a single tool might be unable to describe.

In this paper we explore some nonlinear problems arising when applying dither techniques, also in combination with PI controllers. The approach followed is from a nonlinear point of view, trying to integrate, within the QFT setting, tools from conic analysis, robust nonlinear control, describing function and harmonic balance. The integration is fruitful and feasible due to the use of a common frequencial language, which arises in a natural way within the technique of dither compensation.

1 Introduction

As a motivating example, consider the feedback system in Fig. 1, that can be the simplified model of a D.C. servomotor. The plant equations are $\dot{\theta} = K_t\omega$, $\dot{\omega} = T = T_m - T_f$, $T_m = \gamma(u)$ and $T_f = \phi(\omega)$. Thus, the plant is a double integrator, relating torque T, speed ω and angle θ, affected by memoryless nonlinearities at the input $\gamma(\cdot)$ and in feedback $\phi(\cdot)$. Typically, $\gamma(\cdot)$ models electrical nonlinearities. On the other hand, $\phi(\cdot)$ represents speed-dependent mechanical nonlinearities, typically *friction* producing the friction torque T_f. Although there are more complex friction models (*LuGre*, see [6]), the memoryless uncertain shape for $\phi(\cdot)$ in Fig. 1 works well in practice. It contains the three basic contributions: static friction or *stiction*, constant or Coulomb friction, and the linear-in-ω, viscous component [11].

Both nonlinearities $\gamma(\cdot)$ and $\phi(\cdot)$ can be affected by uncertainty. The source of uncertainty may be parameter drifts due *e.g.* to temperature changes in electronic components and wear and inherent friction uncertainty in the mechanical part. Some of the uncertainties have a dynamic nature (electrical time-constants) and at high frequencies the double integrator is not adequate to model the dynamics. But if one considers the medium-low frequency range, the non-dynamical, memoryless model is adequate in practice.

Open loop experimental data. To check the validity of this model, it will be tested on a d.c. servomotor in our laboratory. Preliminary open loop experiments are shown in Fig.2. The control voltage at the stator windings

F. Colonius, L. Grüne (Eds.): Dynamics, Bifurcations, and Control, LNCIS 273, pp. 117–130, 2002.

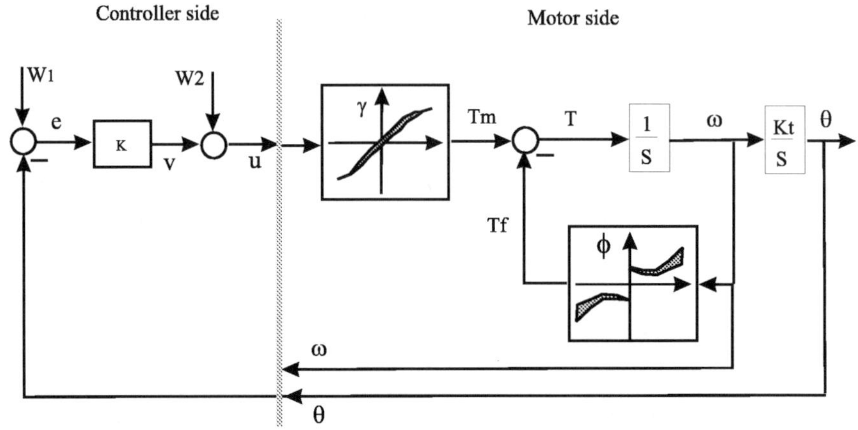

Fig. 1. Block diagram of a d.c. servomotor

is $u(t)$. The motor position $\theta(t)$ and speed $\omega(t)$ can be measured with a potentiometer and tachometer, respectively.

In Fig.2(a) the control input $u(t)$ is a triangular wave between ± 0.6 and period $240 sec$. The motor speed $\omega(t)$ shows clearly the presence of static friction. The acceleration $T(t) = d\omega(t)/dt$ is computed by filtered numeric derivative. Assuming the dependence in Fig.1, put

$$\begin{aligned} T(u,\omega) &= T_m - T_f = \gamma(u) - \phi(\omega) \\ \gamma(u) &= a_0 + a_1 u + a_2 u^2 + a_3 u^3 \\ \phi(\omega) &= b_0^+ + b_1^+ \omega + b_2^+ \omega^2 \text{ for } \omega > 0 \\ \phi(\omega) &= b_0^- + b_1^- \omega + b_2^- \omega^2 \text{ for } \omega < 0. \end{aligned} \tag{1}$$

The coefficients a_j, b_j^+, b_j^- were computed using least-squares, and the estimation of the nonlinearities $\hat{\gamma}, \hat{\phi}$ is shown in Fig.2(b)(c). It can be seen that $\hat{\gamma}$ is not too nonlinear, but the friction $\hat{\phi}$ has a hard, asymmetrical graph.

The residuals or model error, can be considered as an exogenous non-modelled torque $T_e = T(u,\omega) - \left(\hat{\gamma}(u) - \hat{\phi}(\omega)\right)$. But, as the model error is mainly due to the random nature of friction, let us obtain an *uncertain model of the friction*, 'loading' all the uncertainty inside $T_f = \phi(\omega)$. To do this, from data $u(t)$ and $T(t) = d\omega(t)/dt$, compute the time-varying friction $\phi(t) = \hat{\gamma}(u(t)) - T(t)$ and then plot $\phi(t)$ versus $\omega(t)$. This is shown in Fig.2(d).

PI+dither closed loop data. A second set of experiments were performed closing the loop as in Fig.1 and applying a PI controller with *dither* signal. *Dither* [9] [23] [21] is a high-frequency signal used for 'smoothing' the discontinuous effects of friction at low velocities. The advantage of dither for

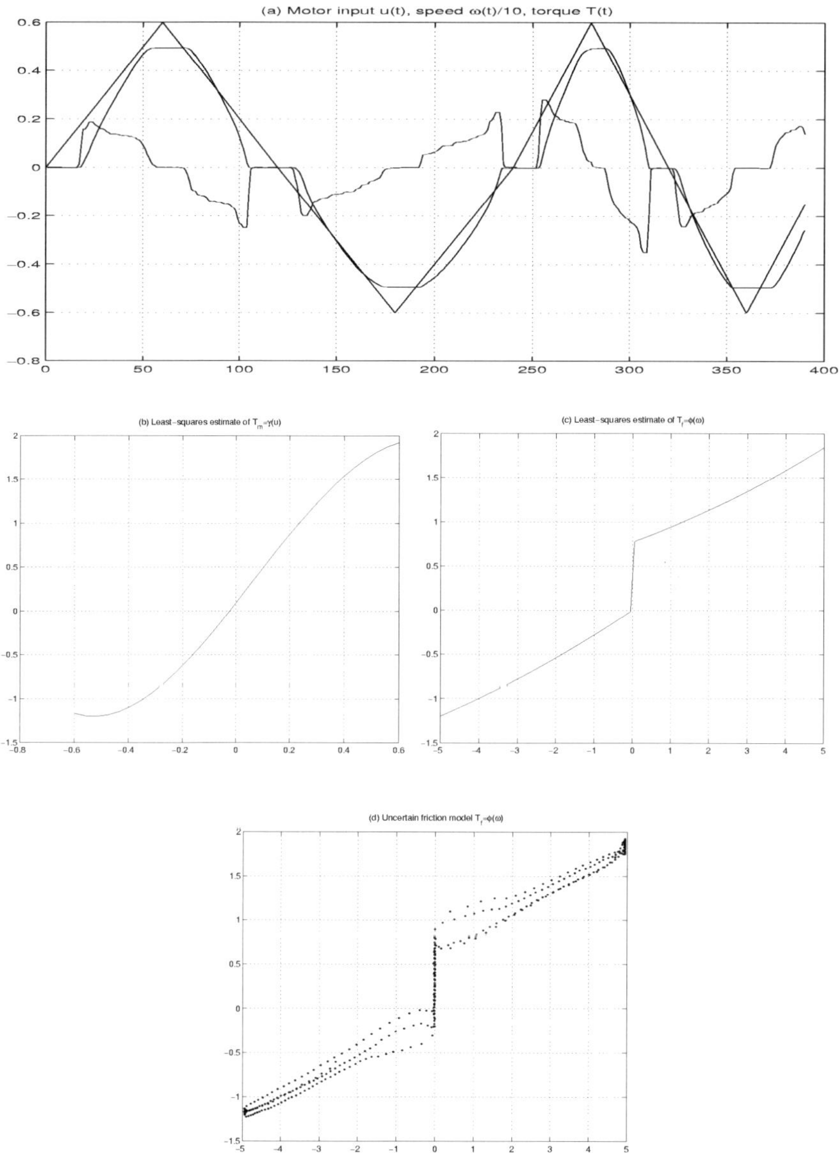

Fig. 2. Experimental data, open loop test

friction compensation is the simplicity of implementation: it does not require complex friction observers, as in the *LuGre* model. It works well in practice and can be used for extinction of undesired limit cycles [23].

In the experiments, the dither signal was applied at $w_2(t)$ (see Fig.1). It was chosen to be a square wave of $50Hz$ (the controller operates at $10Hz$), and amplitude $w_2 = \pm 0.3$. The PI controller was $K(s) = K_p + K_i/s$, with $K_p = K_i = 0.2$, driven by the position error $e(t) = w_1(t) - \theta(t)$. The position setpoint $w_1(t)$ was a chirp signal from $0.1Hz$ to $0.2Hz$. The resulting $\theta(t), \omega(t)$ are shown in Fig.3(a). The error $e(t) = w_1(t) - \theta(t)$ is reasonably small (it was impossible to follow the reference $w_1(t)$ without using dither).

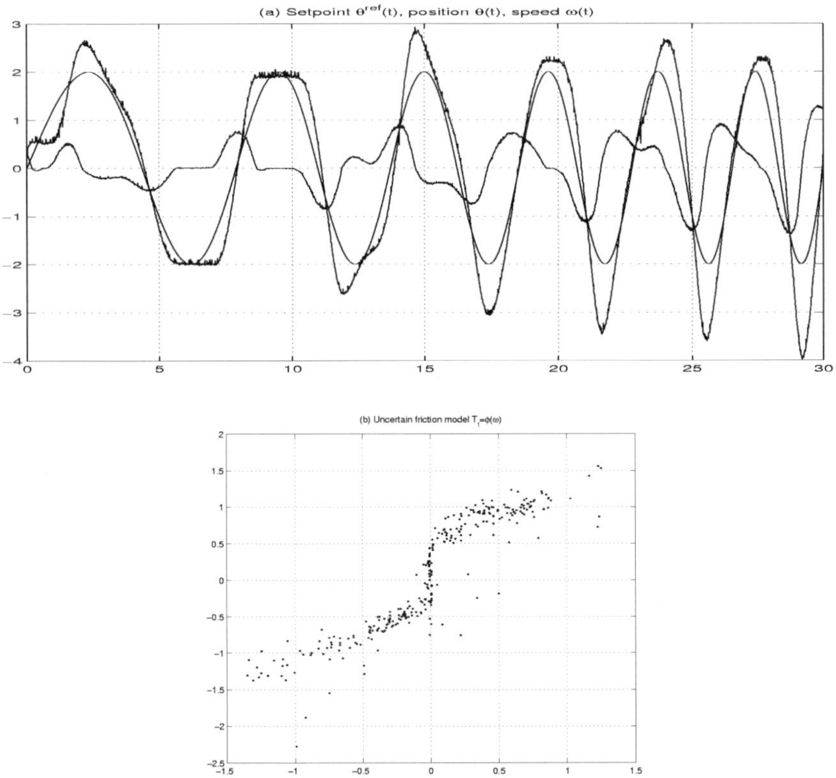

Fig. 3. Experimental data, closed loop test with dither

The same least-squares estimation as in the previous experiment was peformed. But in this case, *linear* models (gains) for $\hat{\gamma}(u) = \hat{K}_\gamma u$ and $\hat{\phi}(\omega) = \hat{K}_\phi \omega$ were obtained. Finally, all the uncertainty (residuals of model) was 'loaded' inside $T_f = \phi(\omega)$, and the resulting uncertain model of dithered-friction is shown in Fig.3(b). Compare Fig.2(d) (actual friction) and Fig.3(b) (equivalent friction usign dither). The advantages of dither are: (i) It works well (ii) The friction smoothing is effective (iii) The conic bounds are more 'tight' and the friction graph is more 'dense' inside this cone.

2 Conic analysis of uncertain friction

Conic analysis is a classical technique for dealing with nonlinear system stability [25] [19] [3] [5]. It provides results (small-gain conditions, circle and Popov criteria, etc.) which guarantee stability in a robust (although conservative) fashion. In this work, conic sector analysis will be combined (in the frequencial QFT setting) with harmonic balance in order reduce conservativeness and improve understanding of the system behaviour.

It can be easily seen, to adopt the conic setting, that the problem of designing the controller $K(s)$ in Fig. 1 can be reduced to the problem of designing the controller $C(s)$ in Fig. 4. This can be accomplished by 'loading' or 'confining' all the nonlinear uncertainty inside a SISO block Δ. For example, one can model γ by a linear, least-squares gain $\hat{K}_\gamma$, and then adopt for the friction $\phi(\cdot) =: \Delta$ an uncertain model like in Fig.2(d) or Fig.3(b). Furthermore, it can be seen that the uncertain block Δ can be 'centered around 0', that is, it can be confined to the sector:

$$\Delta \subseteq \text{Sector } [-\rho,\ \rho], \tag{2}$$

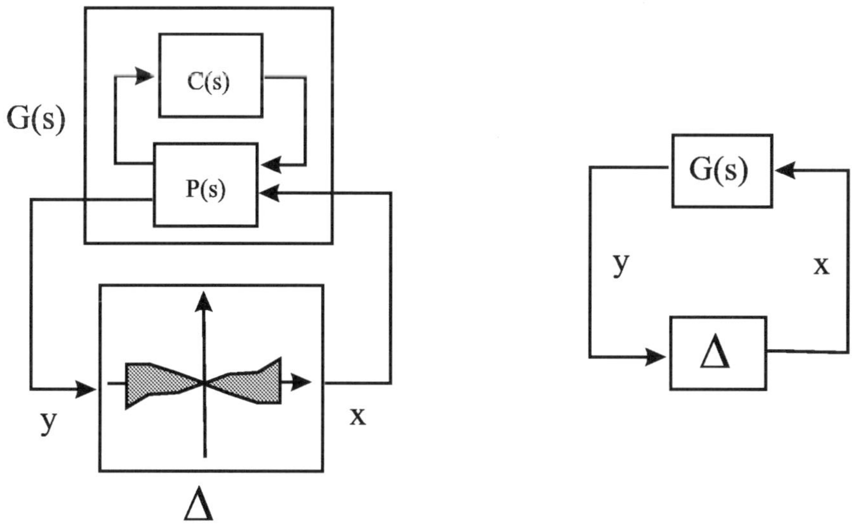

Fig. 4. Normalized loop $\{G(s), \Delta\}$, with $G(s) = \text{LFT}(P(s), C(s))$

where $\rho > 0$ is the cone *aperture* and where Sector$[a, b]$ with $a < b$ stands for the closed region in the graph space bounded between the lower slope $x = ay$ and the upper slope $x = by$. It is standard in conicity to apply *loop transformations* [25] that change any given sector (like $[1, \infty]$, valid for Fig.3b) into the normalized sector (2). The feedforward and feedback loop transformations which achieve this, are included inside the $P(s)$ in Fig.4, together with the motor double integrators. The connection $G(s) = \text{LFT}(P(s), C(s))$ between $P(s)$ and $C(s)$ that gives rise to $G(s)$ is called *'Linear Fractional Transformation'* (LFT), see [5]. From Fig.2(d) and Fig.3(b) and from the previous discussion it is natural to postulate for Δ a nonlinear, time-varying, uncertain nature:

$$\Delta : y(t) \longrightarrow x(t), \text{ such that } x(t) = \delta(t)\, y(t), \text{ and } |\delta(t)| \leq \rho \tag{3}$$

To formulate the problem of global asymptotical stability (GAS), let us define two related stability radii ([8], [14]), the *time-varying radius* $r_t(G(s))$ (aperture tolerance for ensuring stability with cones Δ of *time-varying* gains) and the *complex radius* $r_c(G(s))$ (aperture tolerance for cones Δ of *dynamical* uncertainties). More formally, choose a controllable and observable realization of $G(s) = c(sI - A)^{-1}b$. The matrices (A, b, c) are $n \times n$, $n \times 1$ and $1 \times n$, *resp.* Also, A has to be stable, because Δ contains the zero system $\delta(t) \equiv 0$. Let us denote by $\mathcal{L}$ the set of all the *linear, stable systems* described in this way. Then

$$r_t(G) := \inf\{\rho : \ \dot{x} = (A + b\delta c)x \text{ is not GAS for some } \delta,\ |\delta(t)| \leq \rho\} \tag{4.a}$$

$$r_c(G) := \inf\{\rho : \ \dot{x} = (A + bzc)x \text{ not GAS for some } z \in \mathbf{C},\ |z| \leq \rho\} \tag{4.b}$$

It is known that $r_t(G) \geq r_c(G) = (\max_\omega |G(j\omega)|)^{-1}$, for all $G(s)$, so that

$$\gamma_{tc} := \sup \left\{ \gamma_{tc}(G) = \frac{r_t(G)}{r_c(G)} \ : G \in \mathcal{L} \right\} \geq 1, \tag{4c}$$

but it is not known whether the largest gap γ_{tc}, for all $G(s)$, is finite or infinite, it is stated as an open problem in [18]. If the uncertainty Δ would be *dynamical*, it is well known (circle criterion, small gain condition) that the stability frequencial condition is

$$\rho < r_c(G(s)) = \frac{1}{\|G(s)\|_\infty} = \frac{1}{\max_\omega |G(j\omega)|}, \tag{5}$$

where ρ is the cone aperture. But, in the considered case of time-varying cones, the condition is:

$$\rho < r_t(G(s)). \tag{6}$$

Summarizing, the precise design conditions for the system in Fig. 4 are: given $P(s)$ and $\rho = \|\Delta\|$,

$$\begin{cases} \text{find} & C(s) \\ \text{such that} & \rho < r_t(G(s)) \\ \text{where} & G(s) = \text{LFT}(C(s), P(s)) \end{cases} \tag{7}$$

Recall that the standard approach based on the Circle Criterion is like (7) but using r_c in place of r_t, *i.e.*, given $P(s)$ and $\rho = \|\Delta\|$,

$$\begin{cases} \text{find} & C(s) \\ \text{such that} & \rho < r_c(G(s)) \\ \text{where} & G(s) = \text{LFT}(C(s), P(s)) \end{cases} \tag{8}$$

If the uncertainty is known to be time-varying, and from (4c), it is clear that (7), when compared to (8), is an improved and less conservative approach. However, this is at the cost of an extremely hard computational load, as the following remark explains.

Remark 1. Computation of $r_t(\cdot)$. Discussion on practical viability The computation of $r_t(\cdot)$ can be accomplished using the algorithm in [24]. This implies a bisection search of the aperture ρ, so that at each step of the search one has to solve a Hamilton-Jacobi-Bellman (HJB) equation. For n-dimensional systems, the HJB has to be solved over a mesh obtained by discretization of the unit sphere $S^{n-1} \subset \mathbf{R}^n$. The latter is an $(n-1)$-dimensional mesh, growing exponentially with n, so only for low dimensional systems it could be computed in reasonable time. As r_t is a quantity so hard to compute, it would be interesting to obtain computable bounds for it. If we would know a guaranteed bound $\hat{\gamma}_{tc} \geq \gamma_{tc}$, then we would have:

$$r_c(G) \leq r_t(G) \leq \hat{\gamma}_{tc} \cdot r_c(G) \tag{9}$$

The last upper bound has a great theoretical interest, see [18]. Moreover, (9) provides initial bounds for the bisection search in the algorithm in [24]. However, the upper bound in (9) is not directly useful for (7). Perhaps it might be used in some indirect way, but this requires further research.

Remark 2: On the finitness of γ_{tc}. (i) Experimental evidence. In *all* the examples found in the literature ([8] [14] [24]) it was always true that $r_t(G) \leq 2 \cdot r_c(G)$. It is not known to the authors any example with larger gap. So, all the known practical examples satisfy (by the moment) (9) with $\hat{\gamma}_{tc} = 2$.

(ii) Intuitive arguments. It seems that the systems $G(s)$ in [8] [14] [24] for which the gap $\gamma_{tc}(G)$ takes a large value, have $G(j\omega)$ with a *resonance-peak* at some $\omega = \omega_0$. An intuitive way of exploring the limit of this feature is to consider the narrow-bandpass system $G_0(s)$ given by, say, $G_0(j\omega_0) = \exp(-j\phi)$ and $G_0(j\omega) \approx 0$ for $\omega \neq \omega_0$. The loop $\{G_0(s), \Delta\}$ would be unstable if nonzero signals $x(t), y(t)$ can exist within the loop. It is easy to find a solution for the loop signals given by $y(t) = \cos(\omega_0 t - \phi)$, $\delta(t) = 2\cos(2\omega_0 t - \phi)$

and $x(t) = \delta(t)y(t)$, such that the first harmonic of $x(t)$ is $x_0(t) = \cos\omega_0 t$. So, the loop is made unstable with $|\delta(t)| \leq 2$, then $r_t(G_0) \leq 2$. On the other hand, $r_c(G_0) = 1$, so we have $\gamma_{tc}(G_0) \leq 2$. Then, we reach again to $\gamma_{tc}(G_0) \leq 2$ for a system G_0 for which there are intuitive reasons for giving a large $\gamma_{tc}(G_0)$.

So, the two previous comments suggest that perhaps $\hat{\gamma}_{tc} = 2$. However, this is a difficult open problem and requires further research. Going back to the conic analysis of friction, and after the problems with r_t, we will adopt, instead of the r_t-approach (time-varying uncertainty), the more usual r_c-approach. So, we will consider the loop in Fig. 5, and the friction uncertainty cone will be considered dynamic (or complex). So, the standard results [25] like small-gain, circle criterion, passivity, etc., will be the tools for conic analysis. The way how conicity is applied, in the QFT framework, can not be included for brevity, but details can be seen in [3],[4],[5].

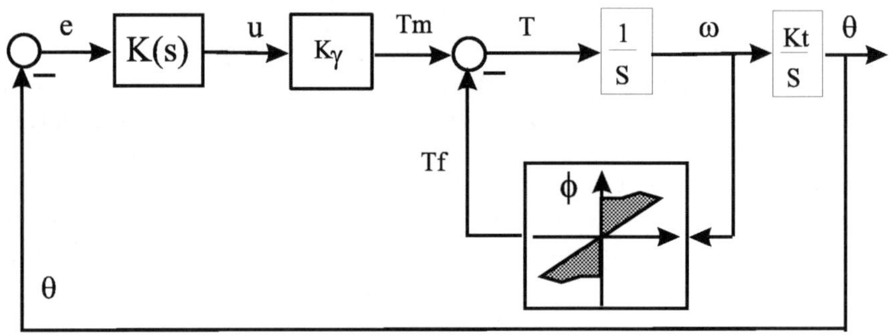

Fig. 5. Block diagram of d.c. motor with uncertain model of friction

3 Harmonic balance

The main drawback of the conic analysis is that usually yields conservative results. This section deals with a complementary method, the harmonic balance, which is approximate but usually gives less conservative results. In this way, it may be used as a guide to search for alternative stability conditions which could relax the conicity conditions. A dynamical system becomes unstable by the loss of local stability of the origin or by the emergence of other limit sets. The loss of local stability can be easily detected by means of linearization techniques. Nevertheless, the emergence of other limit sets is harder to detect. In this paper, an approximate technique, the *harmonic balance*, is used for the detection of limit cycles and undesirable equilibrium points. Therefore, assuming that neither quasi-periodic orbits nor chaos emerge, the method gives approximate bounds for the preservation of global stability.

As the nonlinearities considered in this paper may be asymmetric, the zero-order harmonic must be considered. Neglecting harmonics of order higher than one, the harmonic balance yieds the *dual-input describing function.* In this method, the limit cycles are assumed to be of the form:

$$y_k(t) = a_k^0 + \mathrm{Re}(a_k^1 e^{j\omega t}) \quad \omega > 0, k = 1, \ldots, n \tag{10}$$

Where y_k with $k = 1, \ldots, n$, are the outputs of the linear part, $a_k^0, a_k^1 \in \mathbb{R}$ and $a_k^1 \geq 0$. Notice that while the amplitude of the limit cycles is not necessarily the same for all the outputs, their frequency ω is assumed to be the same. Since $a_k^1 \in \mathbb{R}$ it is assumed that there are no phase shifts between the limit cycles. In a formal analysis these phase shifts should be considered but it can be shown that they do not affect to the final result [1]. This expression covers the case of symmetric limit cycles ($a_k^0 = 0$) and the equilibrium points ($a_k^1 = 0$). The harmonic balance [9] yields the following conditions:

$$(I + G(0)N^0(a^0, a^1, 0))a^0 = 0$$

$$(I + G(j\omega)N^1(a^0, a^1, \omega))a^1 = 0$$

where N^0 and N^1 are respectively the zero-order and first-order describing function matrices [9], and where $a^0 = (a_1^0, \ldots, a_n^0)^T$ and $a^1 = (a_1^1, \ldots, a_n^1)^T$. In order that these equations do not have a trivial solution in a^0, a^1, the following conditions must be fulfilled:

$$\det(I + G(0)N^0(a^0, a^1, 0)) = 0 \tag{11}$$

$$\det(I + G(j\omega)N^1(a^0, a^1, \omega)) = 0 \tag{12}$$

Therefore, Eqs. (11-12) can be used for determining the existence of limit cycles and equilibrium points. One possible use of this result is to perform an optimization of the absolute value of both determinants. In order to deal with uncertainties, the following theorem can be used

Theorem 1. [12] *In the feedback loop of Fig. 6 with a possibly asymmetric nonlinear block, no solution of the harmonic balance equations (11-12) will exist if the following conditions are satisfied:*

1. *There is no solution of Eqs. (11-12), for the nominal closed loop ($\Delta = 0$);*
2. *The model error satisfies one of the following conditions:*

$$\bar{\sigma}(\Delta(0)) < \frac{1}{\Gamma^0(a^0, a^1)} \qquad \forall a^0, a^1 \in \mathbb{R}^n \tag{13}$$

$$\bar{\sigma}(\Delta(j\omega)) < \frac{1}{\Gamma^1(a^0, a^1, \omega)} \quad \forall a^0, a^1 \in \mathbb{R}^n, \forall \omega > 0 \tag{14}$$

with

$$\Gamma^0 = \bar{\sigma}\left(G(0)N^0(I + G(0)N^0)^{-1}\right)$$

$$\Gamma^1 = \bar{\sigma}\left(G(j\omega)N^1(I + G(j\omega)N^1)^{-1}\right)$$

Furthermore, there exists a rational transfer function matrix $\Delta(s)$ *satisfying*

$$\bar{\sigma}(\Delta(0)) \leq \frac{1}{\Gamma^0(a^0, a^1)} \qquad \forall a^0, \forall a^1$$

$$\bar{\sigma}(\Delta(j\omega)) \leq \frac{1}{\Gamma^1(a^0, a^1, \omega)} \qquad \forall a^0, \forall a^1, \forall \omega$$

such that a solution of Eqs. (11-12) exists.

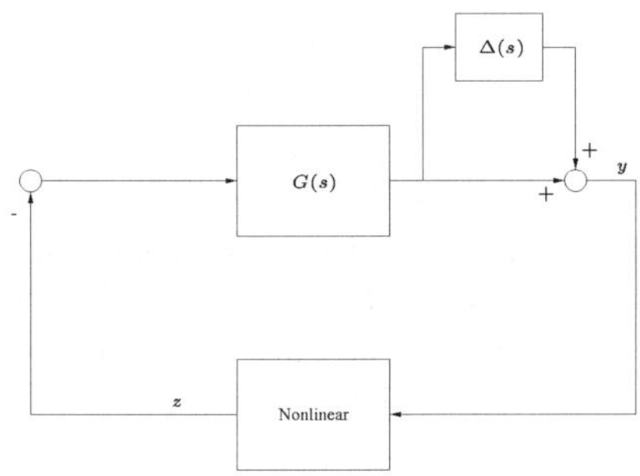

Fig. 6. Normalized loop for robust harmonic balance

The proof is very similar to the one of internal stability for linear systems (see, for example [13], p. 42). In linear systems the number of encirclements of the eigenvalues plot (characteristic loci) of a rational transfer function matrix $G(j\omega)$ around the -1 point changes when the plot crosses the -1 point. The proof in [13] looks for conditions that avoid this intersection. Here, we look for conditions that prevent the same intersection but substituting $G(j\omega)$ by $G(j\omega)N^1$.

In the case considered in this paper, the decomposition between linear and nonlinear parts must be performed carefully. As was pointed in [22] the natural decomposition of Fig. 4 is not useful for the describing function method. The decomposition of Fig. 7 can be used instead. Here, the output of the nonlinear part is the velocity, while its input is the applied torque. With this approach, the harmonic balance works well at the cost of the use of a dynamic nonlinear block. Thus, the describing functions N^0 and N^1 will really depend on the frequency.

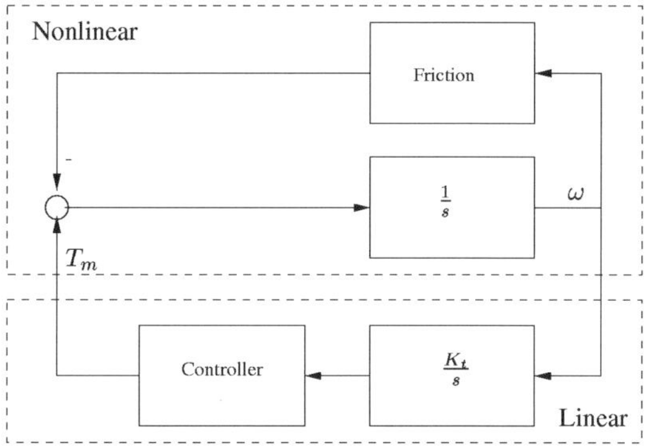

Fig. 7. Loop separation for the problem of motor with friction

4 Frequencial synthesis using QFT

Quantitative Feedback Theory (QFT) provides a natural framework for the integration of frequency domain restrictions in feedback control design. Recently, it has been used for the synthesis of nonlinear stabilizing compensators combining several frequency domain techniques [3],[4],[5], including circle criterion, Popov criterion, harmonic balance and multiplier theory. In this section, some of these results are used to compute stability restrictions, in the form of forbidden regions called *boundaries*, given for several frequencies and represented in the Nichols plane (Angle $(L(j\omega))$ versus $|L(j\omega)|$ in dB).

We will compute the boundaries applying the ideas in sections 2 and 3. After the discussion in section 2, the circle criterion (r_c radius) will be applied, using the robust version in [3]. From the section 3, the harmonic balance in Theorem 1 using the decomposition in Fig 7 will be used to obtain a second set of boundaries. So, consider the Fig. 5, where the friction ϕ is connected to the block $G(s)$ given by:

$$G(s) = \frac{A(s)}{1 + D(s)K(s)},$$

where $A(s) = 1/s$ and $D(s) = K_t K_g/s^2$. Here we have introduced some uncertainty in the parameters K_t and K_g, to allow 10% variations over its nominal values $K_t = K_g = 5$. Direct application of the Robust Circle Criterion [3], modelling the nonlinearity by the Conic Sector $[1, 100]$, results in the bounds in Fig. 8, computed for $L_0(j\omega) = D_0(j\omega)K(j\omega)$, being $D_0(s) = 25/s^2$. The boundaries are given at frequencies $\omega = 0.1, 1, 10, 30$ and 100 rad/s. In Fig. 8, the Nichols diagram of $L_0(j\omega)$ for $K(j\omega)$ given by a PID is also shown. It is given by $K(s) = 0.2(1+s/1.5)/s$, a minor modification of the PID previously used in Section 1. It can be seen how the PID satisfy the bounds for every frequency, thus the closed loop is guaranteed to be stable.

After that, the harmonic balance method of Section 3 is used for deriving bounds over the friction compensator, in order to avoid limit cycles. Obviously, the scalar version of Theorem 1 used. Results for the decomposition of Fig. 7 are given in Fig. 9, where limit cycles are predicted for ≈ 2 rad/s, when the above PID compensator is used.

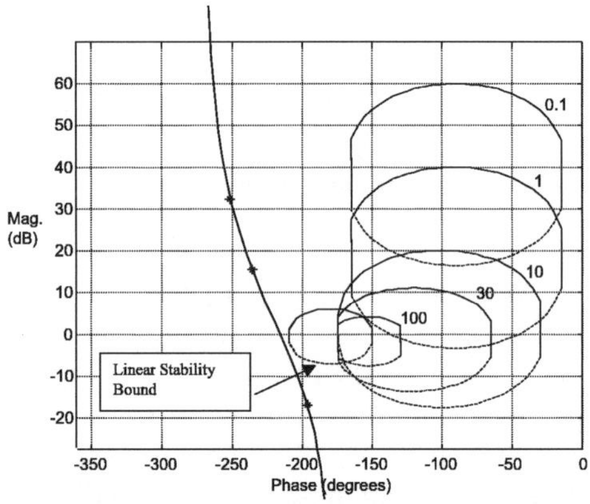

Fig. 8. Boundaries for the open loop $D_0(j\omega)K(j\omega)$ by the Circle Criterion

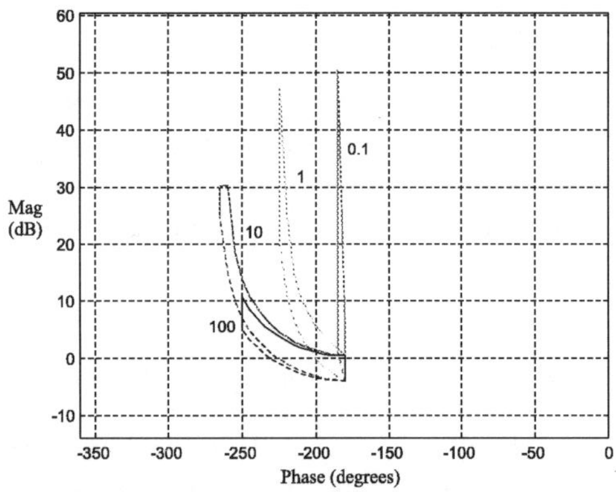

Fig. 9. Boundaries for the open loop $D_0(j\omega)K(j\omega)$ by the Harmonic Balance

5 Discussion

In Armstrong-Hélouvry *et.al.* [2] it is recognized, from a survey of engineers in industry, that the common techniques for friction compensation currently in service are (i) system hardware modification (in particular, lubrication) (ii) high servo gains, integral control (iii) dither (iv) adaptive and/or model-based compensation In this paper we have explored some nonlinear problems arising from (ii) and (iii). The approach followed is from a nonlinear point of view, trying to integrate, within the frequencial QFT setting, tools from conic analysis and harmonic balance. The integration seems fruitful and feasible due to the use of a common frequencial language. Some conclusions and open questions derived from this study are:

- Friction is an elusive, difficult problem in control systems. Due to this, using several techniques is, in principle, a good idea which may help in understanding all the various, related effects, which a single tool might be unable to describe [2].
- From laboratory experiments, we see that friction can be modelled as an uncertain, *time-varying* graph. Then, the natural approach to robustness is computing the 'time-varying radius' r_t. Hard complexity of r_t suggests to use instead the more conservative r_c radius. A practical conjecture on r_t and r_c is postulated.
- Harmonic balance is a key tool in friction control. Dual-input describing function was presented as a tool for determining limit cycles which could arise from bifurcations of equilibria. A theorem was presented which ensures robustness against limit cycles, with possibly asymmetric nonlinearities.
- Finally, Quantitative Feedback Theory (QFT) was introduced as the synthesis technique. QFT is a flexible, transparent tool, with an engineering appeal, for translating and integrating different stability and performance compromises. A particular numerical example, related to our problem, was solved within QFT. It remains for future work to make intensive laboratory experiments to verify the agreement between theory and practice.

Acknowledgements. Work supported by CICYT, Spain, under Project DPI2000-1218 *"Frequency techniques in control systems: effects of hard nonlinearities in electromechanical systems"* (FRENODEL).

References

1. Aracil, J. and Gordillo, F. (Eds.) (2000), *Stability Issues in Fuzzy Control*, Physica-Verlag.
2. Armstrong-Hélouvry B., P. Dupont and C. Canudas de Witt, 1994, "A Survey of Models, Analysis Tools and Compensation Methods for the Control of Machines with Friction", *Automatica*, 30, 7,1083-1138.
3. Baños, A., and Barreiro, A., 2000, "Stability of nonlinear QFT designs based on robust absolute stability criteria", Int. J. Control, 73, 1,74–88.
4. Baños, A., A. Barreiro, F. Gordillo and J. Aracil, "A QFT Framework for Nonlinear Robust Stability", to appear in *Int. Journal of Robust and Nonlinear Control*, I. Horowitz Special Issue, 2002.

5. Barreiro, A. and Baños, A., 2000, "Nonlinear robust stabilization by conicity and QFT techniques", Automatica, Vol. 36, No. 9, pp.1309-1320.
6. Canudas de Wit, C., H. Olsson, K.J. Astrom and P. Lichinsky, "Dynamic friction models and control design", in Proceedings of the 1993 American Control Conference, pp. 1920-1926, San Francisco, 1993.
7. Chait, Y., and Yaniv, O. 1993, "MISO computer-aided control design using the Quantitative Feedback Theory". Int. J. Robust and Nonlinear Control, 3, 47–54.
8. Colonius, F., and W. Kliemann (1990), 'Stability Radii and Lyapunov Exponents,' in *Control of Uncertain Systems*, D. Hinrichsen, B. Martensson, Eds. Birkhauser. pp. 19-55.
9. Cook, P.A., 1986, "Nonlinear Dynamical Systems", Prentice-Hall.
10. Cuesta, F., Gordillo, F., Aracil, J. and Ollero, A., 1999 "Global Stability Analysis of a Class of Multivariable Takagi-Sugeno Fuzzy Control Systems". IEEE Trans. Fuzzy Systems, 7, 5, 508–520.
11. Elhami, M.R. and D.J. Brookfield, (1997). "Sequential identification of Coulomb and Viscous Friction in Robot Drives", *Automatica*, 33, pp. 393-401.
12. Gordillo, F., Aracil J. and Ollero, A. (1999) "Robust Stability Analysis of MIMO Systems with Asymmetric Nonlinearities". Proceedings of the 14th World Congress of IFAC, Vol. E, 123–127.
13. Green, M. and Limebeer D.J.N., 1995, *Linear Robust Control*, Prentice-Hall.
14. Hinrichsen, D., and A.J. Pritchard (1990), 'Real and Complex Stability Radii: A Survey,' in *Control of Uncertain Systems*, D. Hinrichsen, B. Martensson, Eds. Birkhauser. pp. 119-162.
15. Horowitz, I., 1993, Quantitative Feedback Design Theory (QFT), QFT Publications, Boulder, Colorado.
16. Khalil, H.K., 1996, Nonlinear Systems, Prentice Hall, Upper Saddle River, NJ.
17. Mees, A.I.,1981, Dynamics of Feedback Systems, Wiley, New York.
18. Megretski, A.,(1999), 'How conservative is the circle criterion?,' in *Open Problems in Mathematical Systems and Control Theory*, V.D. Blondel, E.D. Sontag, M. Vidyasagar and J.C. Willems, Eds. Springer Verlag. pp. 149-151.
19. Megretski, A. and A. Rantzer, 1997, "System Analysis via Integral Quadratic Constraints", Automatica, 42 , 819–830.
20. Oldak, S., Baril, C., and Gutman, P. O., 1994, "Quantitative design of a class of nonlinear systems with parameter uncertainty", Int. Journal of Robust and Nonlinear Control, 4, 101-117.
21. Oldak, S. and I. Horowitz, "The sensor noise problem in dithered feedback systems", *Automatica* 28, pp. 1021-1025, (1992).
22. Olsson, H., "Describing function analysis of a system with friction", Procceedings of the 4th Conference on Control Applications, pp. 310–315, 1995.
23. Pervozvanski A. and C. Canudas de Wit, "Vibrational smoothing in systems with dynamic friction", Technical report. Laboratoire d'Automatique de Grenoble, 1997.
24. Wirth, F., "On the calculation of the time-varying stability radii", Int. Journal of Robust and Nonlinear Control, 8, 1043-1058 (1998).
25. Vidyasagar, M., 1978, Nonlinear Systems Analysis, Prentice Hall, Englewoods Cliffs.

Time-Optimal Stabilization for a Third-Order Integrator: a Robust State-Feedback Implementation

Giorgio Bartolini, Siro Pillosu, Alessandro Pisano, and Elio Usai

Universitá degli studi di Cagliari
Dipartimento di Ingegneria Elettrica ed Elettronica (DIEE)
Piazza D'Armi, I-09123 Cagliari, Italy

Abstract. This paper deals with the time-optimal control problem for a third-order integrator with saturating control, and a special emphasis is devoted to the state-feedback implementation of the control law and to the purpose of rejecting matched uncertainties. It is known that the time optimal control for the nominal plant is piecewise-constant of the bang-bang type, and is characterized, at most, by two commutations. The idea for constructing a closed-loop stabilizer is well-known since a long time [15], and it is based on backing out of the origin by integrating, with time reversed, the system equation. The aim of this procedure is to derive the control switching surface, i.e. the locus in the state space where the control commutations take place. In particular, in [15], the control switching surface was derived in parametric form, which cannot be used directly for real-time feedback purposes. In this work we derive an analytical form for the switching surface, and, then, we shall prove that it can be successfully used as a sliding manifold for the implementation of a sliding-mode controller; as it can be argued, this give robustness to the closed loop system, and it is demonstrated that global, finite-time, convergence to origin is retained also in presence of unknown drift and gain perturbations acting on the control channel. Further, the effect of control bandwidth limitation is discussed. Simulations confirm the proposed analysis.

1 Introduction

Optimal control is a powerful tool for identifying the "best" performance that can be attained by a given system. It allows for satisfying some specific constraints relevant to the state trajectory and/or to the control effort, and also gives the designer a systematic way to define the control input in accordance with suitable performance criteria involving the transient behaviour, the convergence time, the energy consumption and so on [2,6,9,16,15].

In special cases (e.g. LTI systems with an LQ-type performance index), a static state-feedback solution can be computed, the gain matrix being the solution of the associated Riccati equation [6,15]. The closed-loop form gives the LQ-optimal controller guaranteed robust performance (e.g. guaranteed lower bounds for the stability margins) [5,6].

When considering general control problems, where a closed-loop solution is not available, if the assumption of perfectly known system cannot be met

F. Colonius, L. Grüne (Eds.): Dynamics, Bifurcations, and Control, LNCIS 273, pp. 131–144, 2002.

the actual performance of the open-loop controller designed for the nominal system may be highly unsatisfactory. As a matter of fact, high sensitivity of optimal controllers with respect to modeling errors, and possible lack of stability, have been pointed out in the literature since a long time [5,17], and this constitute an intrinsic limitation to the practical application of the optimal control methodology.

For these reasons, optimal control is actually employed mainly for planning a desired trajectory for the system, deserving to other approaches the associated tracking control problem [12,4]. An on-line generator of time-optimal trajectories has been recently proposed in the literature [14], particularly suited for real-time feedback purposes.

Motivated by the above considerations, some research efforts have been made to develop closed-loop implementations of optimal control techniques, to reduce their sensitivity to modeling errors.

The state-feedback solution to the time-optimal control problem for the double integrator is well known in the literature [6], as well as the idea of using the associated switching surface as a sliding manifold for the implementation of a sliding-mode controller [10]. In [3] it was derived a different form for the time-optimal sliding manifold, which does not require the knowledge of the higher-order phase variable. The proposed solution is sub-optimal in the sense that starting from some region in the phase plane, the resulting control is characterized by two commutations (one more than necessary). It has been shown that the sliding-mode implementation of time-optimal [10] and sub-optimal [3] controllers is robust against unmodelled dynamics [10] and matched uncertainties [3], respectively.

As for the third-order integrator, the closed-loop implementation of the time-optimal stabilizer has been considered in [15] (Chapt. 2, pp. 139-141), where the switching surface was given in parametric form which cannot be directly used for feedback purposes. This work is devoted to extend the analysis in [15], by finding an analytical form for the switching surface, and then to study the properties of the sliding-mode controller which uses, as a sliding manifold, the time optimal switching surface. It is shown that it is always possible to properly set the magnitude of the discontinuous control in order to make attractive and invariant the time-optimal sliding manifold in spite of matched disturbances.

The paper is organized as follows: in the next Section 2, the results in [15] are briefly recalled for the readers' convenience, and an analytic form for the switching surface is derived. In Section 3, the convergence and robustness properties of the associated sliding-mode controller are investigated, and the effect of actuators' bandwidth limitations is discussed. Simulation results are reported in Section 4 to validate the proposed analysis.

2 Closed loop time-optimal stabilization for a third-order integrator

Consider system

$$\begin{cases} \dot{x}_1 = x_2 \\ \dot{x}_2 = x_3 \\ \dot{x}_3 = u \end{cases} \tag{1}$$

subjected to the constraint

$$|u| \leq U \tag{2}$$

where U is a positive constant. It is known that the solution to the time-optimal stabilization problem is a bang-bang control commuting between the extreme values $\pm U$, characterized, at most, by two commutations [6,15].

The class of admissible solutions can be parameterized as follows

$$u(k, T_0, T_1, T_2) = \begin{cases} kU & 0 \leq t < T_0 \\ -kU & T_0 \leq t < T_0 + T_1 \\ kU & T_0 + T_1 \leq t < T_0 + T_1 + T_2 \\ 0 & T_0 + T_1 + T_2 \leq t \end{cases} \tag{3}$$

where $k \in \{-1, 1\}$, $T_i \geq 0$, $i = 0, 1, 2$, and $t_f = T_0 + T_1 + T_2$ is the final time instant, at which the state reaches the origin of the state space. From $t = t_f$ on, the zero input keeps the system at the origin. A possible control law is pictured in the following Fig. 1, where the commutation time instants are explicitly defined as

$$\begin{cases} t_1 = T_0 \\ t_2 = T_0 + T_1 \end{cases} \tag{4}$$

A possible way to find a closed-loop form for the time optimal control is to

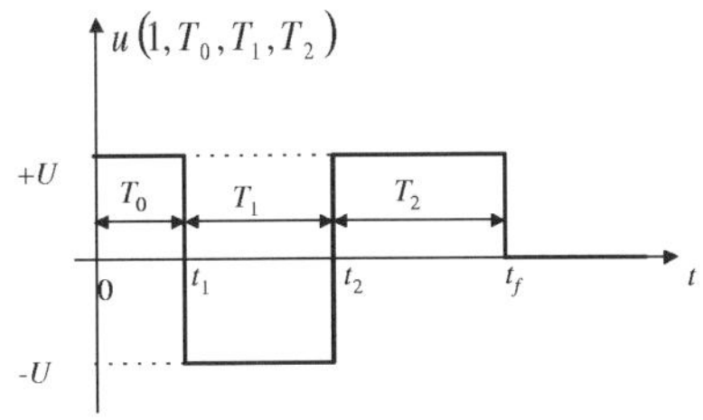

Fig. 1. The optimal control law $u(1, \cdot)$

derive the so-called switching surfaces, i.e. the loci in the state space where the control commutations take place. According to [15], this can be simply done by integrating the system backward in time, starting from the origin of the state space.

It yields that the second commutation ($t = t_2$) occurs along a line (which will be referred to as Γ), while the first one ($t = t_1$) occurs onto a surface (called Σ). Let Γ^+ and Γ^- be the lines associated with the second control commutation, and let $\mathcal{S}^+$ and $\mathcal{S}^-$ be the surfaces on which the first control commutation takes place. Note that the apex '+' refers to the control $u(1;\cdot)$ while the apex '−' corresponds to the symmetrical solution, $u(-1;\cdot)$.

A parametric description of the above loci has been derived in [15] (p. 140), making reference to the special case $U = 1$. We report as follows the parametric equations for arbitrary U, which can be easily derived by using the same procedure in [15]:

$$\Gamma^+ : \begin{cases} x_3 = -UT_2 \\ x_2 = \frac{1}{2}UT_2^2 \\ x_1 = -\frac{1}{6}UT_2^3 \end{cases} \qquad \Gamma^- : \begin{cases} x_3 = UT_2 \\ x_2 = -\frac{1}{2}UT_2^2 \\ x_1 = \frac{1}{6}UT_2^3 \end{cases} \tag{5}$$

$$\Sigma^+ : \begin{cases} x_3 = -U[(T_1 + T_2) - 2T_1] \\ x_2 = \frac{1}{2}U[(T_1 + T_2)^2 - 2T_1^2] \\ x_1 = -\frac{1}{6}U[(T_1 + T_2)^3 - 2T_1^3] \end{cases} \tag{6}$$

$$\Sigma^- : \begin{cases} x_3 = U[(T_1 + T_2) - 2T_1] \\ x_2 = -\frac{1}{2}U[(T_1 + T_2)^2 - 2T_1^2] \\ x_1 = \frac{1}{6}U[(T_1 + T_2)^3 - 2T_1^3] \end{cases} \tag{7}$$

The above parametric descriptions highlight the relationships between the characteristic loci and the parameters defining the control law (obviously $T_1, T_2 \geq 0$).

Eliminating the T_2 parameter from (5), it yields the following analytical expressions for Γ^+ and Γ^-

$$\Gamma^+ \equiv \{(x_1, x_2, x_3) \in R^3 : x_1 = \frac{1}{6U^2}x_3^3\ ,\ x_2 = -\frac{1}{2U}x_3^2\ ,\ x_3 < 0\} \tag{8}$$

$$\Gamma^- \equiv \{(x_1, x_2, x_3) \in R^3 : x_1 = \frac{1}{6U^2}x_3^3\ ,\ x_2 = \frac{1}{2U}x_3^2\ ,\ x_3 > 0\} \tag{9}$$

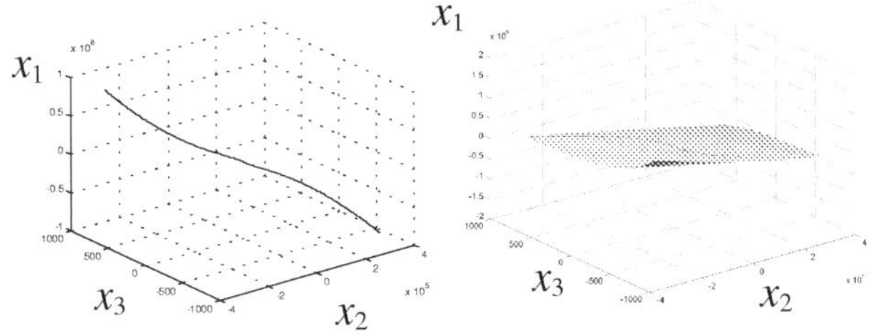

Fig. 2. The line Γ and the surface Σ for $U = 1$.

The lines Γ^+ and Γ^- are smoothly connected at the origin (see Fig. 2), and we can define

$$\Gamma \equiv \Gamma^+ \cup \Gamma^- \cup \{0,0,0\} \tag{10}$$

For further developments, it is useful to consider the projections on the $x_1 = 0$ plane of the lines Γ^+ and Γ^-, which will be referred to as $\mathcal{G}^+$ and $\mathcal{G}^-$, respectively, expressed as follows:

$$\begin{aligned} \mathcal{G}^+ &\equiv \{(x_2, x_3) \in R^2 : \; x_2 = -\tfrac{1}{2U} x_3^2 \;, \; x_3 < 0\} \\ \mathcal{G}^- &\equiv \{(x_2, x_3) \in R^2 : \; x_2 = \tfrac{1}{2U} x_3^2 \;, \; x_3 > 0\} \end{aligned} \tag{11}$$

Eliminating T_1 and T_2 from (6)-(7), we obtain the following analytical form for Σ^+ and Σ^-

$$\begin{aligned} \Sigma^+ &= \left\{(x_1, x_2, x_3) \in R^3 : x_1 = f^+(x_2, x_3) \;,\; (x_2, x_3) \in \mathcal{S}^+\right\} \\ \Sigma^- &= \left\{(x_1, x_2, x_3) \in R^3 : x_1 = f^-(x_2, x_3) \;,\; (x_2, x_3) \in \mathcal{S}^-\right\} \end{aligned} \tag{12}$$

where

$$\begin{aligned} f^+(x_2, x_3) &= -\tfrac{x_3^3}{3U^2} - \tfrac{x_2 x_3}{U} - \tfrac{1}{\sqrt{U}} \left(x_2 + \tfrac{x_3^2}{2U}\right)^{3/2} \\ f^-(x_2, x_3) &= -\tfrac{x_3^3}{3U^2} + \tfrac{x_2 x_3}{U} + \tfrac{1}{\sqrt{U}} \left(-x_2 + \tfrac{x_3^2}{2U}\right)^{3/2} \end{aligned} \tag{13}$$

and

$$\begin{aligned} \mathcal{S}^+ &\equiv \; \{(x_2, x_3) \in R^2 : x_2 > -\tfrac{1}{2U} x_3 |x_3|\} \\ \mathcal{S}^- &\equiv \; \{(x_2, x_3) \in R^2 : x_2 < -\tfrac{1}{2U} x_3 |x_3| \;\} \end{aligned} \tag{14}$$

Σ^+ and Σ^- are smoothly connected along the line Γ (see Fig. 2), then, it makes sense to define the following surface

$$\Sigma \equiv \Sigma^+ \cup \Sigma^- \cup \{0,0,0\} \tag{15}$$

As it was shown in [15], Σ separates R^3 in two regions, called V^+ and V^-, which constitute the part of space above, and below, the surface Σ, respectively (see Fig. 2), i.e.

$$\begin{aligned} V^+ &\equiv \{(x_1, x_2, x_3) \in R^3 : x_1 < f(x_2, x_3)\} \\ V^- &\equiv \{(x_1, x_2, x_3) \in R^3 : x_1 > f(x_2, x_3)\} \end{aligned} \tag{16}$$

where

$$f(x_2, x_3) = \begin{cases} f^+(x_2, x_3) & \text{if } (x_2, x_3) \in \{\mathcal{S}^+ \bigcup \mathcal{G}^-\} \\ f^-(x_2, x_3) & \text{if } (x_2, x_3) \in \{\mathcal{S}^- \bigcup \mathcal{G}^+\} \end{cases} \tag{17}$$

In the following Lemma, a simple algorithm for calculating the open-loop time-optimal control, which relies on the above considerations, is given.

Lemma 1. System (1), with initial condition $\mathbf{x}(0) = [x_1(0), x_2(0), x_3(0)]$, is driven to the origin in minimum time by a control of the type (3), iff the parameters k, T_0, T_1 and T_2 are evaluated by solving the nonlinear algebraic system

$$\begin{cases} x_1(0) = -\frac{1}{6}k[(T_0 + T_1 + T_2)^3 - 2(T_0 + T_1)^3 + 2T_0^3] \\ x_2(0) = \ \ \frac{1}{2}k[(T_0 + T_1 + T_2)^2 - 2(T_0 + T_1)^2 + 2T_0^2] \\ x_3(0) = -k[(T_0 + T_1 + T_2) - 2(T_0 + T_1) + 2T_0] \end{cases} \tag{18}$$

Proof: Relationships (18) are nothing but the parametric description (for $T_0, T_1, T_2 \geq 0$) of the domains V^+ and V^-. They can be easily derived by integrating backward in time, in the time interval $t \in (0, t_1)$, the trajectories of system (1), (3) (with $u = u(1, \cdot)$ and $u = u(-1, \cdot)$, respectively) and initial conditions (6) and (7), respectively. By construction, (18) can be understood in terms a couple of algebraic systems (obtained for $k = 1$ and $k = -1$), only *one* of which can be *uniquely* solved for non-negative values of T_0, T_1, and T_2 to compute the optimal control switching instants. This latter statement relies on the global existence Theorem ([15], Chapt. 2, Th. 8 and 17), and uniqueness Theorem ([15], Chapt. 2, Th. 19, Corollary 4) of time-optimal control law for linear stationary systems. On the basis of such Theorems, it can be claimed that only one between the two algebraic systems (18) admits a solution over the admissible set $(T_0, T_1, T_2) \in (R_0^+ \times R_0^+ \times R_0^+)$, where $R_0^+ \equiv R^+ \cup \{0\}$.

By (16) one can identify a-priori, on the basis of the initial conditions, if the solvable system is either (18) with $k = 1$ or (18) with $k = -1$. In particular, the condition to be checked is the belonging of $(x_1(0), x_2(0), x_3(0))$ to the domain V^+ or to the complementary one V^-. $\triangleleft$

Remark 1. The same analysis has been performed making reference to the n-th order integrator. In that case, the time-optimal control is characterized, at most, by $n-1$ commutations between the extreme values $\pm U$. Thus, the parameters defining the control are the coefficient $k = \pm 1$ and the parameters T_i, $i = 0, 2, \ldots, n-1$, and system (18) can be generalized as

$$x_j(0) = -\frac{(-k)^j}{(n-j+1)!}\left\{\left(\sum_{i=0}^{n-1} T_i\right)^{n-j+1} - 2\sum_{k=2}^{n}\left[(-1)^k\left(\sum_{i=0}^{n-k} T_i\right)^{n-j+1}\right]\right\}$$

$$j = 1, 2, \ldots, n \tag{19}$$

Having expressed the characteristic loci in terms of the system state, the closed-loop time-optimal stabilizer can be expressed as follows:

$$u = \begin{cases} U & \text{if } (x_1, x_2, x_3) \in (V^+ \cup \Sigma^- \cup \Gamma^+) \\ -U & \text{if } (x_1, x_2, x_3) \in (V^- \cup \Sigma^+ \cup \Gamma^-) \\ 0 & \text{if } (x_1, x_2, x_3) = (0, 0, 0) \end{cases} \tag{20}$$

where the domains V^+, V^-, Σ^+, Σ^-, Γ^+ and Γ^- are defined in (8)–(17).

Note that (20) is formally equivalent to the last formula in [15], p. 141, thus we rely on the costructive proof in [15] to claim its effectiveness. In the present treatment, we have expressed the relevant loci depending on the actual system state, thus making the closed-loop controller directly implementable.

The actual behaviour of controller (20) exactly matches that of (3), i.e., after (at most) two commutations system (1) reaches the origin in minimum time. After the origin has been reached, the control is switched off, so that the system state remains confined to the origin.

3 Sliding-mode implementation of the time-optimal controller

In this section we consider the perturbed triple integrator

$$\begin{cases} \dot{x}_1 = x_2 \\ \dot{x}_2 = x_3 \\ \dot{x}_3 = \varphi(\mathbf{x}, t) + g(\mathbf{x}, t)u \end{cases} \tag{21}$$

where $\varphi(\cdot)$ and $g(\cdot)$ are unknown terms satisfying the following inequalities

$$\begin{aligned} &|\varphi(\mathbf{x}, t)| \le F(\mathbf{x}, t) \\ &0 < G_1(\mathbf{x}, t) \le g(\mathbf{x}, t) \le G_2(\mathbf{x}, t) \end{aligned} \qquad \forall(\mathbf{x}, t) \tag{22}$$

$F(\mathbf{x},t)$, $G_1(\mathbf{x},t)$ and $G_2(\mathbf{x},t)$ being known functions.

The aim is to derive a robust terminal (i.e. finite-time) stabilizer for system (21) based on the above developed treatment. The idea is the same as in [10], i.e. that of using the switching surface of time-optimal control as a sliding surface for the implementation of a sliding-mode controller.

The control constraint (2) must be dispensed with, since the control effort has to be large enough to cope with the uncertainties. Moreover, since the reaching time is affected by the uncertainties, time-optimality cannot be ascertained, nevertheless, some degree of near time-optimality for the resulting closed-loop behaviour may be reasonably expected.

The design of a sliding-mode control system consists of two-steps: the first step is the choice of a manifold in the state space such that, once the state trajectory is constrained on it, the system behaviour, described by its zero-dynamics corresponding to the sliding output, exhibits the desired performance. The second step is the design of a control action capable of forcing the system state, after a finite transient time, to evolve permanently on the chosen manifold (accordingly called sliding manifold).

We shall prove that:

A. The manifold Σ in (12)-(15) is a valuable sliding manifold to achieve the terminal stabilization of system (21).

B. The manifold Σ can be made globally attractive and invariant under the action of a control discontinuous across the same manifold.

Item A means that the system dynamics restricted to the manifold Σ, i.e. the system zero-dynamics [8] with respect to the output variable

$$\sigma = x_1 - f(x_2, x_3) \tag{23}$$

where $f(\cdot)$ is defined in (17), (13), (11) and (14), is globally stable and finite-time converging to the origin.

Some care must be taken in defining the system behaviour on the sliding manifold, since after the sliding manifold has been reached the discontinuous control switches at theoretically-infinite frequency, and the system trajectory are described by a differential equation with a *discontinuous right-hand side.* Moreover, we are interested in solving the differential equation exactly on the discontinuity manifold (i.e., the manifold Σ).

Discontinuous differential equations has been thoroughly analyzed by soviet researchers, within the framework of the variable structure systems (VSS) theory, and the seminal work in this field was a book by the russian mathematician A. Filippov [7]. He gave a geometrical interpretation of the solution of a discontinuous differential equation (DDE) along the discontinuity manifold, and a systematic procedure for regularizing DDEs.

In more recent times, a different approach for the regularization of DDEs was provided by V. Utkin [13], formalized in a systematic, analytical, form which makes the proposed approach more effective.

Basically, the method proposed by V. Utkin suggests to evaluate the solution of a DDE, on its manifold of discontinuity, by replacing the discontinuous term (in the actual case the control u) with the continuous one guaranteeing

$$\dot{\pi} = 0 \tag{24}$$

where $\pi = 0$ is the relevant manifold of discontinuity ($\sigma = 0$ for the considered case). Such continuous control action is referred to as the "equivalent control" (and called u_{eq}).

It will be shown that, when constrained onto the manifold Σ by discontinuous control, the "equivalent" system dynamics of the system, is that of a triple integrator subjected to the time-optimal stabilizer. Thus, all the trajectories of the system (21) turn out to converge to the origin in finite-time.

The equivalence of the two methods (the Filippov method and the equivalent control method) was also demonstrated in [13] for nonlinear systems with affine dependence on the control. Equivalence may be lost if more general classes of systems are dealt with.

Proof of step B is a standard application of sliding mode control analysis techniques; it is shown that if the control effort is sufficiently large then the so-called "reaching condition"

$$\dot{\sigma}\sigma \leq -H(\cdot)|\sigma| \quad H(\cdot) > 0 \tag{25}$$

is satisfied for some strictly positive function $H(\cdot)$, thus guaranteeing the finite time convergence of σ to zero and the invariance of the manifold $\sigma = 0$.

The following theorem summarizes the proposed strategy, and give sufficient conditions for the finite-time stabilization of system (21)-(22) through discontinuous feedback.

Theorem 1. *System* (21)-(22) *is globally stabilized in finite-time by the state-feedback control*

$$u = -\frac{U + F(\mathbf{x},t) + \xi^2}{G_1(\mathbf{x},t)}\text{sign}(\sigma) \qquad \xi \neq 0 \tag{26}$$

where $\sigma = \sigma(\mathbf{x}; U)$ *is defined in* (23), (17), (13), (11), (14) *and* U *is a positive arbitrary constant.*

Proof: The proof follows the above-sketched path:

Proof of step A: The system zero-dynamics with respect to the output variable σ is stable and finite-time converging to the origin.

According to previous considerations, in order to compute the system trajectories on the manifold Σ, it must be evaluated the equivalent control to be put in (21) in place of the discontinuous switching control (26). Simple computations show that the derivative of the σ variable is

$$\dot{\sigma} = \Phi(x_2, x_3)\left[\frac{1}{U}(\varphi(\mathbf{x}, t) + g(\mathbf{x}, t)u) + a(x_2, x_3)\right] \tag{27}$$

where

$$\Phi(x_2, x_3) = \begin{cases} \Phi^+(x_2, x_3) & \text{if } (x_2, x_3) \in \{\mathcal{S}^+ \bigcup \mathcal{G}^-\} \\ \Phi^-(x_2, x_3) & \text{if } (x_2, x_3) \in \{\mathcal{S}^- \bigcup \mathcal{G}^+\} \end{cases} \tag{28}$$

$$\begin{cases} \Phi^+(x_2, x_3) = x_2 + \frac{1}{U}x_3^2 + \frac{3}{2\sqrt{U}}x_3\left(x_2 + \frac{1}{2U}x_3^2\right)^{1/2} \\ \Phi^-(x_2, x_3) = -x_2 + \frac{1}{U}x_3^2 - \frac{3}{2\sqrt{U}}x_3\left(-x_2 + \frac{1}{2U}x_3^2\right)^{1/2} \end{cases} \tag{29}$$

$$a(x_2, x_3) = \begin{cases} 1 & \text{if } (x_2, x_3) \in \{\mathcal{S}^+ \bigcup \mathcal{G}^-\} \\ -1 & \text{if } (x_2, x_3) \in \{\mathcal{S}^- \bigcup \mathcal{G}^+\} \end{cases} \tag{30}$$

with domains $\mathcal{S}^+$, $\mathcal{S}^-$, $\mathcal{G}^+$ and $\mathcal{G}^-$ as in (11) and (14).

The definition of the equivalent control (i.e. the solution of $\dot{\sigma} = 0$) yields

$$u_{eq}(t) = -\frac{1}{g(\mathbf{x}, t)}\left[\varphi(\mathbf{x}, t) + Ua(x_2, x_3)\right] \tag{31}$$

Substituting (31) into (21), and considering (30), it yields

$$\begin{cases} \dot{x}_1 = x_2 \\ \dot{x}_2 = x_3 \\ \dot{x}_3 = u_{\Sigma}^{opt} \end{cases} \tag{32}$$

where u_{Σ}^{opt} is

$$u_{\Sigma}^{opt} = \begin{cases} U & \text{if } (x_2, x_3) \in \{\mathcal{S}^- \bigcup \mathcal{G}^+\} \\ -U & \text{if } (x_2, x_3) \in \{\mathcal{S}^+ \bigcup \mathcal{G}^-\} \\ 0 & \text{if } (x_2, x_3) \equiv (0, 0) \end{cases} \tag{33}$$

i.e., nothing but the restriction to the surface Σ of the time optimal control law (20).

Therefore, it is demonstrated that the zero-dynamics of the system, when constrained onto the manifold Σ is that of a triple integrator subjected to the time-optimal stabilizer. Thus, the trajectories of the system (21) converge to the origin in finite-time.

Proof of step B. Control (26) makes globally attractive and invariant the manifold $\sigma = 0$.

The objective is to show that a reaching condition of the type (25) is globally satisfied. Unlike the previous case (see Sect. 2), note that the coefficient U represents no more a pre-specified bound for the control magnitude, but it is actually only an arbitrary positive parameter, entering as a scaling factor in the definition of the sliding manifold.

The derivative of the sliding variable σ is given in (27)-(30). It is plain to verify that $\Phi(x_2, x_3) > 0$ over the whole $x_1 = 0$ plane, therefore, as $|a(x_2, x_3)| = 1$, control (26) sets the sign of $\dot{\sigma}$, and, thus, the following relationship holds

$$\dot{\sigma}\sigma \leq -\frac{1}{U}\Phi(x_2, x_3)\xi^2|\sigma| \tag{34}$$

Since (34) is formally equivalent to (25), the global, finite-time, convergence of σ to zero is guaranteed. ▷

Remark 2: The actual implementation (or even a numerical simulation) of the control law (26) reveals that its switching frequency will be very high but finite, according to the real actuator bandwidth. As a consequence the system trajectories will be constrained within a small neighbour of the sliding manifold (boundary layer, [13]), whose size vanishes as the switching frequency approaches infinity.

Let T be the inverse of the maximum switching frequency (i.e. the minimum time-delay between subsequent control switchings). Simulations show that the real accuracy is proportional to the third power of T, i.e., in the steady-state one has

$$|\sigma| \sim O(T^3) \tag{35}$$

As far as the state variables are considered, it can be derived from (35) that the obtained accuracy is

$$|x_1| \sim O(T^3) \qquad |x_2| \sim O(T^2) \qquad |x_3| \sim O(T) \tag{36}$$

4 Simulation results

To confirm the present analysis, some simulations results are presented hereafter. Consider system (1)-(2) with $U = 1$ and initial conditions

$$\begin{aligned} x_1(0) &= -13/6 \\ x_2(0) &= 1.5 \\ x_3(0) &= -1 \end{aligned} \tag{37}$$

Solving system (18), one can compute the time-optimal control as $u(1,1,1,1)$. It can be concluded from the analysis of Fig. 3, where control law in (20) is depicted, that the response of the proposed closed-loop stabilizer exactly matches the time-optimal control. Fig. 4 reports the state behaviour. In a

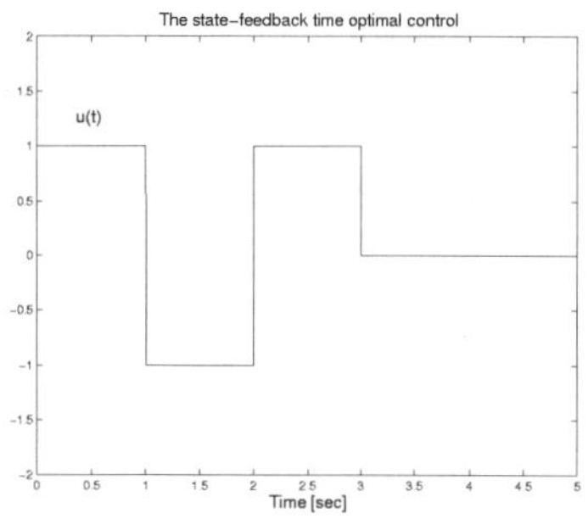

Fig. 3. Nominal system. The state-feedback time-optimal control.

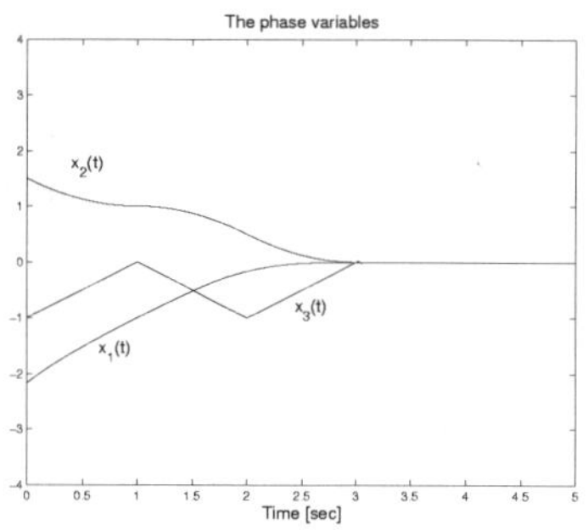

Fig. 4. Nominal system. The phase variables.

second test, the perturbed plant (21) has been considered, where the drift and gain term are given by

$$\begin{aligned} \varphi(\cdot) &= 4 + sin(t) + cos(10t) \\ g(\cdot) &= 1 + 0.5sin(2t) \end{aligned} \tag{38}$$

with initial conditions as in (37). The robust convergence to the origin of the system state despite of the unknown perturbations is highlighted in Fig. 5, while the monotonic convergence to zero of the sliding variable σ is shown in Fig. 6.

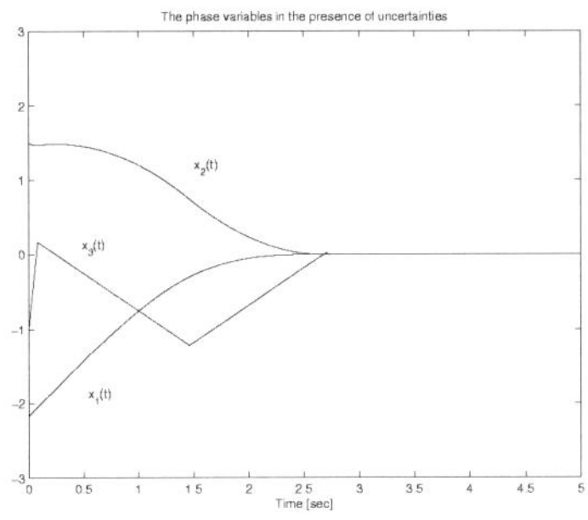

Fig. 5. Perturbed system. The phase variables.

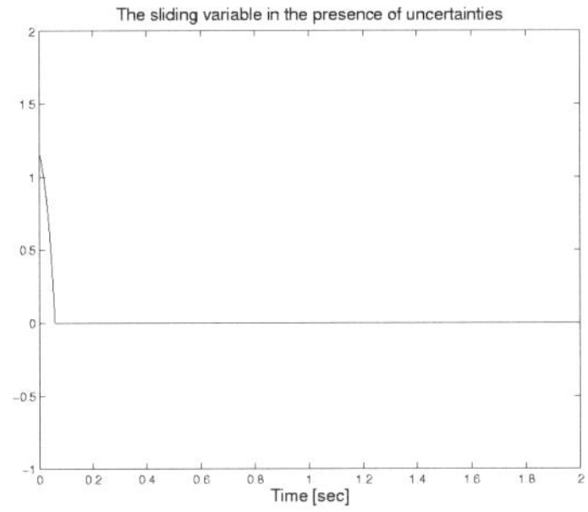

Fig. 6. Perturbed system. The sliding variable $\sigma(t)$.

The dependence of the boundary layer size on the switching delay, quantified by (35), is investigated by performing two simulations with a different sampling step, i.e., $T = 10^{-3}$ and $T = 10^{-4}$. As for the state variable x_1, the expected improvement of accuracy is 1000, and we obtained $|x_1| \leq 10^{-5}$ and $|x_1| \leq 10^{-8}$, respectively, confirming (35).

5 Conclusions

A robust state-feedback implementation of the time-optimal stabilizer for a third-order integrator has been dealt with by means of the sliding mode approach. The effect of model uncertainties has been investigated, and it has been shown that the proposed closed-loop realization of the time-optimal control is robust against matching perturbations, provided that the control effort is sufficiently large. Simulation results have been provided to validate the present analysis.

Acknowledgements

The authors would like to ackowledge Dr. Ing. C. Argiolas, former student of Electrical Engineering at the University of Cagliari, for his useful work on preliminary analysis and simulations.

References

1. Argiolas, C. (2000) Controllo Ottimo a Ciclo chiuso per un triplo integratore, (Closed-loop Optimal Control for a third-order integrator) Laurea Thesis, Univ. of Cagliari (in italian).
2. Athans, M., Falb, P.L. (1966) Optimal Control: An Introduction to the Theory and its Applications, McGraw Hill, NY.
3. Bartolini, G., Ferrara, A., Usai, E. (1997) Application of a sub-optimal discontinuous control algorithm for uncertain second-order systems, Int. J. Rob. Nonlin. Contr., **7**, 299-319
4. Bartolini, G., Sanna, S., Usai, E. (1998) Robust Near-Optimal Control via Unchattering Sliding-Mode Control, in Computing anticipatory systems-CASYS, First Int. Conf., Dubois ed., AIP Conf. Proc., vol. 437, pp. 269-283, American Inst. of Physics, Woodbury, NY.
5. Doyle, J.C. (1978) Guaranteed margins for LQG regulators. IEEE Trans. Aut. Contr., 23, 4, 756-757.
6. Kirk (1970) Optimal Control Theory, Prentice Hall, Englewood Cliffs, NJ.
7. Filippov, A. (1988) Differential Equations with discontinuous right-hand side, Kluwer, Dordrecht, The Netherlands.
8. A. Isidori (1995) Nonlinear Control Systems. Third edition, Springer Verlag, Berlin.
9. Leitmann, G. (1981) The Calculus of Variations and Optimal Control, Plenum Press, NY.
10. Newman, W.S. (1990) Robust Near Time-Optimal Control. IEEE Trans. Aut. Contr., 35, 7, 841-844.
11. Pillosu, S. (2000) Stabilizzazione robusta di sistemi a grado relativo tre (Robust Stabilization of relative-degree-3 systems), Laurea Thesis, Univ. of Cagliari (in italian).
12. Sakawa, A., Shindo, Y. (1982) Optimal Control of container cranes. Automatica, 18, 3, 257–266.
13. Utkin, V.I. (1992), Sliding Modes in Control and Optimization, Springer Verlag, Berlin.
14. Zanasi, R., Guarino Lo Bianco, C., Tonielli , A. (2000) Nonlinear filters for the generation of smooth trajectories, Automatica, 36, 3, 439-448.
15. Lee E.B. and Markus L., (1967) Foundations of Optimal Control Theory, Wiley.
16. Zelikin, M.I., Borisov, V.F. (1994) Theory of Chattering Control with Applications to Astronautics, Robotics, Economics, and Engineering, Birkhauser, Boston.
17. Zinober, A.S.I., Fuller, A.T. (1973) The sensitivity of nominally time-optimal control systems to parameter variation, Int.J. of Control, 17, 673-703.

Stability Analysis of Periodic Solutions via Integral Quadratic Constraints

Michele Basso, Lorenzo Giovanardi, and Roberto Genesio

Università di Firenze, Dipartimento di Sistemi e Informatica, Via di S. Marta 3, I-50139 Firenze, Italy

Abstract. The paper considers stability of periodic solutions in a class of periodically forced nonlinear systems depending on a scalar parameter and subject to disturbances. A result concerning local existence of a family of periodic solutions for such systems is also given. The stability analysis – based on a combined use of linearization techniques and frequency-domain stability criteria expressed via Integral Quadratic Constraints – can be efficiently performed in terms of Linear Matrix Inequalities. An application example is carried out for illustrative purposes.

1 Introduction

Stability analysis of periodic solutions in nonlinear systems is a classical and widely studied subject in nonlinear science [1,2]. A renewed interest in this topic has emerged in recent years, also due to its significant relations with bifurcation and chaos control, where in fact stabilization of periodic solutions is often a satisfactory target [3].

For a single periodic solution, Floquet analysis provides a full characterization of stability in terms of characteristic multipliers. Application of this method requires each considered solution to be completely known, at least in numerical form. On the other hand, in many applications one has to deal with a whole *family* of periodic solutions, as it is the case for periodically forced nonlinear systems in which the amplitude (or frequency) of the forcing signal is a varying parameter. Besides, possible *disturbances* entering the system equations must be taken into account, so that stability robustness has to be guaranteed with respect to both parametric variations and unknown disturbances.

A useful, though conservative tool to tackle this problem is represented by a combined use of linearization techniques and frequency-domain stability criteria. In [4] perturbation methods were used to derive a result concerning local existence of a family of periodic solutions for a general class of periodically forced nonlinear systems. Multivariable circle criterion was subsequently used to develop a general framework for the stability analysis, with a view towards the design of suitable feedback controllers able to extend the family of stable periodic solutions.

In this paper we give an improved existence result which embeds a first-order estimate of the solution as a function of the amplitude of the sinusoidal

F. Colonius, L. Grüne (Eds.): Dynamics, Bifurcations, and Control, LNCIS 273, pp. 145–157, 2002.

forcing input. Moreover, to conduct stability analysis we make use of the more general Integral Quadratic Constraints (IQC) tools [5], which generalize many existing stability results and possess an efficient computational counterpart in terms of Linear Matrix Inequalities (LMI) [6].

The paper is organized as follows. Section 2 introduces a motivating example. Section 3 formulates the problem and gives a preliminary result about the existence of periodic solutions for the considered class of systems. Section 4 develops the stability analysis framework based on linearization and IQC/LMI tools. Section 5 applies the proposed methodology to the example introduced before. Some concluding remarks end the paper in Section 6.

Notation

$\mathbf{Z}$: set of integer numbers;
$\mathbf{R}$: set of real numbers;
$\mathbf{R}^n$: space of n-component real (column) vectors;
$\mathbf{R}^{m\times n}$: space of $(m \times n)$ real matrices;
$'$: transpose operator;
*: transpose conjugate operator;
$|M| := \sqrt{\lambda_{\max}(M'M)}$: norm of matrix $M \in \mathbf{R}^{n\times n}$;
$x(t) \in \mathbf{R}^n$: n-valued time domain signal;
$\|x(t)\| := [x(t)'x(t)]^{1/2}$: euclidean norm of vector $x(t)$;
$\|x\|_\infty := \sup_{t\geq 0} \max_{k\in[1,n]} |x_k(t)|$: ∞-norm of signal $x(t)$;
$\|x\|_2 := [\int_0^\infty \|x(t)\|^2 \, dt]^{1/2}$: 2-norm of signal $x(t)$;
L_2: vector space of signals with finite 2-norm;
$X(j\omega)$: Fourier transform of $x(t) \in L_2$.

2 A motivating example

Figure 1 shows the schematic of a single-mode CO_2 laser in which the cavity losses are sinusoidally modulated by an intracavity electro-optic modulator at the operating frequency $f_0 = 100\,\text{KHz}$. The modulation is intended to produce burst-mode operation, since a high beam intensity is only obtained when the losses are below a certain threshold.

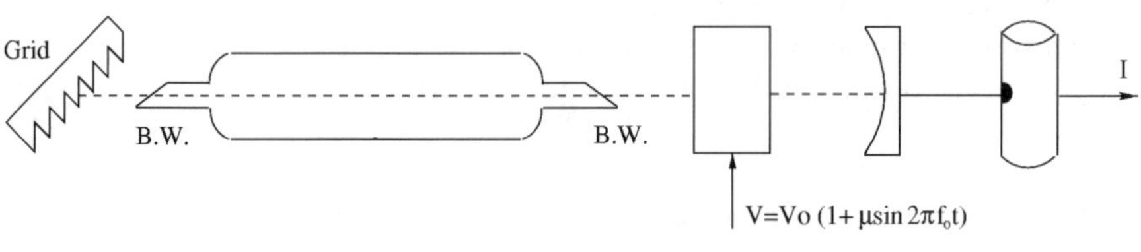

Fig. 1. Schematic of a CO_2 laser with modulated cavity losses

The system behavior can be described with a good accuracy by a two-level model in terms of the following third-order nonlinear differential equation [7]

$$\begin{cases} \dot{I}(t) &= k_0 I(t)\left[N(t) - (1 + \mu \sin 2\pi f_0 t)\right] \\ \dot{N}(t) &= -\Gamma N(t) + \gamma_R M(t) - 2k_0 N(t) I(t) + \gamma_P \\ \dot{M}(t) &= -\alpha M(t) + \beta N(t) \end{cases} \tag{1}$$

where I represents the field intensity and N, M are linear combinations of the populations of the two lasing states and the global populations of the manifolds of rotational levels. The dimensionless parameter μ representing the amplitude of the losses modulation acts on the peak intensity of the bursts, and is here considered as the bifurcation parameter. The remaining physical parameters are set as in Table 1.

Table 1. Parameter values expressed in s^{-1}

k_0	Γ	γ_R	γ_P	α	β
3.18×10^7	7.05×10^6	7.0×10^5	2.8714×10^5	6.767×10^5	6.626×10^6

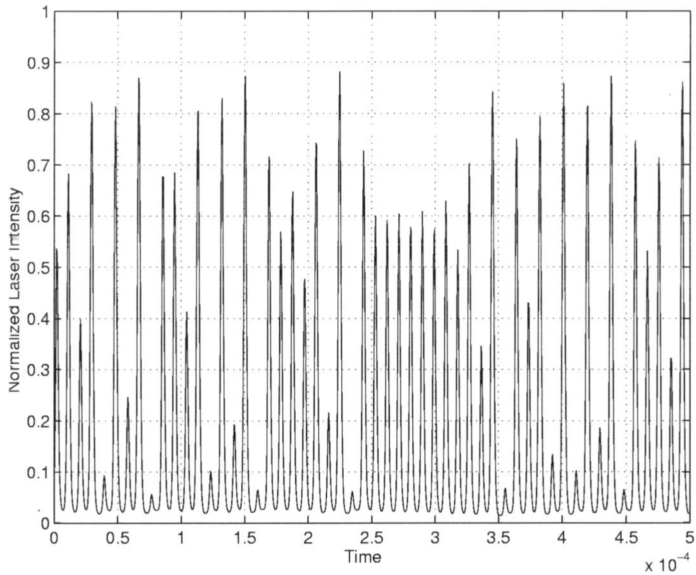

Fig. 2. Laser intensity at the modulation amplitude $\mu = 0.18$

For small values of the amplitude μ the system displays stable periodic solutions with the same period of the forcing input. Increasing μ gives rise to complex behavior which compromises regular operation of the apparatus (see Fig. 2). Such performance degradation is caused by a cascade of period-doubling bifurcations, as shown in the bifurcation diagram of Fig. 3 which approximates rather satisfactorily the experimental bifurcation diagram reported in [7].

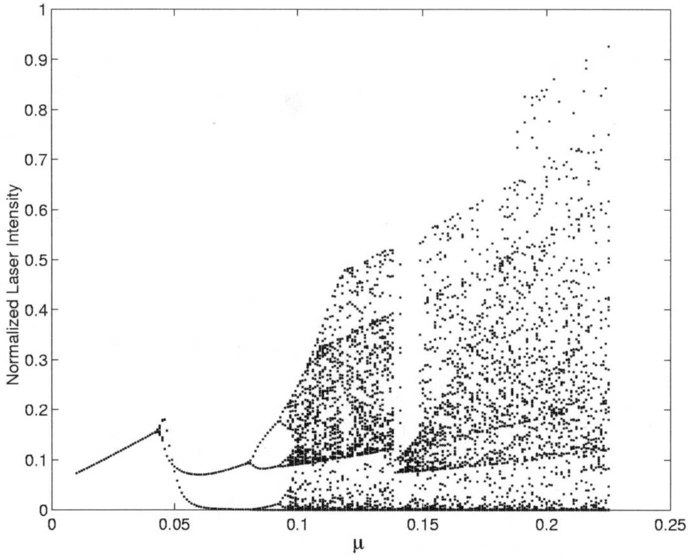

Fig. 3. Bifurcation diagram of the CO_2 laser model (1)

3 Problem formulation and preliminary results

Many application examples, as the one presented in the previous section, motivate the interest in stability analysis of families of periodic solutions in forced nonlinear systems. In particular, it would be useful to compute bounds on the maximal amplitude of the periodic forcing input guaranteeing local stability of the corresponding orbit. The aim of this paper is to tackle this problem by using local stability results derived from absolute stability theory. Despite its inherently conservative nature, this approach does not require much information on the considered solutions. Additionally, it can be suitably exploited for control design aimed at enlarging the range of regular operation [8].

The general class of nonlinear systems Σ under investigation is described by the ordinary differential equation

$$\Sigma : \qquad \dot{x} = f(x, d) + g(x, d)w \tag{2}$$

where $w \in \mathbf{R}$ is the reference scalar signal, $d = (d_1, \ldots, d_q) \in D \subset \mathbf{R}^q$ is the disturbance vector, and $x = (x_1, \ldots, x_n) \in X \subset \mathbf{R}^n$ is the state vector.

We are interested in studying the behavior of system Σ subject to varying-amplitude periodic reference signals w and bounded-energy disturbances d. More specifically, our aim is to investigate existence of state-space periodic solutions of the unperturbed system Σ_u obtained by setting $d = 0$ in (2), i.e.,

$$\Sigma_u : \qquad \dot{x} = f(x, 0) + g(x, 0)w \tag{3}$$

and their stability robustness with respect to disturbances d of bounded energy. In this respect, the exogenous signals in (2) are required to obey the following assumptions.

A1. The forcing input signal w belongs to the linearly parameterized set of periodic signals

$$\mathcal{W} := \{w_\mu(t) = \mu\, \overline{w}(t)\,,\ \mu \geq 0\,,\ \|\overline{w}\|_\infty \leq 1\,,\ \overline{w}(t) = \overline{w}(t+T)\,,\ T > 0\}. \tag{4}$$

A2. The disturbance input signal d belongs to the set

$$\mathcal{D}_\varepsilon := \{d(t) :\ d \in L_2 \ \text{ and } \ \|d\|_\infty \leq \varepsilon\}\,.$$

In the particular case $w = 0$, system (3) reduces to the unforced unperturbed system Σ_{uu}, i.e.,

$$\Sigma_{uu} : \qquad \dot{x} = f(x, 0)\,. \tag{5}$$

Some technical hypotheses on the nonlinear functions $f : (X \times D) \to \mathbf{R}^n$, $g : (X \times D) \to \mathbf{R}^n$ must hold. Moreover, we will assume without loss of generality that the origin is an exponentially stable equilibrium point of system Σ_{uu}. To summarize, the following additional assumptions on the structure of system (2) are required.

A3. X and D are open and connected sets of $\mathbf{R}^n$ and $\mathbf{R}^q$ containing the origin. The functions $f(x,d)$ and $g(x,d)$ are continuously differentiable in the set $(X \times D)$ over which the Jacobian functions can be defined as

$$J_x(x, d, w) := \frac{\partial f(x,d)}{\partial x} + \frac{\partial g(x,d)}{\partial x} w(t)\ , \tag{6a}$$

$$J_d(x, d, w) := \frac{\partial f(x,d)}{\partial d} + \frac{\partial g(x,d)}{\partial d} w(t)\ ; \tag{6b}$$

A4. $f(0,0) = 0 \ ,\quad g(x,0) \neq 0 \ \ \forall\, x \in X;$

A5. The matrix

$$A := J_x(0,0,0) = \left.\frac{\partial f(x,d)}{\partial x}\right|_{(0,0)}$$

has no eigenvalues in the closed right half plane.

As mentioned above, the first issue concerns existence of periodic solutions for system (3) subject to the periodic forcing term defined via A1. Exploiting standard results in [1,2], the following theorem can be obtained (see [9] for a complete proof).

Theorem 1. *Suppose A1–A5 hold.*
Then, there exists a positive constant $\overline{\mu}$ such that for all $\mu \in (0, \overline{\mu})$ system (3) subject to the periodic input $w_\mu \in \mathcal{W}$ possesses a unique non-trivial T-periodic solution $x_\mu(t)$.
Moreover, if $w_\mu(t)$ is a sinusoidal input, i.e. in A1

$$\overline{w}(t) = \cos \omega_0 t \quad , \quad \omega_0 = \frac{2\pi}{T} \, , \tag{7}$$

then, there exists a positive constant ϱ such that

$$\|x_\mu(t) - \mu \, x_L(t)\| \leq \varrho \, \mu^2 \quad , \quad \forall \, t \geq 0 \tag{8}$$

where

$$x_L(t) = \text{Re} \left[(j\omega_0 I - A)^{-1} \, g(0,0) \, e^{j\omega_0 t} \right] \, . \tag{9}$$

This (essentially local) result states that if the origin of the unforced system is an exponentially stable equilibrium point, then there exists a family of T-periodic solutions of the forced system for an interval of the amplitude parameter μ. When the forcing input is further constrained to be sinusoidal, the approximate closed-form expression for the T-periodic solution $x_\mu(t)$ – which is valid for small μ's – is a useful estimate in case the solution is neither analytically nor numerically known.

The second issue concerns stability of the periodic solutions $x_\mu(t)$ with respect to the disturbance set described in A2. In particular, our aim is to provide a lower bound $\hat{\mu}$ on the amplitude range corresponding to stable periodic solutions $x_\mu(t)$. This issue will be dealt with in Sect. 4 by exploiting input-output stability results. Accordingly, we conclude this section by introducing some definitions regarding the L_2-stability setting.

Assume that system Σ is operating in a periodic regime, where (w_μ, x_μ) represents a nominal input-state signal pair of period T, and consider a generic perturbation input d. The following is the standard definition of L_2-stability.

Definition 1.[1] The solution $x_\mu(t)$ of system Σ is L_2-stable if there exists a positive constant γ such that $\|x - x_\mu\|_2 \leq \gamma \|d\|_2 \; \forall \, \delta w \in L_2$.

[1] The definition is sometimes referred to as L_2-stability with finite gain and zero bias [2,10].

The above definition concerns global L_2-stability of a given periodic solution. Here, we are more interested in local stability properties as captured by next definition.

Definition 2. The solution $x_\mu(t)$ of system Σ is small signal L_2-stable if there exist positive constants γ and ε such that $\|x - x_\mu\|_2 \leq \gamma \|d\|_2 \ \forall \ d \in L_2$, $\|d\|_\infty \leq \varepsilon$.

Note that in Definition 2 the perturbation d is required to be uniformly small over the time interval. Hereafter, small signal L_2-stability of the T-periodic solution $x_\mu(t)$ of system Σ will be considered.

4 Sufficient conditions for stability of periodic solutions

The first step in the computation of a lower bound $\hat{\mu}$ ensuring small signal L_2-stability of $x_\mu(t)$, for all $\mu \in [0, \hat{\mu}]$, is to linearize system (2) around $x = x_\mu(t)$, $d = 0$. This leads to the following linear periodic system

$$\Sigma_L : \qquad \delta\dot{x} = (A - \Delta_\mu(t))\, \delta x + \Phi_\mu(t)\, d \tag{10}$$

where

$$\Delta_\mu(t) = A - J_x(x_\mu(t), 0, w_\mu(t)) \ , \tag{11a}$$
$$\Phi_\mu(t) = J_d(x_\mu(t), 0, w_\mu(t)) \tag{11b}$$

are T-periodic bounded matrices and δx is the appropriate perturbed vector. An equivalent feedback scheme representation of (10) is depicted in Fig. 4, where

$$L(s) = (sI - A)^{-1}.$$

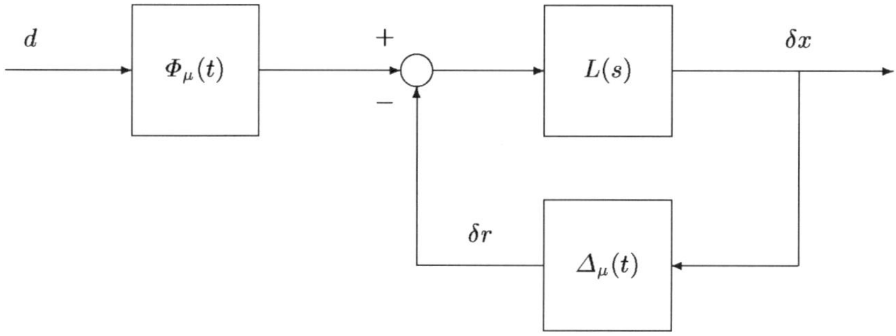

Fig. 4. Linearized system Σ_L

A sufficient condition for small signal L_2-stability of $x_\mu(t)$ can be obtained by investigating L_2-stability of Σ_L, according to the following standard result (see [9] and references therein).

Proposition 1. *If the linear periodic system Σ_L is input-output L_2-stable then the periodic solution $x_\mu(t)$ of system Σ is small signal input-output L_2-stable.*

A powerful tool for L_2-stability analysis of feedback systems is represented by Integral Quadratic Constraints (IQC) [5]. Here, the analysis is based on the ability of describing the relation between δx and $\delta r := \Delta_\mu(t)\delta x$ in system Σ_L by using integral quadratic inequalities, according to the following definition.

Definition 3. Let $\Pi : j\mathbf{R} \to \mathbf{C}^{(2n\times 2n)}$ be a (rational) bounded function. Then, two signals $\delta r \in L_2$ and $\delta x \in L_2$ are said to be related by the IQC defined by Π, if

$$\int_{-\infty}^{+\infty} \begin{bmatrix} \delta R(j\omega) \\ \delta X(j\omega) \end{bmatrix}^* \Pi(j\omega) \begin{bmatrix} \delta R(j\omega) \\ \delta X(j\omega) \end{bmatrix} d\omega \geq 0 \tag{12}$$

where $\delta R(j\omega)$ and $\delta X(j\omega)$ are the Fourier transforms of signals δr and δx, respectively.

Now, consider the linearized system Σ_L in Fig. 4. Since $\Delta_\mu(t)$ is bounded, δr is guaranteed to be square summable if $\delta x \in L_2$. In this case, if (12) holds $\Delta_\mu(t)$ is said to satisfy the IQC defined by Π. The following theorem can be derived by applying the stability result reported in [5, Theorem 1] to the linearized system Σ_L in Fig. 4.

Theorem 2. *Suppose that*

i) A1–A5 hold;
ii) there exist $\hat{\mu} > 0$ and a function Π s.t. $\Delta_\mu(t)$ satisfies the IQC defined by Π, $\forall\, \mu \in [0, \hat{\mu}]$;
iii) there exists $\varepsilon > 0$ such that

$$\begin{bmatrix} L(j\omega) \\ -I \end{bmatrix}^* \Pi(j\omega) \begin{bmatrix} L(j\omega) \\ -I \end{bmatrix} \leq -\varepsilon I \quad \forall\, \omega \in \mathbf{R}\,. \tag{13}$$

Then, the T-periodic solution $x_\mu(t)$ of Σ is small signal L_2-stable $\forall\, \mu \in [0, \hat{\mu}]$.

The above criterion provides a frequency domain tool for assessing small signal L_2-stability of a family of T-periodic solutions of system Σ. When used in conjunction with Theorem 1, which guarantees the existence of a T-periodic solution in a neighborhood of $\mu = 0$, Theorem 2 further provides a useful estimate of the range of stability with respect to the varying parameter μ. In other words, while Theorem 1 does not indicate any estimate of $\overline{\mu}$

concerning the existence, Theorem 2 explicitly does by computing $\hat{\mu}$. It thus follows that $\hat{\mu} \leq \overline{\mu}$.

From a computational point of view, a number of techniques can be employed to numerically solve inequality (13). One of the most efficient is certainly obtained when (13) is expressed in an LMI form [6]. This can be achieved by performing the following spectral factorization of Π

$$\Pi(j\omega) = H^*(j\omega)\, Q\, H(j\omega)\ , \tag{14}$$

and rewriting inequality (13) as the LMI

$$\left[\begin{array}{c|c} \mathcal{A}'P + P\mathcal{A} & P\mathcal{B} \\ \hline \mathcal{B}'P & 0 \end{array}\right] + [\mathcal{C} \quad \mathcal{D}]'\, Q\, [\mathcal{C} \quad \mathcal{D}] < 0\ , \tag{15}$$

where the unknown matrix P is symmetric positive definite and $(\mathcal{A}, \mathcal{B}, \mathcal{C}, \mathcal{D})$ is a state-space realization of the transfer function matrix

$$\mathcal{L}(s) = H(s) \begin{bmatrix} L(s) \\ -I \end{bmatrix}\ . \tag{16}$$

Theorem 2 embeds many existing stability criteria depending on the particular structure of Δ_μ and on the choice of Π according to Definition 3. As a general rule, IQC's for Δ_μ are produced by forming any convex combination of "standard" IQC's derived for its "elementary" subsystems. Therefore, it is evident that a diagonal structure for Δ_μ is highly desirable in most cases, since a valid IQC can be readily obtained by separately considering valid IQC's for the single entries. Most common IQC's for general time-varying and/or nonlinear operators defining the mapping $\delta x \to \delta r$ are reported in [5]. Here, only a few cases of interest are considered for the problem under investigation.

1. **Diagonal bounded $\Delta_\mu(t)$.**
   Assume that $\Delta_\mu(t)$ is diagonal, bounded on $\mu \in [0, \hat{\mu}]$, $t \geq 0$, and compute the diagonal matrices $\underline{\Delta}$ and $\overline{\Delta}$ such that

$$\begin{aligned} \overline{\Delta} - \max_{\mu\in[0,\hat{\mu}]} \max_{t\in[0,T]} \Delta_\mu(t) \geq 0 \\ -\underline{\Delta} + \min_{\mu\in[0,\hat{\mu}]} \min_{t\in[0,T]} \Delta_\mu(t) \geq 0 \end{aligned} \tag{17}$$

   where the max-min operations are componentwise.
   Then, a valid IQC for $\Delta_\mu(t)$ according to Definition 3 is

$$\Pi_D = \left[\begin{array}{c|c} -2\underline{\Delta}\,\overline{\Delta} & \underline{\Delta} + \overline{\Delta} \\ \hline \underline{\Delta} + \overline{\Delta} & -2I \end{array}\right]\ . \tag{18}$$

   Note that $\underline{\Delta}$ and $\overline{\Delta}$ can be easily selected according to (17) even if only some bounds on the T-periodic solutions are known. On the other hand, observe that the above IQC does not explicitly take into account periodicity of the gain Δ_μ. For this particular IQC, Theorem 2 gives rise to the multivariable circle criterion [2].

2. **Harmonic $\Delta_\mu(t)$.**
   Assume that $\Delta_\mu(t) = \mu\Lambda\cos(\omega_0 t + \varphi)$, $\mu \in [0, \hat{\mu}]$, $\Lambda \in \mathbf{R}^{n\times n}$, $\varphi > 0$. Then,

$$\Pi_H(\omega) = \left[\begin{array}{c|c} \hat{\mu}^2\Lambda'[X(\omega+\omega_0) + X(\omega-\omega_0)]\,\Lambda & 0 \\ \hline 0 & -2X(\omega) \end{array}\right] , \tag{19}$$

   where $X(\omega) = X_0 + \dfrac{X_1}{a^2+\omega^2}$, $a \in \mathbf{R}$, $X_0, X_1 \in \mathbf{R}^{n\times n}$.

As already pointed out, most of the available results in multivariable feedback stability theory – and their corresponding IQC counterparts – apply to nonlinear/time-varying/uncertain blocks Δ that are in [block-]diagonal form. For example, the matrix sector conditions pertaining to circle criterion are easily obtainable from the scalar sector conditions of the single entries of Δ only when Δ is diagonal (otherwise, it is still possible to find matrix bounds but at the price of introducing conservativeness).

Several methods can lead to a feedback description, equivalent to that in Fig. 4, in which the block $\Delta(t)$ is diagonal. In principle any interconnection of linear blocks, whatever complicated it might be, can be rearranged in such a way to have a closed subset of them (in this case the TV ones) in a [block-]diagonal form. This is the standard approach in robust control literature.

An alternative procedure was proposed in [4], that starting from Fig. 4 directly obtains a corresponding system in diagonal and "minimal" form. The procedure involves construction of a $n^2{\times}n^2$ diagonal stack matrix obtained by juxtaposing the columns or rows of $\Delta(t)$, and subsequent order reduction via elimination of null diagonal entries.

5 Application example

In this section we apply the presented approach to the laser model introduced in Sect. 2. Equations (1) already fit into the class of systems (2). However, we note that considering the logarithm of the intensity I can simplify the analysis since the first equation becomes linear and, in addition, the field component g does not depend on x. Besides, the equilibrium at $\mu = 0$ needs to be shifted to the origin to accomplish Assumption A4. We therefore consider the change of variables

$$x_1 = \log(I/I_{eq}) \ , \ x_2 = N - N_{eq} \ , \ x_3 = M - M_{eq}$$

where

$$I_{eq} = \eta/2k_0 \ , \ N_{eq} = 1 \ , \ M_{eq} = \beta/\alpha \ , \ \eta = \gamma_P - \Gamma + \gamma_R\beta/\alpha \ ,$$

and a forcing input w belonging to the class of signals (4) with $\overline{w}(t) = \cos \frac{2\pi}{T} t$, $T = 10^{-5}$ s and perturbed by an additive disturbance d which takes into account the noise affecting the intra-cavity modulation. The resulting system Σ, now characterized by

$$f(x,d) = \begin{bmatrix} k_0(x_2 + d) \\ -\Gamma x_2 + \gamma_R x_3 - \eta[(x_2+1)\exp(x_1) - 1] \\ \beta x_2 - \alpha x_3 \end{bmatrix} , \ g(x,d) = \begin{bmatrix} k_0 \\ 0 \\ 0 \end{bmatrix} ,$$

fulfills Assumptions A1–A5. As a consequence Theorem 1 guarantees existence of periodic solutions for some interval $(0, \overline{\mu})$.

The linearization procedure applied to system Σ yields

$$L(s) = \begin{bmatrix} s & -k_0 & 0 \\ \eta & s+\eta+\Gamma & -\gamma_R \\ 0 & -\beta & s+\alpha \end{bmatrix}^{-1} , \ \Delta_\mu(t) = \begin{bmatrix} 0 & 0 & 0 \\ k_1(t) & k_2(t) & 0 \\ 0 & 0 & 0 \end{bmatrix} , \ \Phi_\mu(t) = \begin{bmatrix} k_0 \\ 0 \\ 0 \end{bmatrix}$$

where

$$k_1(t) = \eta\left[(x_{2\mu}(t) + 1)\exp(x_{1\mu}(t)) - 1\right] \ , \ k_2(t) = \eta\left[\exp(x_{1\mu}(t)) - 1\right] \ .$$

When the diagonalization procedure outlined at the end of the previous section is applied to the feedback part of the scheme in Fig. 4, the equivalent diagonalized system $\Sigma_{L_{eq}}$ is as in Fig. 5 with

$$\tilde{L}(s) = \begin{bmatrix} L_{12}(s) & L_{12}(s) \\ L_{22}(s) & L_{22}(s) \end{bmatrix} , \quad \tilde{\Delta}_\mu(t) = \begin{bmatrix} k_1(t) & 0 \\ 0 & k_2(t) \end{bmatrix} . \tag{20}$$

We have applied the IQC methodology outlined in Sect. 4 with the aim of computing the lower bound $\hat{\mu}$ of system (1). All the numerical computations have been performed in MATLAB using the IQCβ Toolbox [11] and its library of IQC's.

Different strategies can be adopted, depending on the assumptions about the a-priori knowledge of the investigated periodic solutions. Here, we briefly summarize the typical case where the available information concerns the min-max bounds of the state variables $x_i(t)$ over the period[2], i.e., the knowledge necessary to plot (or, if the case, obtainable from) a bifurcation diagram as in Fig. 3. This is enough to calculate, via (6a) and (11a), the corresponding

[2] Other cases can be suitably considered, including the one where no knowledge of the periodic solutions is available (see [9] for details) and the one where such knowledge is extended to the whole set of local minima and maxima over the period [12].

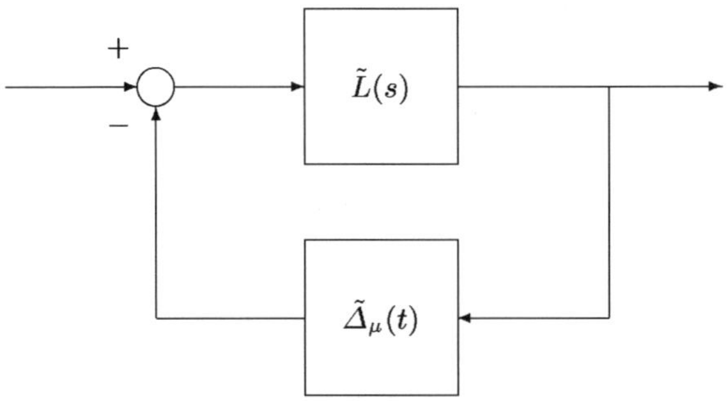

Fig. 5. Equivalent diagonal linearized system $\Sigma_{L_{\rm eq}}$

bounds $\underline{\Delta}$ and $\overline{\Delta}$ on the entries of $\tilde{\Delta}_\mu(t)$. We then exploit Theorem 3 with the IQC valid for $\tilde{\Delta}_\mu(t)$ described in (18). By solving the LMI (15) we get $\hat{\mu} = 8.86\times10^{-3}$, i.e., stability is proved for all forcing amplitudes in the range $\mu \in (0, 8.86\times10^{-3})$.

The result is sharper with respect to the smaller interval $\hat{\mu} = 5.9\times10^{-3}$ obtained applying the standard, classical version of the multivariable circle criterion as done in [4]. This is due to the convex optimization, implicit in the IQC approach, which optimally weights all the scalar sectors relative to the entries of $\tilde{\Delta}_\mu(t)$, and turns out to be equivalent to the use of a "static" (i.e., constant with frequency) positive definite multiplier matrix [13]. Note also that other even less conservative criteria for periodic systems, like Willems criterion [14], cannot be treated numerically via LMI's since their IQC representation involves non-rational (periodic) matrices $\Pi(j\omega)$.

6 Conclusions

The study of the stability properties of periodic solutions in periodically forced nonlinear systems is motivated by several application examples. We have suggested that absolute stability theory can represent a useful tool to conduct such analysis in a wide class of systems. In this context, the use of modern IQC tools can be used to reduce conservatism with respect to existing standard criteria. The proposed approach could be seen as complementary to standard approaches based on Floquet analysis, which however require full knowledge of the considered solution.

References

1. Farkas M. (1994) Periodic Motions. Springer, New York.

2. Khalil H. K. (1996) Nonlinear Systems, 2nd edn. Macmillan, New York.
3. Chen G. (Ed.) (1999) Controlling Chaos and Bifurcations in Engineering Systems. CRC Press, Boca Raton, FL.
4. Basso M., Genesio R., Giovanardi L., Tesi A. (1999) Frequency domain methods for chaos control. In: In Chen G. [3], chapter 9, 175–199.
5. Megretski A., Rantzer A. (1997) System analysis via integral quadratic constraints. IEEE Trans. Automatic Control **42**, 819–830.
6. Boyd S., El Ghaoui L., Feron E., Balakrishnan V. (1994) Linear Matrix Inequalities in System and Control Theory. SIAM, Philadelphia.
7. Stanghini M., Basso M., Genesio R., Tesi A., Ciofini M., Meucci R. (1996) A new three-equation model for the CO_2 laser. IEEE J. Quantum Electron. **32**, 1126–1131.
8. Basso M., Genesio R., Tesi A. (2000) Controller synthesis for stabilizing periodic orbits in forced nonlinear systems. Proc. 39th IEEE CDC, Sydney, Australia.
9. Basso M., Giovanardi L., Tesi A. (2001) Stability analysis of periodic solutions via IQCs. Research Report no. 01/2001, Dipartimento di Sistemi e Informatica, Università di Firenze, Italy.
10. Vidyasagar M. (1992) Nonlinear Systems Analysis, 2nd edn. Prentice-Hall, Englewood Cliffs, NJ.
11. Megretski A., Kao C., Jönsson U., Rantzer A. (2000) A guide to IQCβ: Software for robustness analysis. http://web.mit.edu/ameg/www/index.html
12. Giovanardi L., Basso M., Genesio R., Tesi A. (2000) Lower bounds for the stability degree of periodic solutions in forced nonlinear systems. Internat. J. of Bifurcation and Chaos **10**, 639–653.
13. Schmitt G. (1999) Frequency domain evaluation of discrete-time circle criterion and Tsypkin criterion in the MIMO case. Proc. 5th European Control Conf., Karlsruhe, Germany.
14. Basso M., Genesio R., Tesi A. (1998) Frequency domain methods and control of complex dynamics. Proc. 37th IEEE CDC, Tampa, FL, 1936–1941.

Port Controller Hamiltonian Synthesis Using Evolution Strategies

José Cesáreo Raimúndez Álvarez

Universidad de Vigo, Vigo (Pontevedra) CEP 08544, Spain,
cesareo@uvigo.es

Abstract. Evolution Strategies (ES) are stochastic optimization techniques obeying an evolutionist paradigm, that can be used to find global optima over a response hypersurface. The current investigation focuses on Port Controlled Hamiltonian (PCH) systems stabilization, using the unsupervised learning capabilities of ES's inherited from their evolutionist paradigm. The training process intends to build a complementary Energy Function ($\mathcal{H}_a$) which guarantees local asymptotic stability at the desired equilibrium point.

1 Introduction

The Port Controlled Hamiltonian Systems theory presented in [3] has as the main engineering appeal, the energy modeling through effort-flow ports. A systematic procedure for controller design is already developed, amenable for symbolic tools. The controller energy function synthesis proceeds after a gradient field solution according to a set of conditions, included integrability. For underactuated systems the theory is evolving [4] and seems promising.

In this paper we present a technique to produce a gradient field using a neural net as approximator, avoiding the frequently cumbersome partial equation resolution problem. Under the evolutive paradigm, the problem of control synthesis is reduced to *find the minimum of a function (fitness) over a feasible set of values*. ES are used to find out the solution to that *minimization* problem. The technique is applied to an underactuated hamiltonian system (Ball & Beam) for a fixed point controller.

The content of the paper is as follows. In Section 2 the main results in Port Controlled Hamiltonian Systems are briefly explained. In section 3 the port-controller conditions are presented. In Section 4 is presented an introduction to Evolutive Strategies as well as the evolutionary controller design formulation and the Plant-Controller fitness definition, associated to the minimization search. Section 5 is dedicated to the Evolutionary Formulation. In Section 6 is presented a case study which illustrates the line of implementation. Section 7 is dedicated to conclusions.

The main notation signs used are: $()^\top$ for transpose, `eigen` for eigenvalue, $\propto$ for proportional, $||*||$ for Euclidean norm and C^0 as the set of continuous real functions.

F. Colonius, L. Grüne (Eds.): Dynamics, Bifurcations, and Control, LNCIS 273, pp. 159–172, 2002.

2 Port controlled Hamiltonian systems

According to [3], the energy-conserving lumped-parameter physical systems network modeling with independent storage elements leads to models of the form

$$\Sigma : \begin{cases} \dot{\boldsymbol{x}} = J(\boldsymbol{x})\frac{\partial \mathcal{H}}{\partial \boldsymbol{x}}(\boldsymbol{x}) + \boldsymbol{G}(\boldsymbol{x})\boldsymbol{u} \\ \boldsymbol{y} = \boldsymbol{G}^\top(\boldsymbol{x})\frac{\partial \mathcal{H}}{\partial \boldsymbol{x}}(\boldsymbol{x}) \end{cases} \tag{1}$$

where $\boldsymbol{x} \in \Re^n$ are the energy variables, $C^0 \ni \mathcal{H}(\boldsymbol{x}) : \Re^n \to \Re$ represents the total stored energy and $\boldsymbol{u}, \boldsymbol{y} \in \Re^m$ are the port power variables.

Energy-dissipation is included by terminating some of the ports by resistive elements. Considering instead of $\boldsymbol{G}(\boldsymbol{x})\boldsymbol{u}$ in (1)

$$\left[\boldsymbol{G}(\boldsymbol{x}) \; \boldsymbol{G}_R(\boldsymbol{x})\right]\begin{bmatrix} \boldsymbol{u} \\ \boldsymbol{u}_R \end{bmatrix} = \boldsymbol{G}(\boldsymbol{x})\boldsymbol{u} + \boldsymbol{G}_R(\boldsymbol{x})\boldsymbol{u}_R \tag{2}$$

and extending correspondingly $\boldsymbol{y} = \boldsymbol{G}^\top(\boldsymbol{x})\frac{\partial \mathcal{H}}{\partial \boldsymbol{x}}$ to

$$\begin{bmatrix} \boldsymbol{y} \\ \boldsymbol{y}_R \end{bmatrix} = \begin{bmatrix} \boldsymbol{G}^\top(\boldsymbol{x})\frac{\partial \mathcal{H}}{\partial \boldsymbol{x}} \\ \boldsymbol{G}_R^\top(\boldsymbol{x})\frac{\partial \mathcal{H}}{\partial \boldsymbol{x}} \end{bmatrix} \tag{3}$$

with $\boldsymbol{u}_R = -R\boldsymbol{y}_R$ for some positive semi-definite symmetric matrix R. Incorporating in (1) leads to models with the structure

$$\Sigma : \begin{cases} \dot{\boldsymbol{x}} = [J(\boldsymbol{x}) - R(\boldsymbol{x})]\frac{\partial \mathcal{H}}{\partial \boldsymbol{x}}(\boldsymbol{x}) + \boldsymbol{G}(\boldsymbol{x})\boldsymbol{u} \\ \boldsymbol{y} = \boldsymbol{G}^\top(\boldsymbol{x})\frac{\partial \mathcal{H}}{\partial \boldsymbol{x}}(\boldsymbol{x}) \end{cases} \tag{4}$$

3 Controller design

Considering the closed loop dynamics for the pair plant-controller given by

$$\dot{\boldsymbol{x}} = [J_d(\boldsymbol{x}) - R_d(\boldsymbol{x})]\frac{\partial \mathcal{H}_d}{\partial \boldsymbol{x}} \tag{5}$$

The problem is to find a static feed-back control $\boldsymbol{u} = \boldsymbol{\beta}(\boldsymbol{x})$ in (4) responsible for the new closed loop energy shaping. Defining

$$\begin{aligned} J_d(\boldsymbol{x}) &= J(\boldsymbol{x}) + J_a(\boldsymbol{x}) \\ R_d(\boldsymbol{x}) &= R(\boldsymbol{x}) + R_a(\boldsymbol{x}) \end{aligned}$$

$$\mathcal{H}_d(\boldsymbol{x}) = \mathcal{H}(\boldsymbol{x}) + \mathcal{H}_a(\boldsymbol{x}) \tag{6}$$

and considering

$$\begin{aligned} \dot{\boldsymbol{x}} &= [J_d(\boldsymbol{x}) - R_d(\boldsymbol{x})]\tfrac{\partial \mathcal{H}_d}{\partial \boldsymbol{x}} \\ &= [J + J_a - (R + R_a)]\left(\tfrac{\partial \mathcal{H}}{\partial \boldsymbol{x}} + \tfrac{\partial \mathcal{H}_a}{\partial \boldsymbol{x}}\right) \\ &= [J - R]\tfrac{\partial \mathcal{H}}{\partial \boldsymbol{x}} + [J_a - R_a]\tfrac{\partial \mathcal{H}}{\partial \boldsymbol{x}} + [J_d(\boldsymbol{x}) - R_d(\boldsymbol{x})]\tfrac{\partial \mathcal{H}_a}{\partial \boldsymbol{x}} \\ &= [J - R]\tfrac{\partial \mathcal{H}}{\partial \boldsymbol{x}} + \boldsymbol{G}(\boldsymbol{x})\boldsymbol{\beta}(\boldsymbol{x}) \end{aligned} \tag{7}$$

we conclude

$$[J + J_a - (R + R_a)]\frac{\partial \mathcal{H}_a}{\partial \boldsymbol{x}} + [J_a - R_a]\frac{\partial \mathcal{H}}{\partial \boldsymbol{x}} = \boldsymbol{G}(\boldsymbol{x})\boldsymbol{\beta}(\boldsymbol{x}) \tag{8}$$

which is the basic relationship involving $\boldsymbol{\beta}(\boldsymbol{x})$ and $\mathcal{H}_a$. Considering $\boldsymbol{K}(\boldsymbol{x}) = \frac{\partial \mathcal{H}_a}{\partial \boldsymbol{x}}$, the controller synthesis problem can be stated as:

Given $J(\boldsymbol{x}), R(\boldsymbol{x}), \mathcal{H}, \boldsymbol{G}(\boldsymbol{x})$ and the desired equilibrium to be stabilized $\boldsymbol{x}^* \in \Re^n$, find functions $\boldsymbol{\beta}(\boldsymbol{x}), J_a(\boldsymbol{x}), R_a(\boldsymbol{x})$ and $\boldsymbol{K}(\boldsymbol{x})$ satisfying

$$\begin{aligned} &[(J(\boldsymbol{x}) + J_a(\boldsymbol{x})) - (R(\boldsymbol{x}) + R_a(\boldsymbol{x}))]\, \boldsymbol{K}(\boldsymbol{x}) \\ &= -\left[J_a(\boldsymbol{x}) - R_a(\boldsymbol{x})\right] \frac{\partial \mathcal{H}}{\partial \boldsymbol{x}} + \boldsymbol{G}(\boldsymbol{x})\boldsymbol{\beta}(\boldsymbol{x}) \end{aligned}$$

and such that:

1. *Structure Preservation*

$$\begin{aligned} J(\boldsymbol{x}) + J_a(\boldsymbol{x}) &= -\left[J(\boldsymbol{x}) + J_a(\boldsymbol{x})\right]^\top \\ R(\boldsymbol{x}) + R_a(\boldsymbol{x}) &= \left[R(\boldsymbol{x}) + R_a(\boldsymbol{x})\right]^\top \geq 0 \end{aligned}$$

2. *Integrability* $\boldsymbol{K}(\boldsymbol{x})$ is the gradient of a scalar function:

$$\frac{\partial \boldsymbol{K}}{\partial \boldsymbol{x}}(\boldsymbol{x}) = \left[\frac{\partial \boldsymbol{K}}{\partial \boldsymbol{x}}(\boldsymbol{x})\right]^\top \tag{9}$$

3. *Equilibrium Assignment* $\boldsymbol{K}(\boldsymbol{x})$ at $\boldsymbol{x}^*$ verifies

$$\boldsymbol{K}(\boldsymbol{x}^*) = -\frac{\partial \mathcal{H}}{\partial \boldsymbol{x}}(\boldsymbol{x}^*)$$

4. *Lyapunov Stability* The Jacobian of $\boldsymbol{K}(\boldsymbol{x})$ at $\boldsymbol{x}^*$ satisfies the bound

$$\frac{\partial \boldsymbol{K}}{\partial \boldsymbol{x}}(\boldsymbol{x}^*) \geq -\frac{\partial^2 \mathcal{H}}{\partial \boldsymbol{x}^2}(\boldsymbol{x}^*)$$

Under these conditions, the closed-loop system $\boldsymbol{u} = \boldsymbol{\beta}(\boldsymbol{x})$ will be a Port Controlled Hamiltonian (PCH) system with dissipation of the form (5), where $\mathcal{H}_d(\boldsymbol{x})$ is given by (6) and

$$\frac{\partial \mathcal{H}_a}{\partial \boldsymbol{x}}(\boldsymbol{x}) = \boldsymbol{K}(\boldsymbol{x}) \tag{10}$$

Supposing that such a $\boldsymbol{K}(\boldsymbol{x})$ can be found, the control can be calculated using the formula

$$\begin{aligned}\boldsymbol{\beta}(\boldsymbol{x}) = &\left[\boldsymbol{G}^\top(\boldsymbol{x})\boldsymbol{G}(\boldsymbol{x})\right]^{-1}\boldsymbol{G}^\top(\boldsymbol{x}) \\ &\left\{\left[J_d(\boldsymbol{x}) - R_d(\boldsymbol{x})\right]\boldsymbol{K}(\boldsymbol{x}) + \left[J_a(\boldsymbol{x}) - R_a(\boldsymbol{x})\right]\frac{\partial \mathcal{H}}{\partial \boldsymbol{x}}(\boldsymbol{x})\right\}\end{aligned} \tag{11}$$

4 Preliminaries on evolution strategies

4.1 Evolutionary basics

Evolution Strategies (ES) belongs to the class of stocastic optimization techniques, commonly described as evolutionary algorithms.

Simulated Evolution is based on the collective learning processes within a population of individuals, in the quest for survival [1]. Each individual represents a search point in the space of potential solutions to a given problem.

There are currently three main lines of research strongly related but independently developed in simulated evolution : Genetic Algorithms (GA), Evolution Strategies (ES), and Evolutionary Programming (EP). In each of these methods, the population of individuals is arbitrarily initialized and evolves towards better regions of the search space by means of a stochastic process of selection, mutation, and recombination if appropriate.

These methods differ in the specific representation, mutation operators and selection procedures. While genetic algorithms emphasize chromosomal operators based on observed genetic mechanisms (e.g., cross-over and bit mutation), evolution strategies and evolutionary programming emphasize the adaptation and diversity of behavior from parent to offspring over successive generations.

Evolution is the result of interplay between the creation of new genetic information and its evaluation and selection. A single individual of a population is affected by other individuals of the population as well as by the environment. The better an individual performs under these conditions the greater is the chance for the individual to survive for a longer while and generate offspring, which inherit the parental genetic information.

The main contributions in the evolutionary computation approach are:

- Model regularity independence (Applicable to nonsmooth problems).

- Parallelization to cope with intensive cost fitness computation.
- Population search versus individual search (classical).
- General meta-heuristics.
- Good convergence properties.

Evolutionary algorithms mimic the process of neo-Darwinian organic evolution and involves concepts such as:

t	Time or epoch.
$\{\boldsymbol{\pi}, \boldsymbol{\sigma}\}$	Individual.
	$\boldsymbol{\pi}$ Exogenous parameters. (Search Space).
	$\boldsymbol{\sigma}$ Endogenous parameters. (Adaptation).
P	Population. $P = \{\{\boldsymbol{\pi}_1, \boldsymbol{\sigma}_1\}, \ldots, \{\boldsymbol{\pi}_{n_i}, \boldsymbol{\sigma}_{n_i}\}\}$
$\Phi(P)$	Fitness. $\Phi(P) = (\Phi(\boldsymbol{\pi}_1), \ldots, \Phi(\boldsymbol{\pi}_{n_i}))$ $\Phi(\boldsymbol{\pi}) : \Re^{\lvert\pi\rvert} \rightarrow \Re^+$
operators	(Mutation, Selection, Variation, etc.)

where n_i is the number of individuals in the population.
A simple evolutionary algorithm follows:

```
t ← 0
initialize P
evaluate Φ(P)
while not   terminate
   P' ← variation P
   evaluate Φ(P')
   P ← select (P' ∪ Q)
   t ← t + 1
end
```

Q is a special pot of individuals that might be considered for selection purposes, e.g. $Q = \{\emptyset, P, \cdots\}$. An offspring population P' of size λ is generated by means of variation operators such as recombination and/or mutation from the population P. The offspring individuals $\{\boldsymbol{\pi}_i, \boldsymbol{\sigma}_i\} \in P$ are evaluated by calculating their fitness represented by $\Phi(P)$. Selection of the fittest is performed to drive the process toward better individuals.

In evolution strategies the individual consist on two types of parameters: *exogenous* $\boldsymbol{\pi}$ which are points in the search space, and *endogenous* $\boldsymbol{\sigma}$ which are known too as *strategic parameters*. Variation is composed of *mutation* and *self-adaptation* performed independently on each individual. Thus

$$\{\boldsymbol{\pi}', \boldsymbol{\sigma}'\} \leftarrow \texttt{mutate}(\{\boldsymbol{\pi}, \cdot\}) \cup \texttt{adapt}(\{\cdot, \boldsymbol{\sigma}\}) \tag{12}$$

where mutation is accomplished by

$$\pi_i' = \pi_i + \sigma_i \cdot N(0,1) \tag{13}$$

and adaptation is accomplished by

$$\sigma'_i = \sigma_i \cdot \exp\{\tau' \cdot N(0,1) + \tau \cdot N(0,1)\} \tag{14}$$

where $\tau' \propto (\sqrt{2n_\pi})^{-1}$ and $\tau \propto (\sqrt{2\sqrt{n_\pi}})^{-1}$.

$N(0,1)$ indicates a normal density function with expectation zero and standard deviation 1, and n_π the dimension of the search space $(n_\pi = \mid \boldsymbol{\pi} \mid)$.

Selection is based only on the response surface value of each individual. Among many others are specially suited:

- Proportional. Selection is done according to the individual relative fitness $p_i = \frac{\Phi(\boldsymbol{\pi}_i)}{\sum_k \Phi(\boldsymbol{\pi}_k)}$
- Rank-based. Selection is done according to indices which correspond to probability classes, associated with fitness classes.
- Tournament. Works by taking a random uniform sample of size $q > 1$ from the population, and then selecting the best as a survival, and repeating the process until the new population is filled.
- (λ, μ). Uses a deterministic selection scheme. μ parents create $\lambda > \mu$ offspring and the best μ are selected as the next population $[Q = \emptyset]$.
- $(\lambda + \mu)$. Selects the μ survivors from the union of parents and offspring, such that a monotonic course of evolution is guaranteed $[Q = P]$

4.2 Fitness evaluation

Each individual is characterized by a set of exogenous and endogenous parameter values $\boldsymbol{\pi}$ and $\boldsymbol{\sigma}$ respectively. The exogenous parameters are inherited from the response function. The endogenous parameters also called strategic parameters, they do not influence the fitness measure.

Each individual represents a set of independent paths over the phase space beginning at different initial conditions, under the influence of the same controller. This set of n_s orbits should cover conveniently the phase space and can be randomly generated at the very beginning, being common to all the individuals of the population. Care must be taken in the process of initial conditions generation. The set of initial conditions must spread over the expected attraction basin and to avoid overfitting [9] a minimum must be imposed over n_s and to avoid prohibitive computational costs a maximum must be imposed over n_s.

Given an open-loop system represented as

$$\dot{\boldsymbol{x}} = \boldsymbol{F}(\boldsymbol{x}) + \boldsymbol{G}(\boldsymbol{x})\boldsymbol{u} \tag{15}$$

and a stabilizing controller $\boldsymbol{u} = \boldsymbol{\beta}(\boldsymbol{\pi}, \boldsymbol{x})$ in which $\boldsymbol{\pi}$ represents a set of parameters to be fixed, the controller stabilizing behaviour can be measured taking the set of n_s orbits described by the closed-loop system, beginning at a set of initial conditions spread over a region of interest. Under the stabilizing controller action, the resultant set of orbits should approach the origin considered as an equilibrium point. The fitness should detect and measure this performance to serve as learning factor. Thus being $\boldsymbol{x}^k(0), \{k = 1, \cdots, n_s\}$ initial conditions, each orbit starting at $\boldsymbol{x}^k(0)$ under the influence of the controller parameters $\boldsymbol{\pi}$ can be represented as:

$$\boldsymbol{X}^k(\boldsymbol{\pi}, t) = \boldsymbol{X}(\boldsymbol{\pi}, \boldsymbol{x}^k(0), t) \tag{16}$$

Settling time performance is measured through a function $\rho(\boldsymbol{X}) > 0$ which normally has one of the following structures:

$$\rho(\boldsymbol{X}) = \begin{cases} \max_{\forall t} \|\boldsymbol{X}\| \cdot t^\alpha \\ \int_0^{t_{max}} \|\boldsymbol{X}\| \cdot t^\alpha dt \end{cases} \tag{17}$$

with $\alpha > 1$. A typical fitness measure can be obtained as:

$$f(\boldsymbol{\pi}) = k_1 \sum_k \rho(\boldsymbol{X}^k(\boldsymbol{\pi})) + k_2 \sum_k b(\boldsymbol{X}^k(\boldsymbol{\pi})) + k_3 g(\boldsymbol{\pi}) \tag{18}$$

k_1, k_2 and k_3 being positive scale factors and $g(\boldsymbol{\pi})$ is a measure of closeness from the $\boldsymbol{\pi}$ parameters to the origin, as a means to guarantee regularity [9] in the approximator. Usually $g(\boldsymbol{\pi}) = \|\boldsymbol{\pi}\|$ and $b(\boldsymbol{x}^k(\boldsymbol{\pi}))$ is a barrier function [2] which penalizes unwanted states or control efforts.

5 Evolutionary formulation

Our individual will be represented by a set of exogenous parameters $\boldsymbol{\pi}$ which are the controller parameters. (The endogenous $\boldsymbol{\sigma}$ parameters are related to the search process). As can be seen in (11), the controller depends on $\mathcal{H}_a$, J_a and R_a. In this paper the J_a and R_a values will be heuristically chosen according to Remark 1 later. The evolutionist process will act only in $\mathcal{H}_a$ defined according to (19).

Under the controller action, a set of independent orbits started at previously defined points $\boldsymbol{x}^k(0), \{k = 1, \cdots, n_s\}$ (initial conditions), will reach the equilibrium point $\boldsymbol{x}^*$ and will remain there, assuring asymptotic stability. The asymptotic convergence task is performed by the evolutionist process unsupervised learning capabilities, through a behaviour measure (fitness) minimization. The controller derives from an energy function $\mathcal{H}_a$ whose gradient field is modulated according (10),(11).

Our purpose is then to find a vector field $\boldsymbol{K}_\pi(\boldsymbol{x})$ which is the gradient of an energy function $\mathcal{H}_a^\pi(\boldsymbol{x})$ with structure

$$\mathcal{H}_a^\pi(\boldsymbol{x}) = \boldsymbol{\mu}^\top(\Pi^1, \Pi^2, x)\Pi^0\boldsymbol{\mu}(\Pi^1, \Pi^2, \boldsymbol{x}) \tag{19}$$

where $\boldsymbol{\mu}(\Pi^1, \Pi^2, \boldsymbol{x})$ is a neural net with n inputs, n outputs where $n = \dim\{\boldsymbol{x}\}$, with the structure $\boldsymbol{\mu}(\Pi^1, \Pi^2, \boldsymbol{x}) = \sigma(\Pi^1\sigma(\Pi^2\boldsymbol{x}))$ and $\sigma(x) = 2/(1 + \exp(-2x)) - 1$. Π^0 is a positive definite matrix of weights. $(\Pi^k = \{\pi_{ij}^k\}, k = 0, 1, 2)$ are square matrices, so the vector $\boldsymbol{\pi}$ of design parameters is obtained by putting inside a vector all the independent coefficients of the square matrices Π^0, Π^1, Π^2 of size $n \times n$ each, giving $n_\pi = 3n(n+1)/2$

Adopting for $\mathcal{H}_a^\pi(\boldsymbol{x})$ the structure (19) implies to assume $\boldsymbol{x}^* = \mathbf{0}$. If $\mathtt{rank}(\Pi^k) = n$, $k = 1, 2$ then $\boldsymbol{x} = \boldsymbol{x}^*$ is the only point which obeys $\sigma(\Pi^1\sigma(\Pi^2\boldsymbol{x})) = \boldsymbol{x}^*$

In order to characterize the controller performance, a set of initial conditions $\boldsymbol{x}^k(0), \{k = 1, \cdots, n_s\}$ spread over the desired attraction basin are given, remaining constant during the calculations.

The feasible controllers are those which obey the conditions stated in items 1 to 4. For the feasible controllers, the minimization process involves a measure over the plant-controller behaviour. The better the controller, the smaller the measure (fitness). This measure is achieved through the steps:

- For a given $\boldsymbol{\pi}$ calculate $\boldsymbol{\beta}(\boldsymbol{x}) \equiv \boldsymbol{\beta}(\boldsymbol{\pi}, \boldsymbol{x})$ according to (11) with

$$\boldsymbol{K}(\boldsymbol{x}) \equiv \boldsymbol{K}_\pi(\boldsymbol{x}) = 2\boldsymbol{\mu}^\top(\Pi^1, \Pi^2, \boldsymbol{x})\Pi^0\frac{\partial\boldsymbol{\mu}}{\partial x}(\Pi^1, \Pi^2, \boldsymbol{x})$$

- Integrate for $t \in [0, t_{max}]$ the differential equation

$$\dot{\boldsymbol{x}} = [J - R]\frac{\partial\mathcal{H}}{\partial\boldsymbol{x}} + \boldsymbol{G}(\boldsymbol{x})\boldsymbol{\beta}(\boldsymbol{\pi}, \boldsymbol{x}) \tag{20}$$

- Being $\boldsymbol{x}(0) = \{\boldsymbol{x}^1(0), \ldots, \boldsymbol{x}^{n_s}(0)\}$ the set of n_s initial conditions defining the needed attraction basin, $\boldsymbol{X}^i(\boldsymbol{\pi}, t)$ the closed loop path with initial conditions $\boldsymbol{x}^i(0)$ and $\alpha \geq 1$, calculate the fitness index

$$\rho_i(\boldsymbol{\pi}) = \rho(\boldsymbol{X}^i(\boldsymbol{\pi})) = \max_{t\in[0,t_{max}]} \left\|\boldsymbol{X}^i(\boldsymbol{\pi}, t)\right\| t^\alpha \tag{21}$$

with a suitable norm $\|\cdot\|$ so the fitness is calculated as

$$\rho_\pi = \sum_i^{n_s} \rho_i(\boldsymbol{\pi})) \tag{22}$$

Incorporating the feasibility search in the fitness calculation gives the following procedure:

`function fitness` $\{$

$$f \;=\; \min\left\{\text{eigen}\left(\frac{\partial^2\mathcal{H}}{\partial x_i\partial x_j} + \frac{\partial\mu}{\partial x}^\top \Pi^0\frac{\partial\mu}{\partial x}\right)(\boldsymbol{x}^*), \text{eigen}\left(\Pi^0\right)\right);$$

```
if (f < 0)
    f = k1 + k2|ρ|;
else;
    f = Σ_i^{n_s} ρ_i(π);
end
return f;
}
```

This procedure applies a penalization to $\rho(\boldsymbol{\pi})$ in the case that the desired positive-definiteness conditions fail. k_1 and k_2 are large positive numbers.

6 Case study - ball & beam system

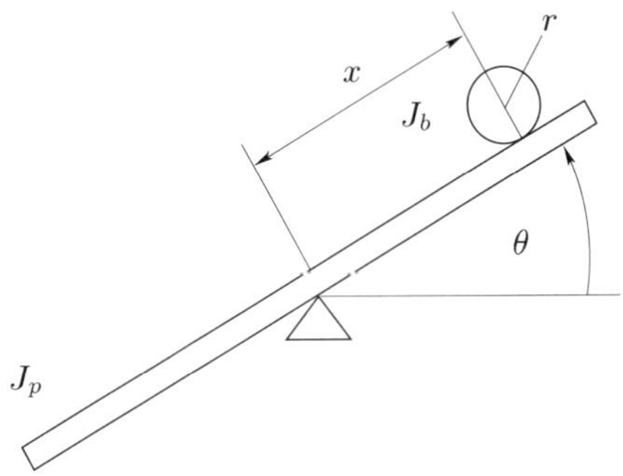

Fig. 1. Ball & Beam diagram

Consider the Ball & Beam plant as can be depicted in Figure 1. The beam is made to rotate in a vertical plane by applying a torque at the center of rotation, and the ball is free to roll along the beam which is one-dimensional. The ball must remain in contact with the beam and the rolling must occur without slipping, which imposes a constraint on the rotational acceleration of the beam as well as in the friction coefficient.

This plant is a well known example of a nonlinear system which is neither feedback linearizable nor minimum phase. Controllers for tracking purposes can be found in [6],[5].

Our goal is to drive the ball to the rest position over a set of initial conditions with values taken on a neighborhood of the origin given by $-0.5 \leq x_i \leq 0.5$, $i = 1, \ldots, 4$ on the phase space, which characterize the needed attraction basin.

Let the moment of inertia of the beam be J_p, the mass and moment of inertia of the ball be m_b and J_b respectively, the radius of the ball be r and the acceleration of gravity be g. Choosing the beam angle θ and the ball position over the beam x as generalized coordinates for the system and according with the above diagram, the kinetic energy is given by

$$T(\theta, x, \dot{\theta}, \dot{x}) = \frac{1}{2}\left[\left(J_p + m_b x^2\right)\dot{\theta}^2 + \left(\frac{J_b}{r^2} + m_b\right)\dot{x}^2\right]$$

and the potential by

$$V(\theta, x) = m_b g x \sin\theta$$

transforming to the hamiltonian formalism in which

$$\begin{aligned} x_1 &= x \\ x_2 &= \theta \\ x_3 &= p_x \\ x_4 &= p_\theta \end{aligned} \tag{23}$$

the total system energy is given by

$$\mathcal{H}(\boldsymbol{x}) = \frac{1}{2}\left(\frac{x_3^2}{J_b/r^2 + m_b} + \frac{x_4^2}{J_p + m_b x_1^2}\right) + m_b g x_1 \sin x_2 \tag{24}$$

and the movement has the description:

$$\begin{pmatrix} \dot{x}_1 \\ \dot{x}_2 \\ \dot{x}_3 \\ \dot{x}_4 \end{pmatrix} = \begin{pmatrix} 0 & 0 & 1 & 0 \\ 0 & 0 & 0 & 1 \\ -1 & 0 & 0 & 0 \\ 0 & -1 & 0 & 0 \end{pmatrix} \begin{pmatrix} \frac{\partial \mathcal{H}}{\partial x_1} \\ \frac{\partial \mathcal{H}}{\partial x_2} \\ \frac{\partial \mathcal{H}}{\partial x_3} \\ \frac{\partial \mathcal{H}}{\partial x_4} \end{pmatrix} + G(\boldsymbol{x})u = \begin{pmatrix} \frac{x_3}{J_b/r^2 + m_b} \\ \frac{x_4}{J_p + m_b x_1^2} \\ \frac{m_b x_1 x_4^2}{(J_p + m_b x_1^2)^2} - m_b g \sin x_2 \\ -m_b g x_1 \cos x_2 + u \end{pmatrix} \tag{25}$$

calculating with the values

$$\begin{aligned} r &= 0.04 \\ J_b &= 0.001 \\ J_p &= 2 \\ m_b g &= 4.9 \end{aligned}$$

Remark 1 The choice of J_a and R_a should accomplish the conditions previously stated within section 3. There are no general rules for their determination. The considered Ball & Beam plant has no dissipation so it is convenient to define $R_a \geq 0$ in order to inject dissipation through the controller action. Concerning J_a, their contribution should be considered for interconnection enhancement in plants which have poor interconnection between states. An example of interconnection enhancement between mechanical and electrical parts using the participation of J_a can be seen in [3]. For our plant we choose

$$J_a = \begin{pmatrix} 0 & 1 & 0 & 1 \\ -1 & 0 & 0 & 0 \\ 0 & 0 & 0 & 0 \\ -1 & 0 & 0 & 0 \end{pmatrix}, \quad R_a = \begin{pmatrix} 4 & 0 & 0 & 0 \\ 0 & 4 & 0 & 0 \\ 0 & 0 & 0 & 0 \\ 0 & 0 & 0 & 0 \end{pmatrix} \tag{26}$$

Processing in a Pentium II at 600Mz after 20 minutes we obtained:

```
PI0 = [
 6.966073 -5.319661 -2.699448 -4.489864
-5.319661  7.976571  3.093717  1.910575
-2.699448  3.093717  1.860503  0.797894
-4.489864  1.910575  0.797894  7.964292 ]';
```

```
PI1 = [
 0.379863 -0.011644 -1.027616 -0.185061
-0.460202 -1.817432 -0.597628 -0.615594
 0.004198 -0.208203 -0.054679  0.459153
 0.172915 -0.018196 -0.085693 -0.274629 ]';
```

```
PI2 = [
-0.882976 -2.350738 -0.035688  0.296927
 0.259639 -0.259648  0.056688  1.268853
-0.547722  0.3397   -0.572971 -1.985099
-1.013208 -0.782236 -0.46785    2.375783 ]';
```

In Figures 2,3 the local potential and kinetic energy modifications introduced by $\mathcal{H}_a$ can be observed. In Figure 4 a set of four orbits with initial conditions over the needed attraction basin are shown characterizing the controller effectiveness.

7 Conclusions

- The main contribution of this technique is associated to the vector field $\boldsymbol{K}_\pi(\boldsymbol{x})$ determination, avoiding the cumbersome resolution of a partial differential equation.
- The saturated nature of neural nets also facilitates the bounding of the control action as practically desired. The vector field $\boldsymbol{K}_\pi(\boldsymbol{x})$ maximum strength is fixed previously, depending on the desired attraction basin, after a judicious effort consideration. The maximum strength is controlled as a barrier on the maximum eigenvalue of Π^0 and limiting a suitable norm over Π^1 and Π^2 (18).
- The controller performance depends on the choice of J_a,R_a which models state interconnections and dissipation and α which models closed-loop settling time.
- The plant model can be described with class C^0 or even piecewise C^0 functions and state space restrictions are also allowed (anti slippage for instance)

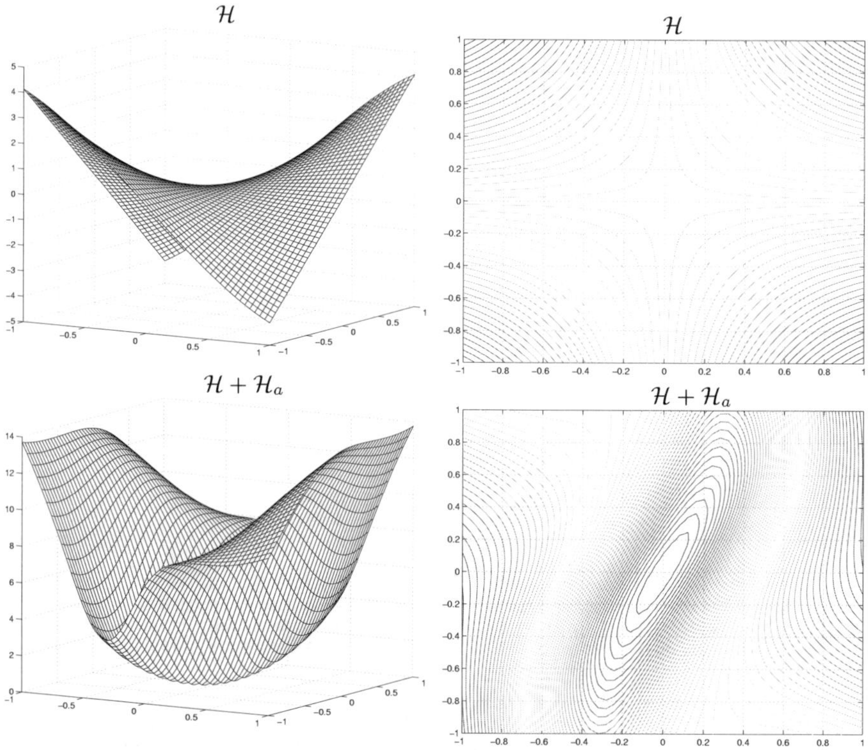

Fig. 2. Comparison of $\mathcal{H}$ and $\mathcal{H}_d = \mathcal{H} + \mathcal{H}_a$ for $\{-1 \leq x_1 \leq 1\}$, $\{-1 \leq x_2 \leq 1\}$, $x_3 = 0$, $x_4 = 0$. (Potential).

As future work we are also introducing the automatic configuration of J_a and R_a using an additional neural net that will evolve jointly with the controller.

Acknowledgements

This work is supported by CICYT, under project TAP99-0926-C04-03.

References

1. Schwefel H.-P., Rudolph, G. (1995), *Contemporary Evolution Strategies.* In: Moran F., Moreno A., Merelo J.J., Chacon, P. (eds.) Advances in Artificial Life. Third International Conference on Artificial Life, vol. 929 of Lecture Notes in Computer Science, 893–907, Springer, Berlin
2. Fiacco A.V., McCormick, G.P. (1968), *Nonlinear Programming: Sequential Unconstrained Minimization Techniques.* John Wiley and Sons, Inc.
3. Ortega R., van der Schaft A.J., Mareels I., Maschke B. (2001), *Putting Energy Back in Control.* Control Systems Magazine, 21, 18–33

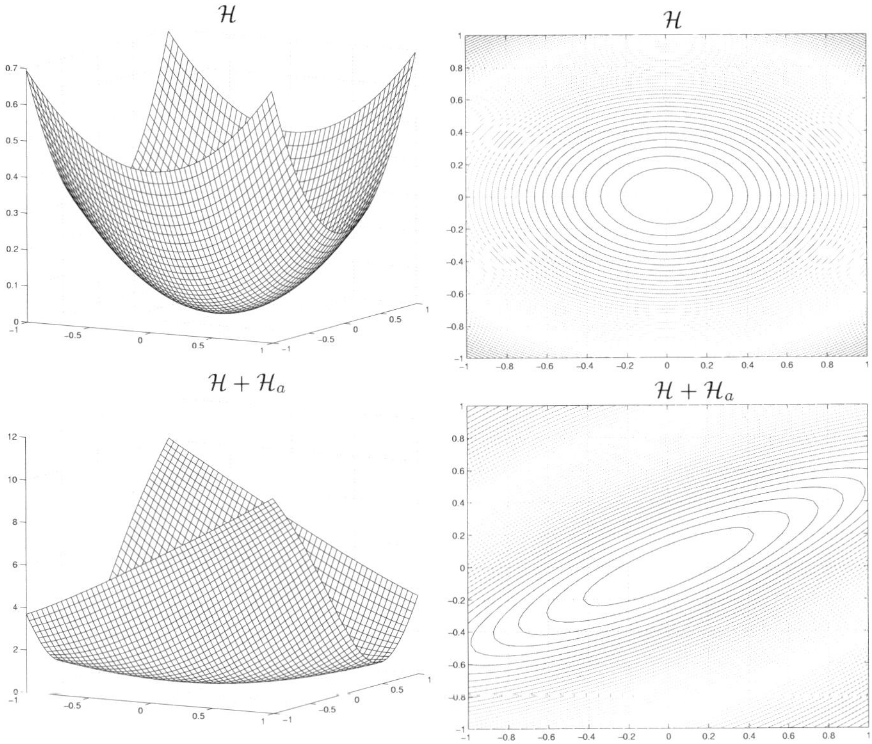

Fig. 3. Comparison of $\mathcal{H}$ and $\mathcal{H}_d = \mathcal{H} + \mathcal{H}_a$ for $x_1 = 0$, $x_2 = 0$, $\{-1 \leq x_3 \leq 1\}$, $\{-1 \leq x_4 \leq 1\}$. (Kinetic).

4. Ortega R., Spong M.W. (2000), *Stabilization of Underactuated Mechanical Systems Via Interconnection and Damping Assignment.* CNRS-SUPELEC, University of Illinois, Proceedings of Lagrangian and Hamiltonian Methods for Nonlinear Control, 1, 69–74
5. Liu P., Zinober A.S.I (1994), *Recursive Interlacing Regulation of Flat and Non-Flat Systems.* School of Mathematics and Statistics, University of Sheffield.
6. Hauser J., Sastry S., Kokotovich P. (1992), *Nonlinear Control Via Approximate Input-Output Linearization: The Ball and Beam Example.* School of Mathematics and Statistics, University of Sheffield, Centre for Systems & Control, University of Glasgow — Internal Report, 37, 392–398
7. van der Schaft A.J. (1996), *$\mathcal{L}_2$ Gain and Passivity Techniques in Nonlinear Control.* Lecture Notes in Control and Information Science, Vol. 218, Springer Verlag, London
8. Bäck T., Hammel U., Schwefel H.P. (1997), *Evolutionary Computation: Comments on the History and Current State.* In: IEEE Trans. on Evolutionary Computation, vol. 1 n° 1
9. Sjöberg J., Ljung L. (1992), *Overtraining, regularization and searching for minimum in neural networks.* In: 4th IFAC Symposium on Adaptive Systems in Control and Signal Processing, Grenoble, France, 669–674

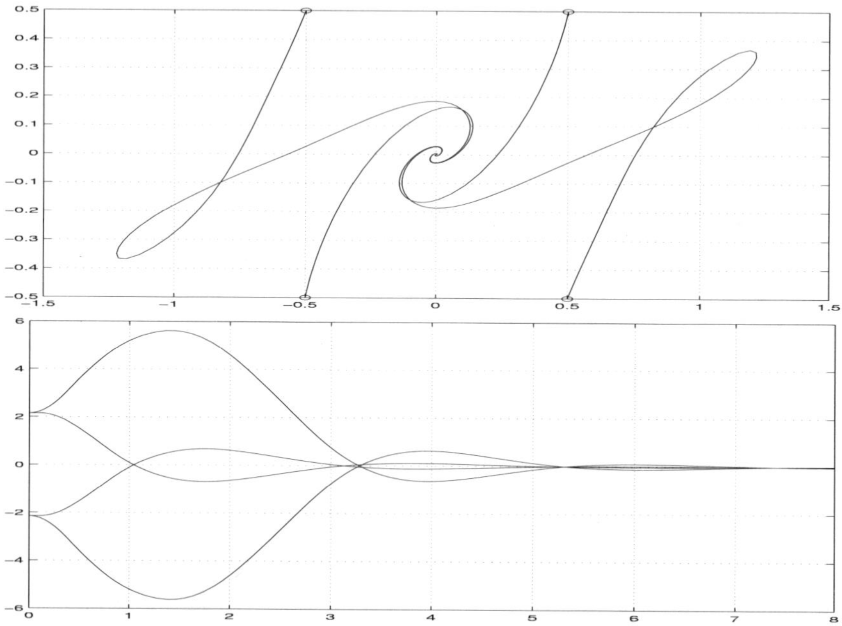

Fig. 4. $x_1 \times x_2$ phase space projection (upper) and $u(t)$ (lower)

Feedback Stabilization and $\mathcal{H}_\infty$ Control of Nonlinear Systems Affected by Disturbances: the Differential Games Approach

Pierpaolo Soravia

Università di Padova, Dipartimento di Matematica Pura e Applicata
via Belzoni, 7, 35131 Padova, Italy

Abstract. In this paper we discuss the differential games approach to stability questions in control systems affected by disturbances. Our main focus will be on nonlinear $\mathcal{H}_\infty$ but we also mention other related problems. We review some existing results on necessary and sufficient conditions based on the existence of a suitably defined Lyapunov function for the system. Being our attention to nonlinear systems, we formulate our conditions allowing nonsmooth Lyapunov functions, using therefore the concept of viscosity solutions for nonlinear partial differential equations. We then present some new results based on an old natural idea recently brought to attention in stabilization of systems by [8], and show how, from semiconcave or just continuous Lyapunov functions one can construct a (discontinuous) feedback that solves the $\mathcal{H}_\infty$ problem in an appropriate sense.

1 Introduction

In this paper we consider a control system affected by disturbances

$$\dot{y} = f(y, u, w), \quad y(0) = x \in \mathbf{R}^N, \tag{1}$$

where the vector field $f : \mathbf{R}^N \times U \times W \to \mathbf{R}^N$ will always be continuous and satisfy enough structural conditions to provide existence and uniqueness of global trajectories $y_x(t, u, w)$ of the system. The control $u(\cdot)$ and the deterministic disturbances $w(\cdot)$ will be taken from suitable admissible classes that we make precise later. The main property that we care about for system (1) here is stability to an equilibrium, to be defined in a suitable way. The approach that we plan to implement is that of worst case disturbances or, equivalently, we want to design stabilizing controllers that are robust in the presence of disturbances. The fact that makes the problem interesting and the differential games approach particularly suitable is that we require such robustness despite all possible disturbances and not only with respect to sufficiently small disturbances according to the data involved and possibly to the controller that we design. Particularly important are such applications as $\mathcal{H}_\infty$ control where disturbances may be unbounded.

We plan to investigate stability as related to the existence of a suitably defined Lyapunov function for the system, i.e. a function satisfying a partial

F. Colonius, L. Grüne (Eds.): Dynamics, Bifurcations, and Control, LNCIS 273, pp. 173–190, 2002.

differential inequality related to the data of the system and the problem. We will define Lyapunov functions relative to some output function

$$z = h(x, u, w), \tag{2}$$

where $h : \mathbf{R}^N \times U \times W \to \mathbf{R}_+$ is at least continuous. This is related to measuring in a quantitative way the rate of decrease of the trajectories of the system relative to the level sets of the Lyapunov function itself. It is then natural to define Lyapunov functions as solutions $V : \mathbf{R}^N \to \mathbf{R}^+$ of the partial differential inequality

$$\min_{w \in W} \max_{u \in U} \{-f(x, u, w) \cdot DV(x) - h(x, u, w)\} \geq 0. \tag{3}$$

The reason why (3) appears can be justified by differential games theory. We are interested in allowing nonsmooth Lyapunov functions, therefore (3) has to be interpreted in a suitable weak sense. We will adopt the concept of viscosity solutions, see e.g. Crandall-Lions [10], and for applications in optimal control the book by Bardi and Capuzzo-Dolcetta [2]. The first goal is to prove that a system is stable if and only if it admits a Lyapunov function. Most of our results will refer explicitly to the so called nonlinear $\mathcal{H}_\infty$ control problem that we define in the next section, although in Section 3 we will mention other related problems where similar ideas can be adapted easily.

In order to be able to implement games theory and to pursue the mentioned equivalence, in a first step we have to admit controllers in larger classes than feedbacks. Thus the second goal becomes the existence of a robust stabilizing feedback. Unfortunately, the classical, formal construction only works in practice when a C^1 Lyapunov function is available and the system is affine (although a few more general results can be obtained). Moreover, even in the case of undisturbed systems, obstructions to existence of continuous stabilizing feedbacks are well known. Thus one is left with the problem of interpreting a closed loop system for a discontinuous feedback. Isaacs' book [18] suggests the use of K-strategies (from S. Karlin) that we rename sampled feedbacks following their recent use in stabilization of nonlinear systems by Clarke-Ledyaev-Sontag-Subbotin [8]. We will show that if a continuous Lyapunov function exists, then we can construct a robust stabilizing sampled feedback.

We finally mention that the theory of differential games starts with the work of Isaacs, contained in some internal reports written in the mid fifties at Rand Corporation. Isaacs' work remained unpublished until his book [18] appeared a few years later. The numerous questions raised by his work motivated the work of many researchers and the theory flourished in the next twenty years. One of the main results of Isaacs work and of key interest here, is the formal derivation of a partial differential equation of Hamilton-Jacobi type satisfied by value functions in differential games. We recall among the main contributions towards the rigorous derivation and understanding of the so called Isaacs equation [14], [16], [12], [23], [4] and the references therein.

A more recent major breakthrough is due to the Crandall-Lions [10] theory of viscosity solutions that made possible to prove rigorously that indeed value functions satisfy the Isaacs equation in the viscosity sense under very general assumptions, see e.g. Barron-Evans-Jensen [3] and Evans-Souganidis [13]. The crucial point of this theory is the fact that viscosity solutions allow uniqueness results in many circumstances.

2 Differential games approach to nonlinear $\mathcal{H}_\infty$ control

In the following, to avoid some technical difficulties, the control set U will be a compact subset of an euclidean space, while the set W where disturbances take their values will be closed but may be unbounded. We define the set of admissible disturbances as $\mathcal{W} = L^2(0,+\infty;W)$. The choice for the exponent is traditional (although irrelevant for exponent $p > 1$) so we will stick to it. The controller will choose open loop controls in $\mathcal{U} = L^\infty(0,+\infty;U)$. More interesting are cases where the controller uses feedback maps $u : \mathbf{R}^N \to U$. When using feedbacks one has to be very careful with existence of trajectories of the closed loop system. In the following we say that a feedback map $u(\cdot)$ is admissible if the closed loop system

$$\dot{y} = f(y, u(y), w), \quad y(0) = x$$

has a unique global, absolutely continuous solution for all $x \in \mathbf{R}^N$ and $w \in \mathcal{W}$. Usually, checking if a feedback is admissible is not straightforward, especially if it is discontinuous.

In order to give sense to our system and choice of disturbances, we need the following set of assumptions on the system. We will henceforth suppose that

$$\begin{cases} |f(x,u,w) - f(y,u,w)| \le L(1+|w|)|x-y|, \\ |f(x,u,w)| \le L(1+|x|+|w|) \quad \text{for all } x,y,u,v, \\ |h(x,u) - h(y,u)| \le L_R|x-y|, \\ \text{for all } |x|,|y| \le R, u \in U,\ R > 0, \end{cases} \tag{1}$$

where $h : R^N \times U \to [0,+\infty[$ is continuous. We can add to the output a lower semicontinuous, extended real valued term to take into account for instance state constraints, see e.g. [22], but we will avoid doing it here for the sake of simplicity.

The nonlinear $\mathcal{H}_\infty$ suboptimal problem with full state information is formulated as follows.

Definition 1. Given a constant $\gamma > 0$ (the disturbance attenuation level), we seek an admissible feedback map $u : \mathbf{R}^N \to U$ and a nonnegative function $G : \mathbf{R}^N \to \mathbf{R}$, null at the origin such that the two following conditions are satisfied:

(i) the feedback $u(\cdot)$ stabilizes the closed loop undisturbed system,

$$\text{(ii)} \quad \int_0^t h(y_x, u(y))ds \le \gamma^2 \int_0^t |w|^2 ds + G(x), \tag{2}$$
$$\text{for all } x \in \mathbf{R}^N,\ t \ge 0,\ w \in \mathcal{W}.$$

Indeed, question (i) turns out to be basically a consequence of (ii), which is therefore the main object of our investigation. Inequality (ii) measures the robustness of the stabilizing feedback.

Remark 1. The $\mathcal{H}_\infty$ control problem was proposed by Zames [33] for linear systems in the frequency domain. The solution of the problem in the linear-quadratic case can be found in [11], and [25]. Namely it was shown that the problem has a solution if and only if a related Riccati equation has a positive definite solution. This result can be extended to completely general systems, as we explain below. Such result is contained in the work by the author, see [28,29] and Appendix B in [2], and holds even for infinite dimensional nonlinear systems, see [21,22] and the references therein. Sufficient conditions for nonlinear-affine systems, based on the existence of a C^1 Lyapunov function, were previously explored in [1], [30] and [31]. The differential games approach is also discussed extensively in the book [5]. We want to mention the interesting connection between stochastic optimal control and $\mathcal{H}_\infty$ control via the risk-sensitive approach, first conjectured by Jacobson [20] and rigorously proved by Fleming-McEneaney [15] by extensively using pde theory and the notion of viscosity solutions .

The connection between the problem as formulated above and differential games is immediate and is derived as follows. It is clear that (2) is equivalent to stating that

$$\sup_{w\in\mathcal{W}} \sup_{t\in\mathbf{R}_+} \int_0^t (h(y_x, u(y_x)) - \gamma^2|w|^2)ds \le G(x) \quad \text{for all } x \in \mathbf{R}^N.$$

We are therefore led to consider the following object

$$\inf_{u(x)} \sup_{w\in\mathcal{W}} \sup_{t\in\mathbf{R}_+} \int_0^t (h(y_x, u(y_x)) - \gamma^2|w|^2)ds. \tag{3}$$

This is what can be related to differential games theory, the so called lower value function and the Hamilton-Jacobi-Isaacs (HJI for short) equation

$$\min_{w\in W} \max_{u\in U} \{-f(x,u,w)\cdot DV(x) - h(x,u) + \gamma^2|w|^2\} = 0. \tag{4}$$

In particular the functional

$$J(x,u,w) = \sup_{t\in\mathbf{R}_+} \int_0^t (h(y_x, u) - \gamma^2|w|^2)ds$$

will be the payoff functional of the game, corresponding to the $\mathcal{H}_\infty$ problem, that the two players, controller and nature (governing disturbances) want to optimize with opposite goals (what is called a zero-sum game). In order to implement games theory some difficulties arise however, that we discuss next. Differential games theory has as a primary goal the existence of value, particularly as a consequence of the existence of saddles for the payoff J. Precisely, given x, a pair (u, w) is said to be a saddle point for J if

$$J(x, u, w) \leq J(x, u, w) \leq J(x, u, w), \quad \text{for all } u, v.$$

Indeed for given x, a saddle (u, w) automatically defines optimal controls for the two players. This general issue is however quite difficult and has to be addressed is some generalized sense. More interesting to us here is the connection with the so called Hamilton-Jacobi-Isaacs equation. In order to derive it, the classical argument by Isaacs uses two main facts, namely the existence of feedback saddles for the payoff, which allows to define a *value* function by setting $V(x) = J(x, u, w)$, and then smoothness of the value function itself. Both required facts are quite restrictive and this is what makes the argument formal in most situations. There is a third, technical but crucial condition in the work by Isaacs, the so called Isaacs condition. Of course in (4) we cannot interchange the roles of *min* and *max* in general, therefore we say that the Isaacs condition holds whenever

$$\begin{aligned} &\min_{w \in W} \max_{u \in U} \left\{-f(x, u, w) \cdot p - h(x, u) + \gamma^2 |w|^2\right\} \\ &= \max_{u \in U} \min_{w \in W} \left\{-f(x, u, w) \cdot p - h(x, u) + \gamma^2 |w|^2\right\}, \end{aligned} \tag{5}$$

for all x, p. To overcome these problems already pointed out by Isaacs, the authors mentioned in the introduction proposed different approaches. All of them give meaning to the Isaacs equation via the dynamic programming principle, without assuming existence of saddles and smoothness of the (upper or lower) value function but only much more reasonable structural conditions on the system. Their work simplifies, extends and is put into a uniqueness theory for Hamilton-Jacobi equations with the introduction of the notion of viscosity solutions. Moreover, a major consequence of using the theory of viscosity solutions is the fact that we can state that all these approaches are equivalent whenever uniqueness of the solution of the HJI equation holds. To complete this part, we now give the definition of viscosity solution for (4).

Definition 2. Let $V : \mathbf{R}^N \to \mathbf{R}$ be a lower semicontinuous function (respectively upper semicontinuous). We say that V is a viscosity supersolution of (4) (resp. subsolution) if whenever $\varphi \in C^2$ and $V - \varphi$ attains a local minimum point (resp. maximum point) at $x_o \in \mathbf{R}^N$ we then have

$$\min_{w \in W} \max_{u \in U} \{-f(x_o, u, w) \cdot D\varphi(x_o) - h(x_o, u) + \gamma^2 |w|^2\} \geq 0 \quad (\text{resp. } \leq 0).$$

In this case we also say that $D\varphi(x_o) \in D^- V(x_o)$ the subdifferential of V at x_o (resp. $D\varphi(x_o) \in D^+ V(x_o)$ the superdifferential). If V is contionuous, we say that it is a viscosity solution of (4) if it is both a subsolution and a supersolution.

What we describe next is maybe the most direct approach to defining the value function for a differential game, as proposed by Varaya-Roxin-Elliott-Kalton, see [12] and the references therein. Instead of looking at (3) directly, it seems more natural to proceed as follows. We say that a functional $\alpha : \mathcal{W} \to \mathcal{U}$ is a strategy, $\alpha \in \Delta$, for the minimizing player if it is nonanticipating (or causal) in the following sense:

$$\text{if } w_1, w_2 \in \mathcal{W} \text{ are such that } w_1 = w_2 \text{ a.e. in } [0,T] \text{ then } \alpha[w_1] = \alpha[w_2] \text{ a.e. in } [0,T].$$

Now we introduce the *lower* value function (in the VREK sense) of the differential game, namely

$$V_\gamma(x) = \inf_{\alpha\in\Delta} \sup_{w\in\mathcal{V}} \sup_{t\in\mathbf{R}_+} \int_0^t (h(y_x, \alpha[w]) - \gamma^2|w|^2)ds. \tag{6}$$

This is the object that we will study. Note that in (6) the roles of the two players are not symmetric. Reversing the roles of the players, i.e. allowing the maximizing player to use strategies and the minimizing one to use open loop controls defines the upper value. This asymmetry is also reflected in the HJI equation (4) in that, as we observed, we cannot interchange *min* and *max* in general. While V_γ can be related to equation (4), the upper value is instead related to the equation where min and max are interchanged. At a first sight, the upper value function does not seem so realistic in practice, thinking of the maximizing player as nature, but in the theory it is important to take it into account, because only when upper and lower values coincide (usually when the Isaacs condition holds), we may say that a value function for the game exists and therefore that an optimal cost for the players can be set precisely. Note that the concept of saddle point for the payoff loses meaning with the VREK approach, and saddles are to be found in some generalized sense.

Remark 2. Even the lower value does not appear too realistic. Indeed, in order to apply a strategy, a player needs to know the control of the opponent. As an example of strategies more reasonable in practical cases than the general ones defined above, we present the following sampled feedbacks (or K-strategies from Isaacs book). Let $u : \mathbf{R}^N \to U$ be a (discontinuous) feedback map and let $(t_n)_n$ be an increasing, unbounded sequence of positive numbers with

$$t_o = 0, \quad 0 < t_{n+1} - t_n \leq \delta.$$

The positive number δ is the sampling rate. For any $w \in \mathcal{W}$, $x \in \mathbf{R}^N$, solve recursively the differential equations

$$\dot{y} = f(y, u(x_n), w(\cdot + t_n)), \quad y(t_n) = x_n,$$
$$x_{n+1} = y(t_{n+1}),$$
$$x_o = x.$$

Then the position $\alpha_\delta[w](t) = u(x_n)$ for $t \in [t_n, t_{n+1}[$ defines a strategy that we call the sampled feedback determined by u and the sequence $(t_n)_n$. Finding strategies in the previous class that solve the $\mathcal{H}_\infty$ problem, will be the highlight of the last section.

Remark 3. The value function V_γ as defined above, can be thought of as the open loop version of the differential game. We observe that the correct open loop version of the game is not in general the simpler

$$\sup_{w \in \mathcal{W}} \inf_{u \in \mathcal{U}} J(x, u, w),$$

because this object lacks an information pattern between the two players completely. This is also called the static lower value because at the time of choosing the optimal control, the minimizing player knows the control chosen by the opponent for all times. In particular the existence of the static value, i.e. asserting that

$$\sup_{w \in \mathcal{W}} \inf_{u \in \mathcal{U}} J(x, u, w) = \inf_{u \in \mathcal{U}} \sup_{w \in \mathcal{W}} J(x, u, w),$$

which is essential in order to be able to derive the HJI equation, is definitely too restrictive to be true in practice.

We will now discuss the solvability of the $\mathcal{H}_\infty$ problem. The value function V_γ gives by construction the best value $G(x)$ to be used in (2), at least if there is a sequence of minimizing feedbacks. Observe that by choosing $0 = t \in \mathbf{R}_+$ in (6), we immediately obtain that $V_\gamma \geq 0$, but this function may be discontinuous and attain the value $+\infty$ at many points. We can expect finiteness of V_γ to be equivalent to having condition (2) satisfied. Below we will suppose that V_γ is at least lower semicontinuous. This can be assured by standard methods, for instance by assuming convexity of the sets $(f(x, U, w), h(x, U))$ for all x, w or by using relaxed controls for the minimizing player.

The following result states the equivalence between the solvability of the $\mathcal{H}_\infty$ problem and that of finding a positive definite solution of the HJI equation, extending to general nonlinear systems the known result holding in the linear quadratic case.

Theorem 1. *([28,29]) The system (1) with output (2) admits a family of strategies $(\alpha_x)_{x \in \mathbf{R}^N}$ such that the L^2 gain condition is satisfied and the undisturbed system is globally asymptotically controllable to the origin with the* *family of controls $(\alpha_x[0])_x$ if and only if the Hamilton-Jacobi-Isaacs equation*

$$\min_{w\in W}\max_{u\in U}\{-f(x,u,w)\cdot DV(x)-h(x,u)+\gamma^2|w|^2\}=0. \tag{7}$$

has a finite, positive definite, lower semicontinuous and continuous at the origin viscosity (super)-solution.

Remark 4. Note that in Theorem 1 equivalence is stated between the existence of a positive definite solution of the HJI equation and the solvability of the open loop version of the problem. In particular the stabilizability of the undisturbed system has been replaced by global asymptotic controllability, and the feedback in (2) by a family of strategies. This is weaker than we sought but essential to obtain an equivalence statement in general. In the final section, we will describe how one can indeed construct a feedback solution.

Remark 5. The statement of Theorem 1 justifies our definition of a viscosity supersolution of (7) to be a Lypaunov function for the $\mathcal{H}_\infty$ problem. This is confirmed by observing that a Lyapunov function V also satisfies

$$\max_{u\in U}\{-f(x,u,0)\cdot DV(x)-h(x,u)\}\geq 0,$$

which is the usual differential inequality characterizing a control Lyapunov function for the undisturbed system.

Idea of the proof of Theorem 1. The necessary part is, at this point, just a matter of putting together differential games and viscosity solutions theories via a dynamic programming principle. More interesting is the sufficiency part. This follows from the equivalence between viscosity supersolutions V of (7) and the optimality principle formula

$$V(x)=\inf_{\alpha\in\Delta}\sup_{w\in\mathcal{W}}\sup_{t\geq 0}\left\{\int_0^t[h(y,\alpha)-\gamma^2|w|^2]ds+V(y(t))\right\}, \tag{8}$$

which is a deep fact. The previous formula describes the way supersolutions propagate along trajectories of the system, which is closely related to the method of characteristics in partial differential equations. ¿From (8) and $V\geq 0$ we then conclude that $V\geq V_\gamma$. On the other hand, choosing $w\equiv 0$ we get

$$V(x)=\inf_{u\in\mathcal{U}}\sup_{t\geq 0}\left\{\int_0^t h(y,u)ds+V(y(t))\right\},$$

where $\dot{y}=f(y,u,0),\quad y(0)=x$. This is the basic equality to conclude global asymptotic controllability of the undisturbed system from the continuity of V at the origin. □

3 Other stability questions

In this section we mention a selection of other notions of stability, that can be used in systems affected by disturbances and discussed via the differential games approach. In the rest of this section also the control set for disturbances W will be bounded, therefore we do not need to worry about unbounded disturbances, and requiring stability despite worst possible disturbances is not too restrictive. In particular, below we refer to a more general output as in (2) and not to the specific one of the previous section. Other assumptions on the system will be as above. We will also refer to a general equilibrium set, namely any closed set $\mathcal{T} \subset \mathbf{R}^N$. We introduce the following quantity

$$t_x(u,w) = \inf\{t \geq 0 : y_x(t,u,w) \in \mathcal{T}\} \, (\leq +\infty).$$

The previous functional is called capture-time in the literature of pursuit-evasion games. We also denote by $d(x) = \operatorname{dist}(x, \mathcal{T})$ the distance function from $\mathcal{T}$.

We now present several notions of stability, the last one in particular is classical in differential games theory and its study was the main motivation for Isaacs' work.

Definition 3. System (1, 2) is:

- stable, if for all open sets $\mathcal{K} \supset \mathcal{T}$ there is an open set $\mathcal{S} \supset \mathcal{T}$ such that: for all $x \in \mathcal{S}$ there is $\alpha \in \Delta$ such that $y_x(t,\alpha,w) \in \mathcal{K}$ for all $w \in \mathcal{W}$ and $t \geq 0$;
- globally asymptotically controllable, if for all $x \in \mathbf{R}^N$ there is $\alpha \in \Delta$ such that $\lim_{t\to+\infty} d(y_x(t,\alpha,w)) = 0$ for all $w \in \mathcal{W}$;
- globally L^p – *stable*, if for all $x \in \mathbf{R}^N$ there are $\alpha \in \Delta$ and $C_x \in \mathbf{R}$ such that

$$\int_0^{+\infty} [d(y_x(t,\alpha,w))]^p dt \leq C_x,$$

  for all $w \in \mathcal{W}$;
- globally terminable, if for all $x \in \mathbf{R}^N$ there are $\alpha \in \Delta$ and $C_x \in \mathbf{R}$ such that $t_x(\alpha,w) \leq C_x$ for all $w \in \mathcal{W}$.

Of course, all notions above have local counterparts, but we will not analyze them explicitly here. Also, all the questions above can be formulated via feedbacks instead of strategies for the controller. In this case all the discussion of the previous and next sections applies as well.

The study of the notions in Definition 3 can be approached by the dynamic programming equation and the use of viscosity solutions. In particular, the Hamilton-Jacobi-Isaacs inequality involved is again of the form

$$\min_{w\in W} \max_{u\in U} \{-f(x,u,w)\cdot DV - h(x,u,w)\} \geq 0, \tag{1}$$

for suitable choices of the output. Possible sufficient results in this direction are as follows. Of course, similarly to Theorem 1 we can obtain necessary conditions as well.

Theorem 2. *([26,27]) Let $V : \mathbf{R}^N \to [0, +\infty[$ be a continuous and positive definite, i.e. vanishing on $\mathcal{T}$ and positive elsewhere, viscosity solution of (1) in $\mathbf{R}^N$. Then:*

- *if $h \equiv 0$, the system is stable;*
- *if h is positive definite, the system is globally asymptotically controllable;*
- *if $h = Cd(x)^p$, the system is L^p-terminable;*
- *if $h(x, u, w) \geq c > 0$ for all x, u, w, the system is globally terminable (here we only need V to solve (1) in $\mathbf{R}^N \backslash \mathcal{T}$).*

In Theorem 2 h positive definite means that $h(x, u, w) = 0$ if $x \in \mathcal{T}$ but $h(x, u, w) > 0$ if $x \notin \mathcal{T}$ for all u, w. For more detailed results on such problems, we refer the interested reader to [26,27] and the references therein.

4 Building a feedback solution for nonlinear $\mathcal{H}_\infty$ control

As noticed above, in order to be implemented, strategies of differential games theory in the VREK sense require the knowledge of the open loop control played by the opponent. This fact makes their use unrealistic in practice. Other notions of value such as that of Fleming [14], Friedman [16], Krassovskii-Subbotin [23] or of Berkovitz [4] implement some sort of discrete time approximation and may appear more natural. In fact, discrete-time games are somewhat easier to discuss. However, we stress the fact that all the notions of value that fit well within the dynamic programming approach are equivalent, as we already observed, when uniqueness of the solution of the Hamilton-Jacobi-Isaacs equation holds in the viscosity sense. Results are available in this direction in many cases, although for instance the equation of $\mathcal{H}_\infty$ control is not included in the uniqueness theory even if we limit ourselves to classical solutions. In this section we want to turn to the construction of ε-optimal feedbacks, implementing the classical idea of the dynamic programming approach proposed by Isaacs. Namely, after giving precise meaning to the Hamilton-Jacobi-Isaacs equation, as we did in the second section, from its solution we build an optimal controller in feedback form.

At this point, however, the Isaacs condition comes into play. If $V \in C^1$ is a solution of

$$\max_{u \in U} \min_{w \in W} \{-f(x, u, w) \cdot DV(x) - h(x, u) + \gamma^2 |w|^2\} \geq 0, \tag{1}$$

then the formula

$$u(x) \in \arg\max_{u \in U} \left\{ \min_{w \in W} \{-f(x, u, w) \cdot DV(x) - h(x, u) + \gamma^2 |w|^2\} \right\} \tag{2}$$

formally provides an answer to our problem. Indeed, if (2) defines an admissible feedback, this solves the $\mathcal{H}_\infty$ problem. This fact is shown for instance in [28] and is rather standard. One can sometimes also handle nonsmooth solutions of (1), see e.g. Xiao-Basar [32]. What makes the argument formal is mainly the fact that we cannot prove existence of an admissible feedback selection of (2) in general. Notice that we are using the upper Isaacs equation instead of the lower value one. Therefore we are requiring a sufficient condition stronger than the one prescribed by Theorem 1, unless the two equations are equivalent, that is the Isaacs condition holds. In general, there might be a gap between the cost that the minimizing player can realize by using feedbacks and the optimal lower value V_γ. Besides the mentioned Isaacs condition, it is not known in which other cases the existence of a Lyapunov function, a supersolution of the lower Isaacs equation, is sufficient to construct an optimal feedback.

In the following we turn to consider nonsmooth Lyapunov functions and to implement the above idea in a weaker sense. We will work with sampled feedback strategies as introduced in Remark 2. For technical reasons, we add a more restrictive structural condition on the system. Precisely, we suppose further that

$$|f(x,u,w)| \leq L(1+|x|),$$

for all x,u,v. Then the following estimate holds for the trajectories

$$\begin{aligned}|y(t)-x| &\leq \int_0^t (L|y(s)-x| + |f(x,u,w)|)ds \\ &\leq L(1+|x|)t + L\int_0^t |y(s)-x|ds.\end{aligned}$$

Hence from Gronwall Lemma we get

$$|y(t)-x| \leq e^{Lt}L(1+|x|)t, \quad \text{for } t \geq 0. \tag{3}$$

What follows next is inspired by the work of Krasovskii-Subbotin [23] as applied to construction of ε-optimal feedbacks in Clarke-Ledyaev-Subbotin [9] and to stabilization of nonlinear systems in [7], [8], [24], see also [19]. Below, $V : \mathbf{R}^N \to \mathbf{R}$ will be at least continuous, positive definite, i.e. $V(0) = 0$, $V(x) > 0$ if $x \neq 0$, and such that $\lim_{x\to\infty} V(x) = +\infty$. From now on V will also be a nonsmooth Lyapunov function, i.e. a viscosity supersolution of the *upper value equation* (1). This means that the inequality (1) is satisfied by all elements of the subdifferential $D^-V(x)$.

We start by considering semiconcave Lyapunov functions in the following sense. We say that V is C-semiconcave if the function $V(x) - C|x|^2$ is concave. It is known that any semiconcave function can be equivalently described, see e.g. [6], as the sum of a concave and a C^1 function, and that it is locally Lipschitz continuous. Therefore semiconcavity is an intermediate regularity between C^1 and Lipschitz. Note that by semiconcavity one deduces the inequality

$$V(y) \leq V(x) + p\cdot(y-x) + C|y-x|^2, \tag{4}$$

for all $x, y \in \mathbf{R}^N$, $p \in D^+V(x)$, by the concavity of the hypograph $\{(x,y) : y \le V(x) - C|x|^2\}$. Here $D^+V(x)$ indicates the superdifferential of V at x, and we recall that for a concave function it coincides with the standard one of convex analysis. Since V is only almost everywhere differentiable in general, we will use the following selection

$$p(x) = \begin{cases} DV(x), & \text{if } DV(x) \text{ exists;} \\ \lim_{x_n \to x} DV(x_n), & \text{for some } x_n \to x \text{ otherwise,} \end{cases}$$

so that $p(x) \in D^+V(x)$ for $x \in \mathbf{R}^N$. This is a basic fact of semiconvex functions, see for instance Proposition II.4.7 in [2]. In particular we may suppose that for suitable C_R we have $|p(x)| \le C_R$ for all $|x| \le R$. In order to solve the $\mathcal{H}_\infty$ control problem, we will use a possibly discontinuous feedback constructed by the relation

$$u(x) \in \arg\max_{u \in U} \left\{ \min_{w \in W} \{-f(x,u,w) \cdot p(x) - h(x,u) + \gamma^2 |w|^2\} \right\}. \tag{5}$$

Therefore by definitions of viscosity solution and of $p(\cdot)$, in particular this means that

$$-f(x,u(x),w) \cdot p(x) \ge h(x,u(x)) - \gamma^2 |w|^2, \tag{6}$$

for all $x \in \mathbf{R}^N$, $w \in W$. Notice that at this stage it is sufficient that V satisfies (1) in the almost everywhere sense. Being discontinuous in general, this feedback will be implemented as a sampled feedback strategy with respect to the sequence $t_n = n\delta$, and a suitable sampling parameter δ to be made precise later. We will denote below by α_δ such a strategy constructed as in Remark 2.

We start by considering the trajectory solving

$$\dot{y} = f(y,u(x),w), \quad y(0) = x \in \mathbf{R}^N, \quad t \in [0,\delta],$$

for any given disturbance $w \in \mathcal{W}$. The following is now easily deduced by combining (3), (4) and (6)

$$\begin{aligned} V(y(t)) &\le V(x) + p(x) \cdot (y(t) - x) + C|y(t) - x|^2 \\ &\le V(x) + p(x) \cdot \textstyle\int_0^t f(y,u(x),w)ds + Ce^{2L\delta}L^2(1+|x|)^2 t^2 \\ &\le V(x) + p(x) \cdot \textstyle\int_0^t f(x,u(x),w)ds \\ &\quad + C_R e^{L\delta} L^2 (1+|x|) t^2/2 + Ce^{2L\delta}L^2(1+|x|)^2 t^2 \\ &\le V(x) - \textstyle\int_0^t [h(x,u(x)) - \gamma^2|w|^2]ds \\ &\quad + (C_R e^{L\delta} L^2(1+|x|)\delta/2 + Ce^{2L\delta}L^2(1+|x|)^2\delta)t \\ &\le V(x) - \textstyle\int_0^t [h(y,u(x)) - \gamma^2|w|^2]ds + (L_{R+1}e^{L\delta}L^2(1+|x|)\delta \\ &\quad + C_R e^{L\delta}L^2(1+|x|)\delta/2 + Ce^{2L\delta}L^2(1+|x|)^2\delta)t \\ &\le V(x) - \textstyle\int_0^t [h(y,u(x)) - \gamma^2|w|^2]ds + tO_R(\delta), \end{aligned} \tag{7}$$

as $\delta \to 0$, for all $|x| \le R$, $w \in \mathcal{W}$, $t \in [0,\delta]$. A consequence of propagating (7) is the following result.

Theorem 3. *Let $0 < r < R$, $K > 0$ and $c \in (0,1)$ be given. Then we can find positive numbers $M(R), m(R), T(r,R,K,c), \delta(r,c), k(r)$ such that $V(x) \le m(r)$ implies $|x| \le r$ and: for all $r \le |x| \le R$ and $w \in \mathcal{W}$, the sampled feedback trajectories constructed above $y_x(\cdot, \alpha_\delta, w)$ satisfy*

(i) $V(y(t)) + \int_0^t [ch(y,\alpha_\delta) - \gamma^2|w|^2]ds \le V(x)$, *for all $t \le n\delta \le T$, where n is the first integer such that $V(y(n\delta)) \le m(r)$;*
(ii) $y(t) \in \{x : V(x) \le M(R) + \gamma^2\|w\|_2^2\}$ *for all $t \ge 0$;*
(iii) $\lim_{R\to 0^+} M(R) = 0$;
(iv) $|y(t)| \le r$, *for all w, $\|w\|_2^2 \le k(r)/\gamma^2$ and all $t \ge T$.*

Remark 6. To comment the statement of Theorem 3, we notice that the sampled feedback α_δ provides a global boundedness condition on the trajectories (ii) which is robust with respect to bounded disturbances (in L^2); uniform attractiveness and stability of the undisturbed system, robust with respect to small disturbances (iii-iv); and the L^2-gain estimate (i) at least as long as the system is not too close to the equilibrium.

Proof. We can define explicitly the positive numbers

$$M(R) = \max\{V(x) : |x| \le R\}, \quad m(r) = \min\{V(x) : |x| \ge r/2\}.$$

Then (iii) holds by continuity of V at the origin. Note that

$$\begin{aligned} h(x,u) - ch(x,u) &= (1-c)h(x,u) \\ &\ge (1-c)\min_{\{z:V(z)\ge m(r)/2\},\, u\in U} h(z,u) = (1-c)\sigma(r) > 0 \end{aligned} \tag{8}$$

holds uniformly for all x, $V(x) \ge m(r)/2$ and $u \in U$. Thus if $V(x) > m(r)$, from (3), (7) and (8) it follows that we can choose a sufficiently small $\delta = \delta(r,c)$ such that, for $t \le \delta$ we have $V(y_x(t,u(x),w)) \ge m(r)/2$ and

$$V(y(t)) + \int_0^t [ch(y,u(x)) - \gamma^2|w|^2]ds \le V(x). \tag{9}$$

If now $V(y(\delta)) > m(r)$, then we can proceed and apply (9) recursively. To obtain (i), we define $\tau = n\delta$, where n is the first integer such that $V(y(n\delta)) \le m(r)$. By (9) we have also

$$V(y(t)) + \sigma(r)t \le \int_0^t \gamma^2|w|^2 ds + V(x), \quad t \le n\delta, \tag{10}$$

in particular such a n exists and $\tau \le T = T(r,R,K,c)$.

Finally, in order to prove (ii) and (iv), note that from (10) it follows $V(y(t)) \le M(R) + \gamma^2\|w\|_2^2$ for $t \le n\delta$. If moreover for some integer m we have $V(y(m\delta)) \le m(r)$, then we may suppose that δ is chosen sufficiently small so that $|y(\delta + t)| < r$ for $t \le \delta$. With a simple induction argument, this choice allows us to conclude that, both (ii) and (iv) are valid. □

Remark 7. The fact that the sampling rate δ, as determined by Theorem 3, depends on the choice of r, R usually means that δ will tend to zero as $r \to 0$ or $R \to +\infty$. We want to observe that there are results in the literature, see e.g. van der Schaft [30], in the case of affine systems with smooth data, that guarantee the existence of a smooth solution of the HJI equation locally around the origin if the $\mathcal{H}_\infty$ problem for the linearized system is solvable. In principle this allows to combine the exact feedback solution in a neighborhood of the origin with a sampled feedback one outside that neighborhood. Together with the global boundedness condition of the trajectories of the closed loop system, this gives us a determined positive sampling rate. An example where it is proved that sampled feedback stabilization occurs with a fixed sampling rate, in the case of homogeneous systems, is the paper by Grüne [17].

We now turn to continuous Lyapunov functions. The idea is to use a semiconcave regularization and then apply the construction above. Semiconcave regularization is obtained through inf-convolution which is defined as follows

$$V_\varepsilon(x) = \inf_{y \in \mathbf{R}^N} \{V(y) + \frac{|x-y|^2}{2\varepsilon^2}\}. \tag{11}$$

For the continuous Lyapunov function V we will denote by $\omega_R(\cdot)$ its modulus of continuity in the ball B_R. The main properties of inf-convolution are contained in the following statement, whose proof is standard but can be found for instance in the books [2], [6].

Proposition 1. *Let $V : \mathbf{R}^N \to [0, +\infty[$ be continuous and positive definite. Then the following properties hold:*

- *V_ε is $\frac{1}{2\varepsilon^2}$ semiconcave;*
- *V_ε is positive definite;*
- *at a point x of differentiability of V_ε we have*

$$DV_\varepsilon(x) = \frac{x - T_\varepsilon x}{\varepsilon^2}$$

$$\varepsilon^2 |DV_\varepsilon(x)|^2 \leq \omega_{R+1}(\sqrt{2V(x)}\varepsilon),$$

for $T_\varepsilon x = \arg\min_y\{V(y) + \frac{|x-y|^2}{2\varepsilon^2}\}$ and all $|x| \leq R$;
- *$V(x) \geq V_\varepsilon(x) \geq V(x) - \omega_{R+1}(\sqrt{2V(x)}\varepsilon)$, for all $|x| \leq R$, hence $V_\varepsilon \to V$ locally uniformly.*

A consequence of the previous properties is that the inf-convolution is an approximate Lyapunov function. This fact is contained in the following Lemma.

Lemma 1. *If V is a continuous Lyapunov function, then we can define a vector function $p_\varepsilon : \mathbf{R}^N \to \mathbf{R}^N$ such that $p_\varepsilon(x) \in D^-V(T_\varepsilon x)$, $p_\varepsilon(x) = DV_\varepsilon(x)$*

at any point of differentiability x of V_ε, and $\varepsilon^2|p_\varepsilon(x)|^2 \le \omega_{R+1}(\sqrt{2V(x)\varepsilon})$ for all $|x| \le R$. Moreover

$$\begin{aligned} &\max_{u\in U} \min_{w\in W}\{-f(x,u,w)\cdot p_\varepsilon(x) - h(x,u) + \gamma^2|w|^2\} \\ &\quad \ge -L\varepsilon^2|p_\varepsilon(x)|^2 - L_R\varepsilon|p_\varepsilon(x)| \ge o_R^\varepsilon(1), \end{aligned} \tag{12}$$

where $o_R^\varepsilon(1) \to 0$ as $\varepsilon \to 0$, uniformly for $|x| \le R$.

Proof. Let $T_\varepsilon x \in \arg\min_{y\in\mathbf{R}^N}\{V(y) + \frac{|x-y|^2}{2\varepsilon^2}\}$, then define $p_\varepsilon(x) = \frac{x-T_\varepsilon x}{\varepsilon^2}$. The result is a consequence of the properties of inf-convolution as in Proposition 1 or in the references above, the definition of viscosity (super)solution and the assumptions on the data. □

Remark 8. Note that, if V is locally Lipschitz continuous, then the inf-convolution V_ε is itself locally Lipschitz continuous with the same Lipschitz constant on bounded sets, uniformly in ε. Therefore in (12) one gets that the right-hand-side of the estimate is in fact $O_R(\varepsilon)$. If V is semiconvex, then classical convex duality shows that V is a sup-convolution. Hence it is known that V_ε is a C^1 function for a sufficiently small ε. Thus in Lemma 1 we can define directly $p_\varepsilon(x) = DV_\varepsilon(x)$. In this case, if an admissible feedback can be selected in (2), as in the case of affine systems, we can work directly with it with no need of using sampled strategies.

Correspondingly with the choice of p_ε as in Lemma 1, we can then define discontinuous feedbacks u_ε as in (5) and derive for the relative sampled feedback strategies $\alpha_\delta^\varepsilon$ estimates similar to (7). We thus get

$$\begin{aligned} V_\varepsilon(y(t)) &\le V_\varepsilon(x) + p_\varepsilon(x)\cdot(y(t)-x) + \tfrac{1}{2\varepsilon^2}|y(t)-x|^2 \\ &\le V_\varepsilon(x) - \textstyle\int_0^t [h(y,u(x)) - \gamma^2|w|^2]ds + (L_{R+1}e^{L\delta}L^2(1+|x|)\delta \\ &\quad + |p_\varepsilon(x)|e^{L\delta}L^2(1+|x|)\delta/2 + \tfrac{1}{\varepsilon^2}e^{2L\delta}L^2(1+|x|)^2\delta + o_R^\varepsilon(1))t \\ &\le V_\varepsilon(x) - \textstyle\int_0^t [h(y,u(x)) - \gamma^2|w|^2]ds + t[O_R(\tfrac{\delta}{\varepsilon^2}) + o_R^\varepsilon(1)], \end{aligned} \tag{13}$$

Remark 9. Note that, if again V is locally Lipschitz continuous (or semiconvex) and so is V_ε uniformly in ε, then we can improve (13) to get that the remainder term of the right-hand-side is of the form $t[O_R(\delta)+O_R(\frac{\delta}{\varepsilon^2})+O_R(\varepsilon)]$, where the singular term is now only due to the semiconcavity constant of V_ε.

At this point we can formulate the result.

Theorem 4. *Let $0 < r < R$, $K > 0$ and $c \in (0,1)$ be given. Then we can find positive numbers $M(R), m(R), T(r,R,K,c), \varepsilon(r,R,K,c), k(r)$ so that $V(x) \le 2m(r)$ implies $|x| \le r$ and for all $0 < \varepsilon \le \varepsilon(r,R,K,c)$ there are $\delta(\varepsilon,r,c)$ and a feedback $u^\varepsilon(\cdot)$ such that: for all $r \le |x| \le R$ and $w \in \mathcal{W}$, the sampled feedback trajectories $y_x(\cdot,\alpha_\delta^\varepsilon,w)$ satisfy*

(i) *$V_\varepsilon(y(t)) + \int_0^t [ch(y,\alpha_\delta^\varepsilon) - \gamma^2|w|^2]ds \le V_\varepsilon(x)$, for all $t \le n\delta \le T$, where n is the first integer such that $V(y(n\delta)) \le 2m(r)$;*

(ii) $y(t) \in \{x : V(x) \leq M(R) + \gamma^2 \|w\|_2^2\}$ *for all* $t \geq 0$*;*
(iii) $\lim_{R \to 0^+} M(R) = 0$*;*
(iv) $|y(t)| \leq r$*, for all* v*,* $\|w\|_2^2 \leq k(r)/\gamma^2$ *and all* $t \geq T$*.*

Proof. We have to implement the approximation and thus we need a little more space. Again we will define explicitly the positive numbers

$$M(R) = 2\max\{V(x) : |x| \leq R\}, \quad m(r) = \min\{V(x) : |x| \geq r/2\}/2.$$

Note that, by Proposition 1, for all sufficiently small ε we can always suppose that

$$\{x : V(x) \leq S\} \subset \{x : V_\varepsilon(x) \leq S\} \subset \{x : V(x) \leq 2S\},$$

for $S = m(r), M(R)/2$, respectively. With this position, the rest of the proof easily follows by adapting the argument of the proof of Theorem 3 with V_ε in place of V and noting that in order to use the estimate (13) we obviously need to choose ε small first and then δ sufficiently small correspondingly. □

Compared to Theorem 3, Theorem 4 has the negative feature that the feedback that we construct will depend on the prefixed region $\{x : r \leq |x| \leq R\}$ where we want to apply it.

References

1. Ball, J.A.; Helton, J.W.; Walker, M.L. (1993) $\mathcal{H}_\infty$ control for nonlinear systems with output feedback. IEEE Trans. Automat. Control **AC-38**, 546–559.
2. Bardi, M.; Capuzzo-Dolcetta, I. (1997) Optimal control and viscosity solutions of Hamilton-Jacobi-Bellman equations, Birkhäuser.
3. Barron, E.N.; Evans, L.C.; Jensen, R. (1984) Viscosity solutions of Isaacs' equations and differential games with Lipschitz controls. J. Diff. Eq. **53**, 213-233.
4. Berkovitz, L.D. (1988) Characterizations of values in differential games. Appl. Math. Optim. **17**, 177–183.
5. Basar, T.; Bernhard, P. (1990) 2nd edition (1995) $\mathcal{H}_\infty$ optimal control and related minimax design problems, Birkhäuser.
6. Cannarsa, P.; Sinestrari C. (2001) Semiconcave Functions, Hamilton-Jacobi Equations and Optimal Control, book in preparation.
7. Clarke, F. H.; Ledyaev, Yu. S.; Rifford, L.; Stern, R. J. (2000) Feedback stabilization and Lyapunov functions. SIAM J. Control Optim. **39**, 25–48.
8. Clarke, F. H.; Ledyaev, Yu. S.; Sontag, E. D.; Subbotin, A. I. (1997) Asymptotic controllability implies feedback stabilization. IEEE Trans. Automat. Control **AC-42**, 1394–1407.
9. Clarke, F. H.; Ledyaev, Yu. S.; Subbotin, A. I. (1997) The synthesis of universal feedback pursuit strategies in differential games. SIAM J. Control Optim. **35**, 552–561.
10. Crandall, M.C.; Lions, P.L. (1983) Viscosity solutions of Hamilton-Jacobi equations. Trans. Amer. Math. Soc. **277**, 1–42.
11. Doyle, J.C.; Glover, K.; Khargonekar, P.P.; Francis, B.A. (1989) State-space solutions to standard $\mathcal{H}_2$ and $\mathcal{H}_\infty$ control Problems. IEEE Trans. Automat. Control **AC-34**, 831–847.

12. Elliott, R.J.; Kalton, N.J. (1972) The existence of value in differential games. Mem. Amer. Math. Soc. **126**.
13. Evans, L.C.; Souganidis, P.E. (1984) Differential games and representation formulas for solutions of Hamilton-Jacobi equations.. Indiana Univ. Math. J. **33**, 773–797.
14. Fleming, W.H. (1961) The convergence problem for differential games. J. Math. Anal. Appl. **3**, 102–116.
15. Fleming, W.H.; McEneaney, W.M. (1995) Risk sensitive control on an infinite time horizon. SIAM J. Control Optim. **33**, 1881–1915.
16. Friedman, A. (1971) Differential games. Wiley.
17. Grüne, L. (2000) Homogeneous state feedback stabilization of homogeneous control systems. SIAM J. Control Optim. **38**, 1288–1314.
18. Isaacs, R. (1965) Differential games. Wiley.
19. Ishii, Hitoshi; Koike, Shigeaki. (2000) On ϵ-optimal controls for state constraint problems. Ann. Inst. H. Poincar Anal. Non Linaire **17**, 473–502.
20. Jacobson, D.H. (1973) Optimal stochastic linear systems with exponential criteria and their relation to deterministic differential games. IEEE Trans. Automat. Control **AC-18**, 121–134.
21. Kocan, M.; Soravia, P. (1999) Nonlinear, dissipative, infinite dimensional systems. In Stochastic Analisys, Control, Optimization and Applications, A volume in honor of W.H. Fleming, W. McEneaney, G.G. Yin, Q. Zhang eds., Birkhäuser, 75–94.
22. Kocan, M.; Soravia, P. (2000) Differential games and nonlinear H-infinity control in infinite dimensions. SIAM J. Control Optim. **39**, 1296–1322.
23. Krasovskii, A.N.; Subbotin, A.I. (1974) Positional differential games (Russian), Mir. Jeux differentiels, French translation Mir 1979; Game theoretical control problems, revised English edition, Springer 1988.
24. Rifford L. (2000) Existence of Lipschitz and Semiconcave Control-Lyapunov Functions. Preprint.
25. Scherer, C. (1989) $\mathcal{H}_\infty$ control by state feedback: an alternative algorithm and characterization of high-gain occurrence. Systems Control Lett. **12**, 382–391.
26. Soravia, P. (1993) Pursuit-Evasion problems and viscosity solutions of Isaacs equations. SIAM J. Control Optim. **31**, 604–623.
27. Soravia, P. (1995) Stability of dynamical systems with competitive controls: the degenerate case. J. Math. Anal. Appl. **191**, 428–449.
28. Soravia, P. (1996) $\mathcal{H}_\infty$ control of nonlinear systems: differential games and viscosity solutions. SIAM J. Control Optim. **34**, 1071–1097.
29. Soravia, P. (1999) Equivalence between nonlinear $\mathcal{H}_\infty$ control problems and existence of viscosity solutions of Hamilton-Jacobi equations. Applied Math. Optim. **39**, 17–32.
30. Van der Schaft, A.J. (1992) L_2 gain analysis for nonlinear systems and nonlinear $\mathcal{H}_\infty$ control. IEEE Trans. Automat. Control **AC-37**, 770–784.
31. Van der Schaft, A.J. (1993). Nonlinear state space $\mathcal{H}_\infty$ control theory. In Perspective in Control, H.L. Trentelman, J.C. Willems eds., Birkhäuser.
32. Xiao, MingQing; Başar, Tamer. (1999) Nonlinear H^∞ controller design via viscosity supersolutions of the Isaacs equation. In Stochastic analysis, control, optimization and applications, A volume in honor of W.H. Fleming, W. McEneaney, G.G. Yin, Q. Zhang eds., Birkhäuser, 151–170.

33. Zames, G. (1981) Feedback optimal sensitivity: model reference transformations, multiplicative seminorms, and approximate inverses. IEEE Trans. Automat. Control **AC-26**, 301–320.

A Linearization Principle for Robustness with Respect to Time-Varying Perturbations *

Fabian Wirth

Zentrum für Technomathematik, Universität Bremen, 28334 Bremen, Germany,
fabian@math.uni-bremen.de

Abstract. We study nonlinear systems with an asymptotically stable fixed point subject to time-varying perturbations that do not perturb the fixed point. Based on linearization theory we show that in discrete time the linearization completely determines the local robustness properties at exponentially stable fixed points of nonlinear systems. In the continuous time case we present a counterexample for the corresponding statement. Sufficient conditions for the equality of the stability radii of nonlinear respective linear systems are given. We conjecture that they hold on an open and dense set.

1 Introduction

A natural question in perturbation or robustness theory of nonlinear systems concerns the information that the linearization of a nonlinear system at a singular point contains with respect to local robustness properties. This question has been treated for time-invariant perturbations in [8] for continuous time, (see the references therein for the discrete time case). The result obtained in these papers was that generically the linearization determines the local robustness of the nonlinear system, where genericity is to be understood in the sense of semi-algebraic geometry (on the set of linearizations).

Specifically, the objects under consideration are the local stability radius of the nonlinear system and the stability radius of the linear system, where as usual the stability radius of a system is the infimum of the norms of destabilizing perturbations in a prescribed class. The question is then, whether these two quantities are equal or, more precisely, when this is case, see also [4, Chapter 11].

In this paper we treat this problem for nonlinear systems subject to time-varying perturbations. Our analysis is based on recent results on the generalized spectral radius of linear inclusions. In particular, we see a surprising difference between continuous and discrete time. While the linearization always determines the robustness of the nonlinear system if the nominal system is exponentially stable this fails to be true for continuous time. On the other hand we are able to give a sufficient condition which guarantees equality between linear and nonlinear stability radius on an open set of systems. As it is known from [9] that the Lebesgue measure of those linearizations for which

* Research supported by the European Nonlinear Control Network.

F. Colonius, L. Grüne (Eds.): Dynamics, Bifurcations, and Control, LNCIS 273, pp. 191–200, 2002.

it is possible that the nonlinear stability radius is different from the linear is zero it seems therefore natural to conjecture that the set of systems where these two quantities coincide is open and dense.

We proceed as follows. In Section 2 we recall the definition of the stability radius for nonlinear systems with time varying perturbations and state some relevant results from the theory of linear inclusions. In particular, we recall upper and lower bounds of the stability radius of the nonlinear system in terms of the stability radius and the strong stability radius of the linearization. In Section 3 we develop a local robustness theory based on the linearization of the system for the discrete time case. It is shown that the two linear stability radii coincide under weak conditions, demonstrating that one need only consider the linearization in order to determine the local nonlinear robustness properties of a system. The continuous time case is treated in Section 4. We first present a counterexample showing that analogous statements to the discrete time case cannot be expected in continuous time. We then present a sufficient condition for the equality of the two linear stability radii on an open set. Concluding remarks are found in Section 5.

2 Preliminaries

Consider nominal discrete and continuous time nonlinear systems of the form

$$x(t+1) = f_0(x(t))\,, \quad t \in \mathbb{N}, \tag{1a}$$

$$\dot{x}(t) = f_0(x(t))\,, \quad t \in \mathbb{R}_+\,, \tag{1b}$$

which are exponentially stable at a fixed point which we take to be 0. By this we mean that there exists a neighborhood U of 0 and constants $c > 1, \beta < 0$ such that the solutions $\varphi(t; x, 0)$ of (1a),(1b) satisfy $\|\varphi(t; x, 0)\| \leq ce^{\beta t}\|x\|$ for all $x \in U$.

As the concepts we will discuss do not differ in continuous and discrete time we will summarize our notation by writing $\mathbb{T} = \mathbb{N}, \mathbb{R}_+$ for the time-scale and $x^+(t) := \dot{x}(t), x(t+1)$ according to the time-scale we are working on.

Assume that (1a),(1b) are subject to perturbations of the form

$$x^+(t) = f_0(x(t)) + \sum_{i=1}^{m} d_i(t) f_i(x(t)) =: F(x(t), d(t))\,, \tag{2}$$

where the perturbation functions f_i leave the fixed point invariant, i.e. $f_i(0) = 0, i = 0, 1, \ldots, m$. We assume that the f_i are continuously differentiable in 0 (and locally Lipschitz in the case $\mathbb{T} = \mathbb{R}_+$). The *unknown* perturbation function d is assumed to take values in $\alpha D \subset \mathbb{R}^m$,

$$d : \mathbb{T} \to \alpha D\,,$$

where in the case $\mathbb{T} = \mathbb{R}_+$ we impose that d is measurable. Here $\alpha > 0$ describes the perturbation intensity, which we intend to vary in the sequel, while the perturbation set D is fixed. Thus structural information about the perturbations one wants to consider can be included in the functions $f_i, i = 1, \dots, m$ and in the set D. For the perturbation set $D \subset \mathbb{R}^m$ we assume that it is compact, convex, with nonempty interior, and $0 \in \operatorname{int} D$. Solutions to the initial value problem (2) with $x(0) = x_0$ for a particular time-varying perturbation d will be denoted $\varphi(t; x_0, d)$.

The question we are interested in concerns the critical perturbation intensity at which the system (2) becomes unstable. The *stability radius* is thus defined as

$$r_{nl}(f_0, (f_i)) := \inf\{\alpha > 0 \mid \exists d^* : \mathbb{T} \to \alpha D : x^+(t) = F(x(t), d^*(t)) \text{ is not asymptotically stable at } 0\}\,. \tag{3}$$

By linearizing the perturbed system in (2) we are led to the system

$$x^+(t) = \left(A_0 + \sum_{i=1}^{m} d_i(t) A_i \right) x(t)\,, \quad t \in \mathbb{T}\,. \tag{4}$$

This is a (discrete or differential) linear inclusion, which is in principle determined by the set

$$\mathcal{M}(A_0, \dots, A_m, \alpha) := \left\{ A_0 + \sum_{i=1}^{m} d_i A_i \;\middle|\; \|d\| \le \alpha \right\}.$$

If the matrices A_i are fixed we will denote this set by $\mathcal{M}(\alpha)$ for the sake of succinctness.

The inclusion (4) is called exponentially stable, if there are constants $M \ge 1, \beta < 0$ such that

$$\|\psi(t)\| \le M e^{\beta t} \|\psi(0)\|\,, \quad \forall t \in \mathbb{T}$$

for all solutions ψ of (4).

Exponential stability is characterized by the number

$$\rho(\mathcal{M}(A_0, \dots, A_m, \alpha)) := \sup_{\psi} \limsup_{t \to \infty} \|\psi(t)\|^{1/t}\,,$$

where the supremum is taken over all solutions ψ of (4). Namely, (4) is exponentially stable iff $\rho(\mathcal{M}(A_0, \dots, A_m, \alpha)) < 1$. Again we will write $\rho(\alpha)$ if there is no fear of confusion.

In the discrete time case the number ρ is known as the joint or the generalized spectral radius. We refer to [2,10] for further characterizations of this number and for further references. In the continuous time case it is more customary to consider the quantity $\kappa(\alpha) := \log \rho(\alpha)$, which is known under the name of maximal Lyapunov exponent, see [4] and references therein.

As in the nonlinear case we now define stability radii by

$$r_{Ly}(A_0, (A_i)) := \inf\{\alpha \geq 0 \mid \rho(\alpha) \geq 1\},$$
$$\bar{r}_{Ly}(A_0, (A_i)) := \inf\{\alpha \geq 0 \mid \rho(\alpha) > 1\}.$$

The relation between the linear and the nonlinear stability radii is indicated by the following result which is contained in [3] for the continuous and in [7] for the discrete time case.

Lemma 1. *Let* $\mathbb{T} = \mathbb{N}, \mathbb{R}_+$ *and consider system* (2) *and its linearization* (4), *then*

$$r^{\mathbb{T}}_{Ly}(A_0, (A_i)) \leq r^{\mathbb{T}}_{nl}(f_0, (f_i)) \leq \bar{r}^{\mathbb{T}}_{Ly}(A_0, (A_i)).$$

It is the aim of this paper to obtain further results on the information the linear stability radii contain for the nonlinear system.

The following set of matrix sets will play a vital role in our analysis. Recall that a set of matrices $\mathcal{M}$ is called irreducible if only the trivial subspaces of $\mathbb{R}^n$ are invariant under all $A \in \mathcal{M}$.

We define

$$I(\mathbb{R}^{n\times n}) := \{\mathcal{M} \subset \mathbb{R}^{n\times n} \mid \mathcal{M} \text{ compact and irreducible}\}.$$

Note that this set is open and dense in the set of compact subsets of $\mathbb{R}^{n\times n}$ endowed with the usual Hausdorff metric.

The proof of the following statements can be found in [10]. They are the foundation for our analysis of linearization principles.

Theorem 1. *(i) The generalized spectral radius is locally Lipschitz continuous on* $I(\mathbb{R}^{n\times n})$.
(ii) The maximal Lyapunov exponent is locally Lipschitz continuous on $I(\mathbb{R}^{n\times n})$.

Furthermore in the discrete time case a strict monotonicity property can be shown to hold, under the assumption that the following condition can be satisfied. Given $A \in \mathbb{R}^{n\times n}$ we denote by P_A the reducing projection corresponding to the eigenvalues $\lambda \in \sigma(A)$ with $|\lambda| = r(A)$.

Property 1. The set $\mathcal{M} \subset \mathcal{K}(\mathbb{R}^{n\times n})$ is said to have Property 1 if $n = 1, 2$ or if there exists an $A \in \operatorname{conv} \mathcal{M}$ such that

$$r(A) < \rho(\mathcal{M}), \quad \text{or} \quad \operatorname{rank} P_A \neq 2, \quad \text{or} \quad \sigma((I - P_A)A) \neq \{0\}.$$

It is easy to construct a set $\mathcal{M}$ that does not possess Property 1 (just take a set of matrices with entries $A_{11} = A_{22} = 1$ and $A_{12} = c$ and zero elsewhere). The interesting question, however, is whether it is possible to do this in a way so that $\mathcal{M}$ is irreducible. We would assume that this is not the case, but this matter remains unresolved for the moment. In any case, the negation of Property 1 is highly nongeneric, as it requires that the spectrum of all $A \in \mathcal{M}$ are contained in $\{0\} \cup \{z \in \mathbb{C} \mid |z| = \rho(\mathcal{M})\}$.

Modulo this point we now state a monotonicity property of the generalized spectral radius valid in the discrete time case. In the following we denote the affine subspace generated by a set $\mathcal{M} \subset \mathbb{R}^{n\times n}$ by $\operatorname{aff}\mathcal{M}$ while $\operatorname{int}_{\operatorname{aff}\mathcal{M}}Y$ denotes the interior of Y with respect to this affine subspace.

Proposition 1. *Let $\mathcal{M}_1, \mathcal{M}_2 \in I(\mathbb{R}^{n\times n})$ satisfy $\mathcal{M}_1 \neq \mathcal{M}_2$ and*

$$\mathcal{M}_1 \subset \operatorname{int}_{\operatorname{aff}\mathcal{M}_2}\operatorname{conv}\mathcal{M}_2 \,. \tag{5}$$

Assume that $\mathcal{M}_1$ has Property 1 then

$$\rho(\mathcal{M}_1) < \rho(\mathcal{M}_2)\,.$$

3 The discrete time case

In discrete time the situation turns out to be particularly simple. In fact, if Property 1 holds then we can immediately conclude the following linearization principle.

Theorem 2. *Let $\mathbb{T} = \mathbb{N}$ and consider the discrete-time system* (1a) *and the perturbed system* (2) *along with its linearization* (4)*. If for some $\alpha^* < r_{Ly}(A_0,(A_i))$ the set $\mathcal{M}(\alpha^*)$ is irreducible and satisfies Property 1, then*

$$r_{Ly}(A_0,(A_i)) = r_{nl}(f_0,(f_i)) = \bar{r}_{Ly}(A_0,(A_i))\,.$$

Proof. The assumptions guarantee that the map $\alpha \mapsto \rho(\alpha)$ is strictly increasing on $[\alpha^*,\infty)$ by Proposition 1. This implies $r_{Ly}(A_0,(A_i)) = \bar{r}_{Ly}(A_0,(A_i))$. The assertion now follows from Lemma 1.

The situation simplifies even more if we assume that the unperturbed system (1a) is exponentially stable. We can use this natural assumption to replace the somewhat awkward condition concerning Property 1. The reason for this is simple. Exponential stability implies that $r(A_0) < 1$. The stability radius $r_{Ly}(A_0,(A_i))$ equals α only if $\rho(\mathcal{M}(A_0,A_1,\dots,A_m,\alpha)) = 1 > r(A_0)$. These two things enforce that $\mathcal{M}(A_0,A_1,\dots,A_m,\alpha)$ has Property 1.

Corollary 1. *Let $\mathbb{T} = \mathbb{N}$ and consider the discrete-time system* (1a) *and the perturbed system* (2) *along with its linearization* (4)*. If the point $x^* = 0$ is exponentially stable for the unperturbed system*

$$x(t+1) = f_0(x(t))$$

then

$$r_{Ly}(A_0,(A_i)) = r_{nl}(f_0,(f_i)) = \bar{r}_{Ly}(A_0,(A_i))\,.$$

Proof. There exists a similarity transformation T such that all A_i, $i = 0, \dots, m$ are similar to matrices of the form

$$TA_iT^{-1} = \begin{bmatrix} A^i_{11} & A^i_{12} & \dots & \dots & A^i_{1d} \\ 0 & A^i_{22} & A^i_{23} & \dots & A^i_{2d} \\ 0 & 0 & A^i_{33} & & \vdots \\ \vdots & & \ddots & \ddots & \vdots \\ 0 & & \dots & 0 & A^i_{dd} \end{bmatrix},$$

where each of the sets $M_j := \{A^i_{jj} \mid i = 0, \dots, m\}, j = 1, \dots, d$ is irreducible. It holds that $\rho(\alpha) = \max_{j=1,\dots,d} \rho(M_j(\alpha))$.

Thus it is sufficient to consider the blocks individually to determine r_{Ly}, resp. $\bar{r}_{Ly}$. Under the assumption of exponential stability we have $r(A_0) < 1$. Hence for each j we have $r(A^0_{jj}) < 1$ and the set $\mathcal{M}_j(\alpha)$ has Property 1 for all $\alpha > 0$ such that $\rho(\mathcal{M}_j(\alpha)) > r(A_0)$. Now the result follows from Theorem 2.

Corollary 2. *Let $\mathbb{T} = \mathbb{N}$. The stability radius of linear systems with respect to time-varying perturbations r_{Ly} is continuous on the set*

$$\{(A_0, \dots, A_m) \in (\mathbb{R}^{n\times n})^{m+1} \mid r(A_0) \neq 1\}.$$

Furthermore, the set

$$\{(A_0, \dots, A_m) \in (\mathbb{R}^{n\times n})^{m+1} \mid r_{Ly}(A_0, \dots, A_m) \neq \bar{r}_{Ly}(A_0, \dots, A_m)\}$$

is contained in a lower dimensional algebraic set.

Proof. It was shown in [7] that $r_{Ly}, \bar{r}_{Ly}$ are upper respectively lower semicontinuous on $(\mathbb{R}^{n\times n})^{m+1}$. The preceding Corollary 1 shows that these two functions coincide if $r(A_0) < 1$, which shows continuity in this case. If $r(A_0) > 1$ the statement is obvious as both functions are equal to 0.

The second statement now follows because a necessary condition for the condition $r_{Ly}(A_0, \dots, A_m) \neq \bar{r}_{Ly}(A_0, \dots, A_m)$ is $r(A_0) = 1$. The latter condition defines a lower dimensional algebraic set.

The result for the linear stability radii extends to the case of nonlinear systems as follows. First, denote by $C^1(\mathbb{R}^n, \mathbb{R}^n, 0)$ the set of continuously differentiable maps from $\mathbb{R}^n$ to itself satisfying $f(0) = 0$. This space may be endowed with the C^1 topology inherited from the topologies on the space $C^1(\mathbb{R}^n, \mathbb{R}^n)$, (see [6, Chapter 17]).

Corollary 3. *Given $n, m \in \mathbb{N}$, the set $\mathcal{W}$ of functions $(f_0, f_1, \dots, f_m) \in C^1(\mathbb{R}^n, \mathbb{R}^n, 0)^{m+1}$ for which*

$$r_{nl}(f_0, (f_i)) = r_{Ly}(A_0, (A_i)) \tag{6}$$

contains an open and dense subset of $C^1(\mathbb{R}^n, \mathbb{R}^n, 0)^{m+1}$ with respect to both the coarse and the fine C^1 topology.

Proof. This is immediate from the definition of the C^1 topology.

4 Continuous time

A natural question is if statements similar to those of Theorem 2 and Corollary 1 hold in continuous time. The fundamental tool for this results is the monotonicity property given by Proposition 1. This statement is unfortunately in general false in continuous time, as any subset $\mathcal{M}_1$ of the skew-symmetric matrices generates a linear inclusion whose system semigroup is a subset of the orthogonal group and for which the maximal Lyapunov exponent is therefore equal to 0. Taking a set $\mathcal{M}_2$ which contains $\mathcal{M}_1$ in its interior (with respect to the skew-symmetric matrices) does not yield a Lyapunov exponent larger than zero, so that the strict monotonicity property fails to hold. This example leaves still some hope that maybe a statement corresponding to Corollary 1 remains true in continuous time. The following example shows that even such expectations are unfounded.

Example 1. Consider the matrices

$$A(d) := \begin{bmatrix} 0 & d \\ -d & -2+d \end{bmatrix}.$$

It is easy to see that $A^*(d) + A(d) \leq 0$ for all $d \in (-\infty, 2)$. Hence for $D \subset (-\infty, 2)$ it is immediate that $\kappa(D) \leq 0$ as the Euclidean unit ball is forward invariant under the associated time-varying linear system. On the other hand while for the spectral abscissa $\gamma(A(0)) = 0$, we have $\gamma(A(d)) < 0$ for all $d \in (0, 2)$, see Figure 1.

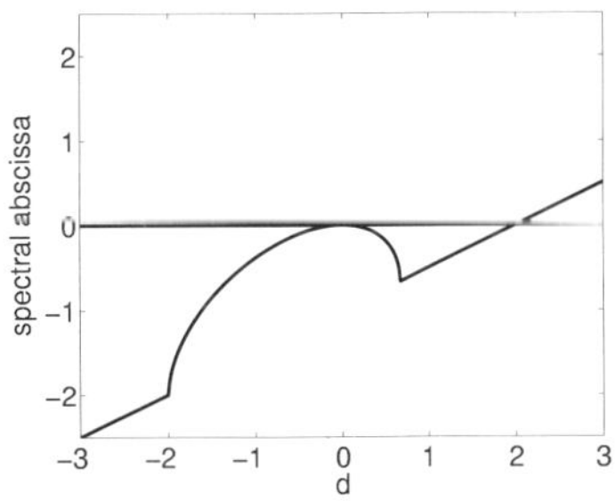

Fig. 1. The spectral abscissa of $A(d)$ in dependence of d.

The consequence of this is the following. If we define $A_0 = A(1/2)$ and

$$A_1 := \begin{bmatrix} 0 & 1 \\ -1 & 1 \end{bmatrix},$$

then

$$0 < r_{Ly}(A_0, A_1) \leq \frac{1}{2} < \frac{3}{2} = \bar{r}_{Ly}(A_0, A_1),$$

because at least $A_0 - 1/2A_1 = A(0)$ is not asymptotically stable. While on the other hand for $\alpha < 3/2$ the perturbation set is a strict subset of $(-\infty, 2)$ and $\gamma(A_0) = -(3+\sqrt{5})/4 \approx -0.191$ so that the unperturbed system is exponentially stable.

While this example shows that we cannot expect a continuous time counterpart to the discrete-time results of Section 3 we are able to show that the property that the stability radius of the linearization determines the nonlinear stability radius is true on certain open sets. We even conjecture that it is true on an open and dense set, but this point remains open for the moment. The following theorem strengthens the result obtained in [9, Theorem 3.1 (i)]. Here the local Lipschitz continuity property of the maximal Lyapunov exponent will play a vital role, as it will allow the application of the implicit function theorem for Lipschitz continuous functions. To this end we will need the Clarke subdifferential of a function g, which we denote by $\partial_{\mathrm{Cl}} g(x)$. Here we will not need the most general definition. For our purposes it is sufficient to know that if we assume that $g : \mathbb{R}^p \to \mathbb{R}$ is locally Lipschitz continuous then

$$\partial_{\mathrm{Cl}} g(x) = \mathrm{conv} \left\{ c \in \mathbb{R}^p \;\middle|\; \exists x_k \to x : c = \lim_{k\to\infty} \nabla g(x_k) \right\}, \tag{7}$$

see [5, Theorem II.1.2], where we tacitly assume that the gradient ∇g exists in x_k if we write $\nabla g(x_k)$. Note that Lipschitz continuity of g implies that it is differentiable almost everywhere by Rademacher's theorem. For further details we refer to [5].

The following lemma ensures that the theory of the Clarke generalized gradient is applicable in our case.

Lemma 2. *The map*

$$(A_0, \dots, A_m, \alpha) \mapsto \kappa(A_0, \dots, A_m, \alpha) := \kappa\left(\left\{ A_0 + \sum_{i=1}^m d_i A_i \;\middle|\; \|d\| \le \alpha \right\}\right)$$

is locally Lipschitz continuous on the set $I(\mathbb{R}^{n\times n}) \times \mathbb{R}_{>0}$.

Proof. Note that the map

$$(A_0, \dots, A_m, \alpha) \mapsto \left\{ A_0 + \sum_{i=1}^m d_i A_i \;\middle|\; \|d\| \le \alpha \right\}$$

is Lipschitz continuous. As the composition of Lipschitz continuous maps is again Lipschitz continuous the claim follows from Theorem 1 (ii).

Proposition 2. *Let* $n, m \in \mathbb{N}$. *Fix* $\{A_0^*, \dots, A_m^*\} \in I(\mathbb{R}^{n\times n})$ *and let*

$$r_{Ly}(A_0, (A_i)) < \infty\,.$$

Consider the map $\kappa : (A_0, \ldots, A_m, \alpha) \mapsto \kappa(\mathcal{M}(\alpha))$ *and denote*

$$\partial_{\mathrm{Cl},\alpha}\kappa(z) := \left\{ c \in \mathbb{R} \mid \exists p' \in (\mathbb{R}^{n\times n})^{m+1} : (p', c) \in \partial_{\mathrm{Cl}}\kappa(z) \right\} .$$

If

$$\inf \partial_{\mathrm{Cl},\alpha}\kappa(A_0^*, \ldots, A_m^*, r_{Ly}(A_0^*, (A_i^*))) > 0 , \tag{8}$$

then $r_{Ly} = \bar{r}_{Ly}$ *on a neighborhood of* $(A_0, \ldots, A_m) \in (\mathbb{R}^{n\times n})^{m+1}$ *and on this neighborhood* r_{Ly} *is locally Lipschitz continuous.*

Proof. By Lemma 2 and (8) we may apply the implicit function theorem for Lipschitz continuous maps [5, Theorem VI.3.1] which states that for every $(B_0, \ldots, B_m)$ in a suitable open neighborhood of $(A_0, \ldots, A_m) \in (\mathbb{R}^{n\times n})^{m+1}$ the map

$$\alpha \mapsto \kappa(\mathcal{M}(B_0, \ldots, B_m, \alpha))$$

has a unique root and this root is a locally Lipschitz continuous function of $(B_0, \ldots, B_m)$. In other words, this means that on this neighborhood the functions r_{Ly} and $\bar{r}_{Ly}$ coincide and are locally Lipschitz continuous.

A complete characterization of the cases where Proposition 2 is applicable is not yet available. In view of on the results in [9, Theorem 3.1 (i)], where it is shown that the set $\{(A_0, \ldots, A_m) \mid r_{Ly}(A_0, (A_i)) \neq \bar{r}_{Ly}(A_0, (A_i))\}$ has Lebesgue measure zero, the following conjecture seems reasonable.

Conjecture 1. Let $\mathbb{T} = \mathbb{R}_+$. For fixed $m \geq 1$ the set $\mathcal{L} \subset (\mathbb{R}^{n\times n})^{m+1}$ given by

$$\{(A_0, \ldots, A_m) \mid r_{Ly}(A_0, (A_i)) = \bar{r}_{Ly}(A_0, (A_i))\}$$

contains an open and dense set. Furthermore, the Lebesgue measure of the complement $\mathcal{L}^c$ is 0.

5 Conclusion

In this paper it was shown that linearization at singular points can provide information about the stability radius of a nonlinear system with respect to time-varying perturbations. In discrete time this information is complete if the nominal system is exponentially stable, while this is false in continuous time.

The fundamental difference between discrete and continuous time lies in the fact that the perturbation in discrete time is on the level of the systems semigroup, whereas in continuous time the perturbations act on the level of the Lie algebra of the system. This at least gives an indication that some differences are to be expected.

We conjecture that also in continuous time the linearization provides sufficient information at least on an open and dense set of systems. If Conjecture 1 can be proved to hold it is clear how to formulate results for the continuous time case analogous to Corollaries 2, 3.

References

1. Barabanov N. E., (1988) Absolute characteristic exponent of a class of linear nonstationary systems of differential equations. Sib. Math. J. 29(4):521–530
2. Berger M. A., Wang Y., (1992) Bounded semigroups of matrices. Lin. Alg. Appl. 166:21–27
3. Colonius F., Kliemann W., (1995) A stability radius for nonlinear differential equations subject to time varying perturbations. 3rd IFAC Symposion on Nonlinear Control Design NOLCOS95, Lake Tahoe, NV, 44–46
4. Colonius F., Kliemann W., (2000) The Dynamics of Control. Birkhäuser, Boston
5. Demyanov V. F., Rubinov A. M., (1995) Constructive Nonsmooth Analysis. Verlag Peter Lang, Frankfurt Berlin
6. Dieudonné J., (1972) Treatise of Modern Analysis, volume 3. Academic Press, New York
7. Paice A. D. B., Wirth F. R., (1997) Robustness of nonlinear systems subject to time-varying perturbations. In: Proc. 36th Conference on Decision and Control CDC97, San Diego, CA, 4436-4441
8. Paice A. D. B., Wirth F. R., (1998) Analysis of the Local Robustness of Stability for Flows. Math. Control, Signals Syst. 11(4):289-302
9. Paice A. D. B., Wirth F. R., (2000) Robustness of nonlinear systems and their domain of attraction. in: Colonius F., et al., (eds) Advances in Mathematical Systems Theory, Birkhäuser, Boston
10. Wirth, F. (2001) The generalized spectral radius and extremal norms. Lin. Alg. Appl. to appear

Part III

Global Dynamics of Control Systems

On Constrained Dynamical Systems and Algebroids

Jesús Clemente-Gallardo[1,3], Bernhard M. Maschke[2], and Arjan J. van der Schaft[1]

[1] Faculty of Mathematical Sciences
Department of Systems, Signals and Control
University of Twente P.O. Box 217
7500 AE Enschede The Netherlands
[2] Laboratoire d'Automatique et de Genie des Procedes UCB Lyon 1
UFR Genie Electrique et des Procedes - CNRS UMR 5007
Universite Claude Bernard- Lyon 1
Villeurbanne France
[3] Control Laboratory, Faculty of Information Technology & Systems
Delft University of Technology P.O. Box 5031
2600 GA Delft The Netherlands

Abstract. In 1994, van der Schaft and Maschke defined a(n) (almost) Poisson structure for the study of constrained port controlled Hamiltonian systems as systems obtained by reduction. This note intends to provide a geometrical framework that justifies such construction, based on the use of Lie algebroids, and which extends the work presented in [3].

1 Introduction: Constrained Hamiltonian systems

The purpose of this note is to present a geometrical construction which allows us to obtain an unconstrained description for a controlled dynamical system with holonomic constraints. What we do is to include the constraints in the geometrical structure which defines the dynamics, defining a description restricted to the constraint submanifold. This structure is, for the case of a holonomically constrained system, a Lie algebroid.

From the geometrical point of view, the description of constrained dynamical system is achieved usually by using Lagrange multipliers in, for instance, a Lagrangian formalism. In such a case, we take the Lagrangian function $L \in C^\infty(TM)$. Representing the constraints with the matrix of one forms A, the corresponding Euler-Lagrange equations (a general reference for this and all the geometrical mechanical topics throughout the paper can be found in [9]) turn out to be, written in coordinates $(q^i, \dot{q}^i)$ and in matrix notation:

$$\begin{aligned} &\frac{d}{dt}\left(\frac{\partial L}{\partial \dot{q}}\right) - \frac{\partial L}{\partial q} = A(q)\lambda \\ &A^T(q)\dot{q} = 0 \ . \end{aligned} \tag{1}$$

F. Colonius, L. Grüne (Eds.): Dynamics, Bifurcations, and Control, LNCIS 273, pp. 203–216, 2002.

The usual procedure to solve the equations starts eliminating the Lagrangian multipliers from the equations above. The geometrical meaning of this procedure is that we are restricting the velocities to belong to the kernel of A^T, i.e. we are admitting only those velocities which belong to the foliation defined by A.

If we assume that the Lagrangian (equivalently the Hamiltonian) is regular [1] the corresponding Hamiltonian formalism is written, in the canonical coordinates (q^i, p_i), and in matrix notation as:

$$\frac{dq}{dt} = \frac{\partial H}{\partial p} \tag{2}$$

$$\frac{dp}{dt} = -\frac{\partial H}{\partial q} - A(q)\lambda \tag{3}$$

$$A^T \frac{\partial H}{\partial p} = 0 \ , \tag{4}$$

for $A(q)\lambda$ the corresponding constraint forces, and the Hamiltonian defined as:

$$H(q,p) = p_i \dot{q}^i(q^i, p_i) - L(q, \dot{q}^i(q^i, p_i)) \ .$$

The geometrical effect of the constraints on the cotangent bundle is to define also a submanifold where the dynamics takes place, but, in general, it will not have a well defined structure (i.e. in the most general case it will not be a subbundle of T^*M). It is defined by the last of the equations above, i.e. we define the constraint submanifold of the cotangent bundle as:

$$M_c = \{(q,p) \in T^*M | A^T \frac{\partial H}{\partial p} = 0\} \ . \tag{5}$$

The elimination of the Lagrange multipliers in this case leads to a Hamiltonian vector field X_H which is tangent to M_c, but the procedure may be difficult depending on the constraints and the Hamiltonian. In [15] two of us introduced a Poisson structure adapted to M_c which allows a description of the system on M_c treated as an unconstrained system. In [3] this construction was analysed from the geometrical point of view, by using the recently formulated Lagrangian and Hamiltonian formalisms on Lie algebroids [11,10,16]. The main idea is that, for many important cases, the subbundle of the tangent bundle defined by the velocities which satisfy the constraint equation contains a geometrical structure which admits an extension of the usual Lagrangian and Hamiltonian formalisms. It encodes the constraints directly in the geometrical structure and allows us to treat the system as an unconstrained one.

[1] in general this implies that the Legendre transformation is a local diffeomorphism what for the systems we are going to study means that the mass matrix is invertible; see [9] for a general reference of the geometric concepts involved

The present work extends this treatment of constrained mechanical system to the case of control systems, i.e. we consider that we act on the system described by (1) by using a set of control forces $\{F_k\}$ thus defining a dynamical system, that in the Hamiltonian version takes the form:

$$\frac{dq}{dt} = \frac{\partial H}{\partial p} \tag{6}$$

$$\frac{dp}{dt} = -\frac{\partial H}{\partial q} - A(q)\lambda + Fu \tag{7}$$

$$A^T \frac{\partial H}{\partial p} = 0 . \tag{8}$$

The extension is done in the framework of Dirac structures and Port Controlled Hamiltonian systems since the geometrical formulation of both is perfectly adapted to the present case.

The paper is organised as follows: section 2 introduces the basic notions of the theory of Lie algebroids and the corresponding Lagrangian and Hamiltonian formalisms and section 3 contains a brief introduction to Dirac structures and Port Controlled Hamiltonian systems, with special emphasis in the case we are interested in. Section 4 introduces briefly the main results presented in [3] and section 5 presents the main result of the paper.

2 What is a Lie algebroid?

2.1 Generalities

The concept of Lie algebroids has been used in the last fifty years in the algebraic-geometrical framework, under different names (see [8,12,7]); but the first proper definition, from the point of view of Differential Geometry, is due to Pradines [13]. The interested reader may find a more detailed description in [3] and some applications to Mechanics and Control Theory in [2].

Definition 1. A **Lie algebroid** on a manifold M is a vector bundle $E \to M$, in whose space of sections we define a Lie algebra structure $(\Gamma(E), [\cdot,\cdot]_E)$, and a mapping $\rho : E \to TM$ which is a homomorphism for this structure in relation with the natural Lie algebra structure of the set of vector fields $(\mathfrak{X}(M), [\cdot,\cdot]_{TM})$. We have therefore:

$$\rho([\eta,\xi]_E) = [\rho(\eta),\rho(\xi)]_{TM} \quad \forall\, \eta,\xi \in \Gamma(E),$$

and

$$[\eta, f\xi]_E = [\eta,\xi]_E + (\rho(\eta)f)\,\xi \quad \forall\, \eta,\xi \in \Gamma(E), \quad \forall\, f \in C^\infty(M) .$$

The definition may be summarised in the following diagram:

$$\begin{array}{ccc} E & \xrightarrow{\rho} & TM \\ {\scriptstyle\tau}\downarrow & \swarrow{\scriptstyle\pi} & \\ M & & \end{array}$$

For simplicity in the notation we often omit the subscripts of E and TM of the commutator.

If we take coordinates in the base manifold $\{x^i\}$ and a basis of sections in the bundle $\{e_\alpha\}$, we can consider $\{(x^i, \lambda^\alpha)\}$ to be the coordinates for E, with respect to which we write the expression of the anchor mapping as

$$\rho(e_\alpha) = \rho^i_\alpha \frac{\partial}{\partial x^i},$$

and the expression of the Lie bracket in this base as

$$[e_\alpha, e_\beta] = C^\gamma_{\alpha\beta} e_\gamma.$$

The main idea we have to keep in mind is that a Lie algebroid is a geometrical object very similar to a tangent bundle, that has a more general Lie algebra structure on the space of sections as its main characteristic.

For the problem we are going to deal with now, there is one property of Lie algebroids which is particularly useful: the dual of any Lie algebroid is a Poisson manifold [2] , as happens in the case of the tangent bundle, with its dual, the cotangent bundle.

Theorem 1. *For any Lie algebroid E the dual bundle E^* is a Poisson manifold.*

Proof. See [16] (direct construction), or [11].

If we take a basis of sections of E^* as the dual of the basis $\{e_\alpha\}$ of sections of E, and denote the corresponding coordinates as (x^i, μ_α), the expression of the Poisson bracket above turns out to be:

$$\{x^i, x^j\} = 0 \quad \{x^i, \mu_\alpha\} = \rho^i_\alpha \quad \{\mu_\alpha, \mu_\beta\} = C^\gamma_{\alpha\beta}\mu_\gamma \tag{9}$$

We can remark in this expression the similarities with the usual Lie-Poission structure on the dual of a Lie algebra, which is an example of Lie algebroid where the base manifold is a single point. Once the manifold is presented, it is clear that the best way of thinking on Lie algebroids is just to consider them as a generalised tangent bundle. Since tangent bundles are the natural framework for the usual Lagrangian mechanics, a natural question arises: is it possible to define a kind of Lagrangian mechanics on these generalised tangent

[2] i.e. it is a differentiable manifold where a Poisson structure is defined, see [9]

bundles? The question was first stated (and answered) by A. Weinstein in [16]. The main ingredient of his construction is the natural Poisson structure that admits the dual bundle $E^* \to M$ and that we saw above. Define now the Lagrangian of the system as a function $L \in C^\infty(E)$, and let us consider the analogous to the usual Legrendre transform. We define it as the derivative on the fibre of the bundle, i.e. $\mathcal{F}L : E \to E^*$ is such that:

$$\mathcal{F}L(q^i, \lambda^\alpha) = (q^i, \frac{\partial L}{\partial \lambda^\alpha}) . \tag{10}$$

As we said before, this transformation defines a local diffeomorphism when the Lagrangian is regular (and global when it is hyperregular) from E to E^*, and hence we can use this diffeomorphism to pull back the Poisson structure from E^* to E, thus defining a Poisson structure on E which depends on the Lagrangian. The dynamics is then defined on the variables $\{q^i, \frac{\partial L}{\partial \lambda}^\alpha\}$ using the analogue of the energy function on the tangent bundles, which here is written as:

$$E_L = \sum_\alpha \lambda^\alpha \frac{\partial L}{\partial \lambda^\alpha} - L .$$

Finally, we can propose a dynamics for the algebroid, considering the equations:

$$\begin{aligned} \frac{dq^i}{dt} &= \{q^i, E_L\} \\ \frac{d}{dt}\left(\frac{\partial L}{\partial \lambda^\alpha}\right) &= \left\{\frac{\partial L}{\partial \lambda^\alpha}, E_L\right\} , \end{aligned} \tag{11}$$

which can be considered as the generalisation of the analogous procedure that can be built up for the tangent bundle case.

2.2 The algebroid structure of an integrable subbundle of a tangent bundle

Let us consider a differentiable manifold M and an integrable subbundle $\mathcal{D} \subset TM$, i.e., a bundle whose sections define a subalgebra in $(\mathfrak{X}, [\cdot,\cdot])$. We consider this subbundle as a separate object, and we denote it as $\mathcal{A}_\mathcal{D}$.

The algebroid components are defined as follows:

- The vector bundle is the bundle $\mathcal{A}_\mathcal{D}$, and its base is the base manifold M.
- The Lie algebra structure on the sections is defined by using the Lie algebra structure of vector fields on the tangent bundle. The fact that the subbundle is integrable means that we have a subalgebra of the algebra of vector fields, and hence, on $\mathcal{A}_\mathcal{D}$ we have a relation of the type:

$$[e_\alpha, e_\beta] = C_{\alpha\beta}^\gamma e_\gamma \tag{12}$$

where $\{e_\mu\}$ defines a basis of sections of $\mathcal{A}_\mathcal{D}$, and $C^\gamma_{\alpha\beta}$ are the structure functions defining the Lie algebra structure on this basis.

- Finally, the anchor mapping is defined as the natural inclusion of the elements of $\mathcal{A}_\mathcal{D}$ in $\mathcal{D} \subset TM$:

$$\rho : \mathcal{A}_\mathcal{D} \hookrightarrow TM \tag{13}$$

Regarding the dual bundle $\mathcal{A}^*_\mathcal{D}$, we have a new vector bundle, which is not a Lie algebroid but which is a Poisson manifold as we saw above, and will be the natural framework for a Hamiltonian dynamical description. Note that the construction of $\mathcal{A}^*_\mathcal{D}$ is purely geometrical, and no dynamical considerations are taken into account. Another important property of the construction is the canonical mapping from T^*M to $\mathcal{A}^*_\mathcal{D}$ defined by the dual of the anchor mapping:

$$\rho^* : T^*M \to \mathcal{A}^*_\mathcal{D} .$$

3 Dirac structures and Port Controlled Hamiltonian systems

3.1 Dirac structures

The use of Dirac structures in the description of dynamical systems is quite recent: the first results appeared in the eighties in [5,6] by Courant and Weinstein and Dorfman respectively. The main difference of this approach with the best known symplectic or Poisson ones lies in the use of pair of elements made up by differential forms and vector fields defined on a differential manifold N, where the dynamics of our system is defined. A very good summary for the theory of Dirac manifolds can be found in [4], where the differential geometric exposition of the topic, as well as its main interesting properties are exposed (though for the finite dimensional case, while Dorfman's approach was developed for the general infinite dimensional case), or to [1] for a more recent one. In this short introduction we will just introduce the general issues, and we refer the reader to the aforementioned references for a more detailed presentation.

Constant Dirac structures The simplest example of Dirac structure is defined on vector spaces. Let V be a vector space(which in applications will be called flow space) and consider also its dual space V^* (in applications, effort space) with respect to the duality product $\langle \cdot, \cdot \rangle$. With this product we define the symmetrised one:

$$\langle (a_1, a_2), (b_1, b_2) \rangle_+ = \langle a_1, b_2 \rangle + \langle a_2, b_1 \rangle \quad \forall (a_1, a_2), (b_1, b_2) \in V \times V^*$$

We can consider a subset D of the space $V \times V^*$ which is maximally isotropic with respect to $\langle \cdot, \cdot \rangle_+$, i.e. $D = D^\perp$ where

$$D^\perp = \{(w_1, w_2) \in V \times V^* | \langle v_1, w_2 \rangle + \langle w_1, v_2 \rangle = 0 \quad \forall (v_1, v_2) \in D \subset V \times V^*\}$$

We will say that this subspace D is a **constant Dirac structure**. Any subspace $D \subset V \times V^*$ of dimension $n = \dim V$ such that:

$$\langle v, v^* \rangle = 0 \quad \forall (v, v^*) \in D \tag{14}$$

will define a constant Dirac structure. In practical applications we will use the notation f for elements in $V = \mathcal{F}$ (the flow space) and e for elements in $V^* = \mathcal{E}$ (the effort space).

General Dirac structures The definition of a constant Dirac structure can also be generalised to the non constant case, when we consider that this vector space V is the tangent space to a manifold N in one point. A constant Dirac structure on the point $p \in N$ is defined on T_pN as a subspace $D(p) \subset T_pN \times T_p^*N$ which satisfies the condition above ($D(p)^\perp = D(p)$). We define a global structure by making this $D(p)$ to be the value at the point $p \in N$ of a geometric object D, for which now the vectors become vector fields as well as the covectors become one forms. The product now is the natural pairing $\langle \cdot, \cdot \rangle : \mathfrak{X}(N) \times \Lambda(N) \to C^\infty(N)$ (we denote by $\mathfrak{X}(N)$ the set of vector fields on N and by $\Lambda(N)$ the set of one forms) and the definition of $D^\perp$ becomes now:

$$D^\perp = \{(Y, \beta) \in TN \oplus T^*N | \langle X, \beta \rangle + \langle Y, \alpha \rangle = 0 \ \forall (X, \alpha) \in D \subset TN \oplus T^*N\}$$

The condition to be satisfied by the set D (actually it must be a subbundle of $TN \oplus T^*N$) is still the same:

Definition 2. A subbundle $D \subset TN \oplus T^*N$ is said to be a **generalised Dirac structure** defined on a manifold N if and only if it is maximally isotropic with respect to the duality product:

$$D^\perp = D$$

Th geometrical characterisation of a Dirac structure is done in terms of two distributions and two codistributions:

- $\mathcal{G}_0 = \{X \in \mathfrak{X}(N) | (X, 0) \in D\}$
- $\mathcal{G}_1 = \{X \in \mathfrak{X}(N) | \exists \alpha \in \Lambda^1(M) \text{ such that } (X, \alpha) \in D\}$
- $\mathcal{P}_0 = \{\alpha \in \Lambda^1(N) | (0, \alpha) \in D\}$
- $\mathcal{P}_1 = \{\alpha \in \Lambda^1(N) | \exists X \in \mathfrak{X}(M) \text{ such that } (X, \alpha) \in D\}$

which are related by the relations

$$\mathcal{P}_0 = \mathrm{Ann}\mathcal{G}_1$$
$$\mathcal{G}_0 = \mathrm{Ker}\mathcal{P}_1$$

These definitions allow us to define some representations of Dirac structures, particularly useful in what follows. Given a generalised Dirac structure on N, with the corresponding distributions and codistributions defined as above, we have that:

- If $\mathcal{G}_1$ is constant dimensional, there exists a skew-symmetric linear map $\hat{\omega} : \mathcal{G}_1 \to \mathcal{G}_1^* \subset T^*N$ such that the pairs of the Dirac structure are defined as its graph, modulo the elements of Ann$\mathcal{G}_1$, i.e.:

$$D = \{(X, \alpha) | \alpha - \hat{\omega}(X) \in \mathrm{Ann}\, \mathcal{G}_1 \forall X \in \mathfrak{X}(N)\} . \tag{15}$$

This is the clear generalisation of the usual symplectic structure (see [9]).

- If $\mathcal{P}_1$ is constant dimensional, there exists a skew-symmetric linear map $\hat{J} : \mathcal{P}_1 \to \mathcal{P}_1^* \subset TN$ such that the pairs of the Dirac structure are defined as its graph, modulo the elements of Ker$\mathcal{P}_1$, i.e.:

$$D = \{(X, \alpha) | X - \hat{J}(\alpha) \in \mathrm{Ker}\, \mathcal{P}_1 \forall \alpha \in \Lambda^1(N)\} , \tag{16}$$

which generalises the Poisson case, and will be used later in the definition of Implicit Hamiltonian Systems.

3.2 Dynamics on Dirac structures

The definition of dynamics on Dirac structures generalises the Hamiltonian formalism defined on symplectic and Poisson manifolds.

Let $D \subset TN \oplus T^*N$ be a Dirac structure in some differentiable manifold N. Let $H \in C^\infty(N)$.

Definition 3. A curve $\gamma(t)$ on N is a solution of the Hamiltonian dynamics defined by the function H if and only if the tangent vector at each point satisfies that

$$(\dot{\gamma}(t), dH(\gamma(t))) \in D .$$

3.3 Application: Mechanical system with constraints

It is simple to implement the framework of Dirac structure to the case of a constrained mechanical system defined on a cotangent bundle. Let us apply Definition 3 to the case of a cotangent bundle, i.e. $N = T^*M$ and the Dirac structure defined by the graph of the canonical symplectic two form ω and the set of constraints (1). Such a Dirac structure is defined by the following set of distributions:

$$\mathcal{G}_1 = \{(X, \hat{\omega}(X) | X \in \mathfrak{X}(N)\}$$

$$\mathcal{G}_0 = \mathrm{Ker} A$$

where $\hat{\omega}$ is the isomorphism defined as:

$$\hat{\omega}(X)(Y) = \omega(X, Y) \quad \forall X, Y \in \mathfrak{X}(M). \tag{17}$$

Definition 3 written in terms of the representation (16), takes the form:

$$(X, \alpha) \in D \Leftrightarrow \begin{cases} dH - \hat{\omega}(X) = \lambda A \\ A(X) = 0 \end{cases}$$

But this is precisely the expression of the dynamics (1). In the following, the expression of the dynamics will be written in canonical coordinates (q^i, p_i) for T^*M as:

$$\begin{aligned} \begin{pmatrix} \dot{q} \\ \dot{p} \end{pmatrix} &= \begin{pmatrix} 0 & I_n \\ -I_n & 0 \end{pmatrix} \begin{pmatrix} \frac{\partial H}{\partial q} \\ \frac{\partial H}{\partial p} \end{pmatrix} + \begin{pmatrix} 0 \\ A(q) \end{pmatrix} \lambda \\ 0 &= \left(0 \; A^T(q)\right) \begin{pmatrix} \frac{\partial H}{\partial q} \\ \frac{\partial H}{\partial p} \end{pmatrix} . \end{aligned} \tag{18}$$

3.4 Port controlled Hamiltonian systems

The main application of Dirac structures in the context we are interested in are (implicit) port controlled Hamiltonian systems, introduced by Maschke and van der Schaft in the early nineties, and which have proved to be a very useful tool in the study of system and control theory. In the context of port controlled Hamiltonian systems we find different examples of the types of Dirac structures we described above. The main characteristic of these systems is that they model very successfully power conserving dynamical systems. Given a manifold M, with coordinates $\{x^i\}$, and vector spaces $\mathcal{E}$ and $\mathcal{F}$ (where $\mathcal{E}$ is considered to be dual to $\mathcal{F}$, $\mathcal{F}^* \sim \mathcal{E}$), the dynamics of the simplest port controlled Hamiltonian we are interested in is:

$$\dot{x} = J(x)\frac{\partial H}{\partial x} + g(x)f \tag{19}$$

$$e = g^T(x)\frac{\partial H}{\partial x} , \tag{20}$$

where $J(x)$ is an antisymmetric matrix representing a Poisson structure, the variables $e \in \mathcal{E}$ and $f \in \mathcal{F}$ are known as efforts and flows respectively, and model the elements with which the dynamical system interacts, and the matrix $g(x)$ represents the input vector fields. The paradigmatic physical system described by these equations is a LC electrical circuit with independent elements, where the external variables are voltages and currents. In order to describe more complicated systems, with energy-depending elements, more general types of Dirac structures must be used. These new systems where general Dirac structures are involved are known as **implicit port-controlled Hamiltonian systems**. The mathematical description mixes now the manifold and the ports and then we consider the generalised Dirac structure defined on the manifold $M \times \mathcal{F}$, which means that it is a bundle whose fibre at the point (x, f) is:

$$D_{(x,f)} \subset T_{(x,f)}(M \times \mathcal{F}) \oplus T^*_{(x,f)}(M \times \mathcal{F}) \sim T_xM \times \mathcal{F} \oplus T^*_xM \times \mathcal{E} \tag{21}$$

which does not depend on $f \in \mathcal{F}$ but only on $x \in M$ (i.e. the geometry depends only on the space state variables, but not on the external variables).

Considering it globally, we define a subbundle $D \subset T(M \times \mathcal{F}) \oplus T^*(M \times \mathcal{F})$ which defines the *implicit port controlled Hamiltonian system.* The dynamics is introduced using a function on the manifold M, called *Hamiltonian* and stating that the curves:

$$\gamma(t) = \{(x(t), f(t), e(t)) \in M \times \mathcal{F} \times \mathcal{E}\}$$

are a solution of the dynamical system defined by the Hamiltonian H if and only if the following property holds:

$$(\dot{x}(t), f(t), -dH(x(t)), e(t)) \in D(x(t)) \forall t \in I \ , \tag{22}$$

where $I \subset \mathbb{R}$ defines a range of valid times.

The case we will be most interested in will be the implicit port controlled Hamiltonian systems described by the equations:

$$\dot{x} = J(x)\frac{\partial H}{\partial x} + g(x)f + b(x)\lambda \tag{23}$$

$$e = g^T(x)\frac{\partial H}{\partial x} \tag{24}$$

$$0 = b^T(x)\frac{\partial H}{\partial x} \ , \tag{25}$$

where we can recognise the structure we mentioned in the previous section, but applied to the manifold $M \times \mathcal{F}$, i.e. the points are pairs (p, f) with coordinates (x, y) in such a way that the Dirac structure becomes:

$$\begin{aligned} D = \{(X, f, \alpha, -e) \in T(M \times F) \oplus T^*(M \times \mathcal{F})| \\ X(x, y) - J(x)\alpha(x, y) - g(x)f(x, y) \in B(x) \\ e(x, y) = g^T(x)\alpha(x, y), \forall (x, y) \in M \times \mathcal{F}, \alpha \in \mathrm{Ann} B\} \end{aligned}$$

where we use B to denote the constraint distribution. The characteristic distribution $\mathcal{G}_0$ will be then given by:

$$\mathcal{G}_0(x, y) = \mathcal{G}_0(x) = \mathrm{Im} \begin{pmatrix} g(x) & b(x) \\ \mathbb{I}_n & 0 \end{pmatrix} .$$

The presence of the constraints, which define $\mathcal{G}_0$ or/and $\mathcal{P}_0$, is the reason why the resulting dynamics is obtained in an implicit form. The projection of this dynamics into a submanifold which incorporates the constraints and on which the dynamics is explicit is a very interesting construction from the practical point of view; but in order to preserve the nice properties that Dirac structures exhibit for the description of control systems, we need to construct a new Port Controlled Hamiltonian system on the submanifold. The following sections present a simple case in which this construction is possible.

Application: controlled constrained mechanical systems To incorporate controls into the Dirac structure we introduced above in 3.3 is an easy task. We have to consider that the system has two external ports, the flow is represented by the control force and the effort by the velocity of the system. The previous equation takes the form for this case:

$$\begin{aligned}
\begin{pmatrix} \dot{q} \\ \dot{p} \end{pmatrix} &= \begin{pmatrix} 0 & I_n \\ -I_n & 0 \end{pmatrix} \begin{pmatrix} \frac{\partial H}{\partial q} \\ \frac{\partial H}{\partial p} \end{pmatrix} + \begin{pmatrix} 0 \\ F(q) \end{pmatrix} f + \begin{pmatrix} 0 \\ A(q) \end{pmatrix} \lambda \\
0 &= \left(0 \; A^T(q)\right) \begin{pmatrix} \frac{\partial H}{\partial q} \\ \frac{\partial H}{\partial p} \end{pmatrix} \\
e &= \left(0 \; F^T(q)\right) \begin{pmatrix} \frac{\partial H}{\partial q} \\ \frac{\partial H}{\partial p} \end{pmatrix} ,
\end{aligned}$$

where $\left(0 \; F^T(q)\right)$ represents the input vector fields.

This is clearly an implicit port controlled Hamiltonian system.

4 Constrained mechanical systems and algebroids

In [15] it was proved that constrained mechanical systems admit a Hamiltonian description as an unconstrained system by using the Poisson structure defined by the restriction of the canonical Poisson bracket of T^*M to the constraint manifold. We denote this Poisson structure by P_c. In [3] the geometrical nature of this Poisson structure was clarified. It is proved that is the Lie algebroid defined by the subbundle of admissible velocities defined by the constraints which provides the Poisson structure defined by van der Schaft and Maschke. The main results are as follows:

First of all, it is proved that, from the geometrical point of view, the constraint manifold is equivalent to the dual of the Lie algebroid defined by the admissible velocities:

Theorem 2 (C-G.M.vdS). *Let the mechanical system with constraints be defined on a cotangent bundle as before. Consider the natural set of coordinates* $\{q^i, p_i\}$. *Assume that the Hamiltonian* $H \in C^\infty(T^*M)$ *is regular and quadratic in momenta. Consider the algebroid* $\mathcal{A}_\mathcal{D}$ *defined in the previous section. The constraint manifold* $M_c = \{(q,p) \in T^*M | A\frac{\partial H}{\partial p} = 0\}$ *is diffeomorphic to the dual bundle* $\mathcal{A}^*_\mathcal{D}$, *i.e. there exists a diffeomorphism*

$$\Psi : M_c \to \mathcal{A}^*_\mathcal{D}. \tag{26}$$

And then it is also proved that the Poisson structure is the image of the canonical Poisson structure on $\mathcal{A}^*_\mathcal{D}$ by this transformation:

Theorem 3 (C-G.M.vdS). *The transformation* Ψ *above is a Poisson morphism.*

Therefore, the Poisson structure which in [15] was defined only in coordinates is actually representing the natural Poisson tensor of the dual of the Lie algebroid. The diffeomorphism Ψ allows us to transfer the Hamiltonian from M_c to $\mathcal{A}_{\mathcal{D}}^*$ defining

$$H^{\mathcal{A}_{\mathcal{D}}^*} = (\Psi^{-1})^* H_c,$$

and hence, the consequence at the level of the Hamiltonian systems is immediate:

Corollary 1. *The Hamiltonian systems* (M_c, P_c, H_c) *and* $(\mathcal{A}_{\mathcal{D}}^*, P^{\mathcal{A}_{\mathcal{D}}^*}, H^{\mathcal{A}_{\mathcal{D}}^*})$ *are equivalent.*

Therefore, the natural geometrical restriction of the original Dirac structure (18) is naturally mapped on the Dirac structure which $P^{\mathcal{A}_{\mathcal{D}}^*}$ defines on $\mathcal{A}_{\mathcal{D}}^*$:

$$D^{\mathcal{A}_{\mathcal{D}}^*} \subset T\mathcal{A}_{\mathcal{D}}^* \oplus T^*\mathcal{A}_{\mathcal{D}}^* = \{(P^{\mathcal{A}_{\mathcal{D}}^*}(\alpha), \alpha) | \alpha \in \Lambda^1(\mathcal{A}_{\mathcal{D}}^*)\} \tag{27}$$

which exhibits an explicit dynamics. The mapping will be denoted by $\hat{\Psi}$:

$$\hat{\Psi} : D \subset T(T^*M) \oplus T^*(T^*M) \to D^{\mathcal{A}_{\mathcal{D}}^*} \subset T\mathcal{A}_{\mathcal{D}}^* \oplus T^*\mathcal{A}_{\mathcal{D}}^* \tag{28}$$

which because of the corollary maps solutions of the dynamics of the Hamiltonian $H \in C^\infty(T^*M)$ defined on T^*M into solutions of the Hamiltonian $H^{\mathcal{A}_{\mathcal{D}}^*}$ on $\mathcal{A}_{\mathcal{D}}^*$.

5 Control of constrained mechanical systems

Now we intend to apply the construction before to the controlled case. As we saw above, controls are included in the Port Controlled Hamiltonian system as forces which act on the ports of the mechanical system.

The dynamics including control forces has been seen to be:

$$\begin{aligned}
\begin{pmatrix} \dot{q} \\ \dot{p} \end{pmatrix} &= \begin{pmatrix} 0 & I_n \\ -I_n & 0 \end{pmatrix} \begin{pmatrix} \frac{\partial H}{\partial q} \\ \frac{\partial H}{\partial p} \end{pmatrix} + \begin{pmatrix} 0 \\ F(q) \end{pmatrix} f + \begin{pmatrix} 0 \\ A(q) \end{pmatrix} \lambda \\
0 &= \begin{pmatrix} 0 & A^T(q) \end{pmatrix} \begin{pmatrix} \frac{\partial H}{\partial q} \\ \frac{\partial H}{\partial p} \end{pmatrix} \\
e &= \begin{pmatrix} 0 & F^T(q) \end{pmatrix} \begin{pmatrix} \frac{\partial H}{\partial q} \\ \frac{\partial H}{\partial p} \end{pmatrix} .
\end{aligned}$$

It is necessary to extend the mapping $\hat{\Psi}$ (28) in order to include the ports. From the geometrical point of view, the only obstruction is an possible non-tangentiality of the control forces to the constraint manifold M_c. From the physical point of view this is not a problem, since we know that the constraint forces would compensate any not tangential force. But we need a geometrical procedure to obtain the desired projection, in a canonical way. The constraint forces A defined in (1) can define such a projection in the following way: take the one forms of the codistribution A and pull-back them to forms defined on T^*M by using the canonical projection of the cotangent bundle:

$$\tilde{A} = \pi^*(A) \ .$$

This is a semibasical one form defined on the cotangent bundle (i.e. it vanishes when saturated with vertical vector fields). Take now the transformation (17) and define a vertical distribution on T^*M:

$$\hat{A} = (\hat{\omega})^{-1}(\tilde{A})$$

Lemma 1. *At each point of the constraint manifold $(q,p) \in M_c \subset M$, $\hat{A}$ defines a subspace of the tangent space of TM that is complementary to the tangent space of the constraint manifold, i.e.:*

$$T_{(q,p)}(T^*M) = \hat{A}_{(q,p)} \oplus T_{(q,p)}M_c \tag{29}$$

Proof. This is precisely the choice of basis on the tangent space $T_{(q,p)}T^*M$ which corresponds to the choice of coordinates $\{p^{(1)}, p^{(2)}\}$ proposed in [15] for the fibre of T^*M. Since it defines a basis for the cotangent bundle, its lifted version must do it also for the corresponding tangent bundle.

This decomposition provides the desired projection for the control forces in our system. First of all, we restrict the codistributon F to the constraint submanifold:

$$\tilde{F}_c = F|_{M_c}$$

In order to eliminate the component of the control forces in the direction of the constraint forces, we take the image of $\tilde{F}_c$ by the transformation $\hat{\omega}^{-1}$:

$$\hat{\tilde{F}}_c = \hat{\omega}^{-1}(\tilde{F}_c),$$

and decompose the corresponding distribution according to (29):

$$\hat{\tilde{F}}_c = \hat{F}^A + \hat{F}_c \ .$$

Finally, we map the distribution $\hat{F}_c$ back to $T^*(T^*M)$ and take its image as the control forces on the constraint manifold:

$$F_c = \hat{\omega}(\hat{F}_c)$$

This decomposition also allows us to incorporate the control forces to the explicit system defined by the Dirac structure $D_c \subset TM_c \oplus T^*M_c$ defined by the Poisson structure P_c (or alternatively to the equivalent system defined by (27)). From the physical point of view, constraint forces will compensate in any case the non tangential part of the control forces. Therefore only the tangential part, F_c is meaningful for the dynamics on M_c.

But hence we have proved that:

Theorem 4. *Under the assumptions of Theorem 2, the Port controlled Hamiltonian system defined by* (M_c, D_c, H_c, F_c) *is equivalent to the original one* (M, D, H, F).
Alternatively, we can also formulate it as $(\mathcal{A}^*_{\mathcal{D}}, D^{\mathcal{A}^*_{\mathcal{D}}}, H^{\mathcal{A}^*_{\mathcal{D}}}, F^{\mathcal{A}^*_{\mathcal{D}}})$ *where we represent by* $F^{\mathcal{A}^*_{\mathcal{D}}}$ *the image of* F_c *under the diffeomorphism (26).*

Acknowledgements: For two of us (J. C-G and A. J. vdS) this work was supported by the European Union through the TMR network in Nonlinear Control (Contract ERB FMRXCT-970137).

References

1. G. Blankenstein. *Implicit Hamiltonian systems:symmetry and interconnection.* PhD thesis, University of Twente, 2000.
2. J. Clemente-Gallardo. Applications of lie algebroids in mechanics and control theory. In *Nonlinear Control in the year 2000*, volume 1, pages 299–314. Springer, 2000.
3. J. Clemente-Gallardo, B. Maschke, and A.J. van der Schaft. Kinematical constraints and algebroids. *Rep. Math. Phys.*, **47** 3: 413-429, 2001
4. T.J. Courant. Dirac manifolds. *Trans. of the AMS*, 319(2):631–661, June 1990.
5. T.J. Courant and A. Weinstein. Beyond poisson structures. Technical report, UCB, 1986.
6. I. Dorfman. Dirac structures of integrable evolution equations. *Physics Letters*, A 125, 1987.
7. R. Hermann. Analytic continuation of group representations, IV. *Comm. Math. Phys.*, 5:131–156, 1967.
8. J.-C. Herz. Pseudo-algèbres de Lie. *C. R. Acad. Sci. Paris, Série A*, 236:1935–1937, 1953.
9. P. Libermann and Ch-M Marle. *Symplectic Geometry and Analytical Mechanics.* Reidel, 1987
10. E. Martínez. Hamiltonian mechanics on lie algebroids. Technical report, Universidad de Zaragoza, 1999.
11. E. Martínez. Lagrangian mechanics on lie algebroids. *Acta Applicandae Mathematicas*, To appear, 2000.
12. R. S. Palais. *A global formulation of the Lie theory of transformation groups*, volume 22 of *Mem. Amer. Math. Soc.* American Mathematical Society, Providence, R. I., 1957.
13. J. Pradines. Théorie de Lie pour les groupoïdes différentiables. Relations entre propriétés locales et globales. *C. R. Acad. Sci. Paris Sér. I Math.*, 263:907–910, 1966.
14. A.J. van der Schaft and M. Dalsmo. On representations and integrability of mathematical structures in energy-conserving physical systems. *SIAM J. Control and Optimization*, 1999.
15. A.J. van der Schaft and B. M. Maschke. On the Hamiltonian formulation of nonholonomic mechanical systems. *Rep. Math. Phys.*, 34(2):225–233, 1994.
16. A. Weinstein. Lagrangian mechanics and groupoids. *Fields Institut Com.*, pages 207–231, 1996.

On the Classification of Control Sets

Fritz Colonius[1] and Marco Spadini[2]

[1] Institut für Mathematik, Universität Augsburg, 86135 Augsburg, Germany
[2] Dipartimento di Matematica Applicata 'G. Sansone', Università di Firenze, Via S. Marta 3, 50139 Firenze, Italy

Abstract. The controllability behavior of nonlinear control systems is described by associating semigroups to locally maximal subsets of complete controllability, i.e., local control sets. Periodic trajectories are called equivalent if there is a 'homotopy' between them involving only trajectories. The resulting object is a semigroup, which we call the *dynamic index* of the local control set. It measures the different ways the system can go through the local control set.

1 Introduction

The purpose of this paper is to contribute to the qualitative study of the controllability behavior of nonlinear control systems. We classify the behavior within locally maximal subsets of complete controllability, *local control sets* as introduced in [4]. To each local control set a semigroup is associated which is constructed from periodic trajectories in the local control set. We take inspiration from the classical construction of the (first) homotopy group in algebraic topology, but use periodic orbits instead of closed loops. Two periodic orbits are equivalent, if they can be connected via a homotopy involving only periodic trajectories. This leads to some technical difficulties as it is necessary to make such an equivalence compatible with the natural composition of orbits with the same initial point. The resulting object is a commutative semigroup; in general it is not a group. This is due to the fact that for many nonlinear control systems the equivalence classes of periodic trajectories need not admit an inverse. We stress the fact that, even allowing trajectories followed backward in time, we would not in general obtain a group. In fact a neutral element would still be lacking.

The so-constructed semigroup is called *dynamic index* of the local control set. It measures the "different" ways in which the system can go through the local control set. It turns out, that for linear systems with controllable (A, B) and admissible control range U the index is always trivial. This remains true for small nonlinear perturbations. If the control range is small enough, we can also show that for a local control set around an attracting periodic solution of the uncontrolled system the index is isomorphic to the additive semigroup of natural numbers $\mathbb{N}$.

Compare also San Martin and Santana [11], where the homotopy type of Lie semigroups and invariant control sets is studied. We remark that in our construction the direction of the trajectories plays a crucial role. This is a

F. Colonius, L. Grüne (Eds.): Dynamics, Bifurcations, and Control, LNCIS 273, pp. 217–231, 2002.

decisive difference of our semigroup from homotopy groups. Katok and Hasselblatt [8, p. 117] briefly discuss other constructions of topological invariants using trajectories of dynamical systems. But perhaps closest in spirit to our paper are the papers [12,13] by A. Sarychev. He studied homotopy properties of the space of trajectories. However, he was interested in the case, where the systems are completely controllable or, in our terminology, where the control set coincides with the whole state space. Furthermore, his main result concerns systems without drift where trajectories can be reversed.

After some basic definitions in Section 2 we define in Section 3 the key technical notion for the construction of the index, the so-called 'strong inner pairs', and show some of their relevant properties. Section 3 is devoted to the construction of the index and two simple examples are provided. Section 4 presents the explicit computation of the index in the case of the control set which arises, for a small control range, around an attracting periodic orbit of the uncontrolled system.

2 Basic definitions

We will consider the following control-affine system in $\mathbb{R}^d$

$$\dot{x}(t) = f\big(x(t), u(t)\big) := f_0(x(t)) + \sum_{i=1}^{m} u_i(t) f_i(x(t)), \tag{1}$$
$$u \in \mathcal{U} = \{u \in L_\infty(\mathbb{R}, \mathbb{R}^m),\ u(t) \in U \text{ for almost all } t \in \mathbb{R}\}.$$

with sufficiently smooth vector fields f_i, $i = 0, 1, ..., m$, on $\mathbb{R}^d$ and a compact convex neighborhood U of the origin in $\mathbb{R}^m$. We assume that for every control $u \in \mathcal{U}$ and every initial condition $x(0) = x_0 \in \mathbb{R}^d$ there exists a unique trajectory which we denote by $\varphi(t, x, u)$, $t \in \mathbb{R}$. Our results will also hold–with some technical modifications–for systems on manifolds. Note that for control affine systems, the trajectories $\varphi(t, x, u)$ depend continuously on (t, x, u), uniformly on bounded time intervals; here $\mathcal{U}$ will be considered in the weak* topology inherited from the inclusion $\mathcal{U} \subset L_\infty(\mathbb{R}, \mathbb{R}^m) = \big(L_1(\mathbb{R}, \mathbb{R}^m)\big)^*$. Notice that $\mathcal{U}$ is in this topology a compact and separable metrizable space (see, e.g., Dunford/Schwartz [6]); an appropriate metric will be fixed throughout and denoted by 'd'; compare e.g. [1, Chapter 4] for this setting.

At some places we will consider for $0 \le \rho \le 1$, the control range $\rho U = \{\rho \cdot x : x \in U\}$. Then we denote by $\mathcal{U}^\rho$ the set of all $L_\infty(\mathbb{R}, \mathbb{R}^m)$ control functions taking values in ρU.

The following definitions specify subsets of complete approximate controllability, which are our primary concern in this paper.

Definition 1. A subset D with nonempty interior of the state space $\mathbb{R}^d$ is a precontrol set if for all $x, y \in D$ and every $\varepsilon > 0$ there exist $T > 0$ and $u \in \mathcal{U}$ such that

$$\varphi(t, x, u) \in D \text{ for all } t \in [0, T] \quad \text{and} \quad |\varphi(T, x, u) - y| < \varepsilon.$$

A precontrol set D of $\mathbb{R}^d$ is a local control set if there exists a neighborhood V of cl D such that for every precontrol set D' with $D \subset D' \subset V$ one has $D' = D$.

Thus a local control set is a locally maximal precontrol set. Note also that control sets (with nonvoid interior) as discussed in [1] are globally maximal precontrol sets. The sets of reachable points from x and controllable to $x \in \mathbb{R}^d$ in time $T > 0$ are denoted by

$$\mathcal{O}^+_{\leq T}(x) = \{y \in \mathbb{R}^d, \text{ there are } 0 \leq t \leq T \text{ and } u \in \mathcal{U} \text{ with } y = \varphi(t, x, u)\}$$

and

$$\mathcal{O}^-_{\leq T}(x) = \{y \in \mathbb{R}^d, \text{ there are } 0 \leq t \leq T \text{ and } u \in \mathcal{U} \text{ with } x = \varphi(t, y, u)\},$$

respectively. Throughout this paper we require local accessibility, that is, $\mathcal{O}^+_{\leq T}(x)$ and $\mathcal{O}^-_{\leq T}(x)$ have nonvoid interiors for all $x \in \mathbb{R}^d$ and all $T > 0$. Then in the interior of a local control set exact controllability holds. Recall also that local accessibility is guaranteed by the following *accessibility rank condition*:

$$\dim \Delta_{\mathcal{L}}(x) = d \text{ for all } x \in \mathbb{R}^d, \tag{2}$$

where $\mathcal{L}$ denotes the Lie algebra generated by the vector fields $f_0, ..., f_m$, and $\Delta_{\mathcal{L}}(x)$ is the subspace of the tangent space (identified with $\mathbb{R}^d$) generated by the vector fields in $\mathcal{L}$.

3 Strong inner pairs

In this section we specify the subclass of periodic trajectories which will be used for the construction of the dynamic index.

Definition 2. A pair $(u, x) \in \mathcal{U} \times \mathbb{R}^d$ is called a strong inner pair, if the control u is piecewise constant with $u(t) \in \operatorname{int} U$ for all $t \in \mathbb{R}$ and there is $\delta > 0$ such that for all $\tau > 0$, small enough, and for all $y \in \mathbb{R}^d$ with $|x - y| < \delta$ the following property holds:
For all $0 < t \leq \tau$ there are neighborhoods $N_t^{\pm}(y)$ of $\varphi(\pm t, y, u)$ such that for any curve $\lambda \mapsto z_\lambda^+ \in N_t^+(y)$ and $\lambda \mapsto z_\lambda^- \in N_t^-(y)$, with $z_0^{\pm} = \varphi(\pm t, y, u)$, there are continuous maps

$$\lambda \mapsto (\pm t_\lambda^{\pm}, u_\lambda^{\pm}) : [0, 1] \to (0, T) \times \mathcal{U},$$

with $u_\lambda^{\pm}$ piecewise constant for $\lambda \in [0, 1]$, and

$$(\pm t_0^{\pm}, u_0^{\pm}) = (\pm t, u) \text{ and } \varphi(\pm t_\lambda^{\pm}, y, u_\lambda^{\pm}) = z_\lambda^{\pm}.$$

The following remarks show that strong inner pairs are abundant provided that local accessibility holds. Here we write $e^{tX}x$, with $X = f(\cdot, u)$, in place of $\varphi(t, x, u)$ provided that u is a constant control.

Remark 1. Assume that for some $\varepsilon > 0$ there exist $s_i^\pm \in (0, \varepsilon)$ and $u_1^\pm, \ldots, u_d^\pm \in \operatorname{int} U$ such that the two maps

$$(t_d, ..., t_1) \mapsto e^{\pm s_d X_d^\pm} \cdots e^{\pm s_1 X_1^\pm} x,$$

where $X_i^\pm := f(\cdot, u_i^\pm)$, have full rank on $(0, \varepsilon) \times ... \times (0, \varepsilon)$. On the interval $(-\sum_{i=1}^d s_i^-, \sum_{i=1}^d s_i^+]$, define

$$u(t) = \begin{cases} u_i^+ \text{ for } t \in (\sum_{j=1}^i s_j, \sum_{j=1}^{i+1} s_j], \\ u_i^- \text{ for } t \in (-\sum_{j=1}^{i+1} s_j, -\sum_{j=1}^i s_j]. \end{cases}$$

Then (u, x) is a strong inner pair. To see that, it is sufficient to notice that the rank condition holds for any y in a neighborhood of x and that neighborhoods of $\varphi(\pm t, y, u)$ are of the form

$$\{e^{\pm t_d X_d} ... e^{\pm t_1 X_1} x, \text{ with } t_1, ..., t_d \in (0, \varepsilon)\}.$$

Hence the required continuous families can be obtained by appropriately changing the times t_i.

Strong inner pairs can also be easily obtained when the linearized control system is controllable. Here it is convenient–also for later use in Section 4–to consider ρ-dependent control ranges. Recall that for two vector fields X, Y one defines $\mathrm{ad}_X^0 Y = Y$ and for $k = 1, 2, ...$ one defines $\mathrm{ad}_X^k Y$ as the Lie bracket $\mathrm{ad}_X^k Y := [X, \mathrm{ad}^{k-1} Y]$.

Proposition 1. *Let $x \in \mathbb{R}^d$ and assume that*

$$\operatorname{span} \{\mathrm{ad}_{f_0}^k f_i(x),\ i = 1, ..., m,\ k = 0, 1, ...\} = \mathbb{R}^d. \tag{3}$$

Then for $\rho > 0$, small enough, each $(u, y) \in \mathcal{U}^\rho \times \mathbb{R}^d$ with u piecewise constant and $u \in \mathcal{U}^{\rho'}$, for some $\rho' < \rho$ and $|y - x| < \rho'$, is a strong inner pair for the ρ-system.

Proof (Sketch of the proof). If $\rho > 0$ and $T > 0$ are small enough, assumption (3) clearly holds for $\varphi(T, y, u)$ with y in a small neighborhood of x and $\|u\|_\infty < \rho$. This, for all $0 < \tau \leq T$, guarantees controllability for the control system linearized along $\varphi(t, y, u)$ (with unbounded controls). Then a standard result in nonlinear control theory, see, e.g., [1, Theorem A.4.11 and Remark A.4.12] guarantees that the nonlinear control system with controls in $\mathcal{U}^\rho$ is locally controllable about the trajectory $\varphi(t, y, u)$, provided that $u \in \mathcal{U}^{\rho'}$ for some $\rho' < \rho$. This is based on an application of the inverse function theorem, which also provides the existence of neighborhoods $N_t^\pm(y)$ as in Definition 2.

The following proposition shows, in particular, that the interior of a local control set corresponds to strong inner pairs.

Proposition 2. *Let D be a local control set for (1) and assume that the accessibility rank condition holds in D. Then, for any $x, y \in \operatorname{int} D$, there are $T > 0$ and a T-periodic control function $u \in \mathcal{U}$ such that (u, x) is a strong inner pair and $y \in \varphi([0, T], x, u)$.*

Proof. By the accessibility rank condition, as in the proof of Krener's Theorem (compare [9]), it follows that there exist $u_1, \ldots, u_d \in \operatorname{int} U$ and $\delta > 0$ such that,

$$N^+ = \operatorname{int}\left\{e^{t_d X_d} \cdots e^{t_1 X_1} x : 0 \leq t_i \leq \delta,\, i = 1, ..., d\right\} \neq \emptyset,$$
$$N^- = \operatorname{int}\left\{e^{t_d X_d} \cdots e^{t_1 X_1} x : -\delta \leq t_i \leq 0,\, i = 1, ..., d\right\} \neq \emptyset,$$

where $X_i = f(\cdot, u_i)$ for $i = 1, ..., d$.

Take $x^+ \in N^+$. Since in the interior of D approximate controllability holds, one can find a control function v_0 and a time S_0 such that $x^- := \varphi(S_0, x^+, v_0) \in N^-$. By continuous dependence we can assume that v_0 is a piecewise constant function with values in $\operatorname{int} U$. Let $v^+, v^- \in \mathcal{U}$ and $S^+, S^- > 0$ be such that

$$x^+ = \varphi(S^+, x, v^+) \quad \text{and} \quad x = \varphi(S^-, x^-, v^-).$$

Concatenating v^-, v^+ and v_0, and taking $T = S^+ + S_0 + S^-$ one gets a T-periodic trajectory driven by some T-periodic piecewise constant control function u. One can also construct u as a control function which connects x^+ to y and y to x^-, in a way that essentially follows the line of the first part of the proof.

4 The dynamic index

In this section we construct a dynamic index for local control sets.

We consider a local control set D for (1) and assume throughout that the accessibility rank condition holds. Define the set

$$\mathcal{P}(D) = \left\{ (T, u, x) \in (0, \infty) \times \mathcal{U} \times \mathbb{R}^d : \begin{array}{c} (u, x) \text{ is a } T\text{-periodic} \\ \text{strong inner pair, } T > 0, \text{ and} \\ \varphi(t, x, u) \in D,\, \forall t \in [0, T] \end{array} \right\},$$

endowed with the metric topology given by

$$\operatorname{dist}\big((T, u, x), (S, v, y)\big) = |T - S| + \|x - y\|_{\mathbb{R}^d} + \mathrm{d}(u, v).$$

Below, when no confusion can possibly arise, we shall omit the explicit dependence on the base set D.

Let us now introduce a relation on $\mathcal{P}$.

Definition 3. $(T,u,x) \sim (S,v,y)$ in $\mathcal{P}$ if there are $k+1$ elements (T_0,u_0,x_0), $\ldots$, (T_k,u_k,x_k) in $\mathcal{P}$ with the following properties:

(i) $(T_0,u_0,x_0) = (T,u,x)$ and $(T_k,u_k,x_k) = (S,v,y)$;
(ii) for $i = 0,...,k$ there are

$$0 = \tau_i^0 < ... < \tau_i^{k_i} = T_i \text{ and } 0 = \sigma_{i+1}^0 < ... < \sigma_{i+1}^{k_i} = T_{i+1},$$

such that $\varphi(\tau_i^j, x_i, u_i) = x_i$ and $\varphi(\sigma_{i+1}^j, x_{i+1}, u_{i+1}) = x_{i+1}$ for all i and all j;
(iii) there are continuous maps $H_i^j : [0,1] \to \mathcal{P}$ such that for $i = 0,...,k$ and $j = 0,...,k_i - 1$

$$H_i^j(0) = \left(\tau_i^{j+1} - \tau_i^j, u_i(\tau_i^j + \cdot), x_i\right), \quad \text{and}$$
$$H_i^j(1) = \left(\sigma_{i+1}^{j+1} - \sigma_{i+1}^j, u_{i+1}(\sigma_{i+1}^j + \cdot), x_{i+1}\right).$$

In other words, (T_i,u_i,x_i) and $(T_{i+1},u_{i+1},x_{i+1})$ are chopped into k_i periodic pieces of period $\tau_i^{j+1} - \tau_i^j$ and $\sigma_{i+1}^{j+1} - \sigma_{i+1}^j$ respectively, and the corresponding pieces are homotopic via trajectories.

This definition makes the natural operation of concatenation of trajectories 'compatible' with the relation '$\sim$'. To be more precise consider two continuous maps H and H' from $[0,1]$ to $\mathcal{P}$. Clearly they establish 'homotopies' between $H(0) = (T,u,x)$ and $H(1) = (S,v,y)$, and between $H'(0) = (T',u',x)$ and $H'(1) = (S',v',y)$ respectively. Define $u \circ u'$ as the concatenation on $[0,T+T']$ of u and u' extended $(T+T')$-periodically to $\mathbb{R}$. According to Definition 3 $(T+T', u \circ u', x)$ is related to $(S+S', v \circ v', y)$; whereas there might not exist any continuous function $F : [0,1] \to \mathcal{P}$ with the property that $F(0) = (T+T', u \circ u', x)$ and $F(1) = (S+S', v \circ v', y)$.

Notice that the relation introduced above is an equivalence relation. Then, consider on $\mathcal{P}/\sim$, the set $\mathcal{Q}$ of all the formal (juxtaposition) products, i.e., the free semigroup on $\mathcal{P}/\sim$. (See, e.g., Howie [7] for some general facts about the algebraic theory of semigroups.) Usually, we shall set

$$[T,u,x]^n = \underbrace{[T,u,x]\cdots[T,u,x]}_{n \text{ times}},$$

for any $n \geq 0$. Here the square parentheses denote the equivalence classes.

Clearly $\mathcal{Q}$ is a semigroup which, besides its non-commutativity, is far too large for being of any use. Below we factorize it over the congruence induced by two families of equations among the elements of $\mathcal{Q}$. Recall that a congruence on a semigroup $(S,\cdot)$ is an equivalence relation '$\equiv$' such that

$$a \equiv a' \text{ and } b \equiv b' \text{ imply } a \cdot b \equiv a' \cdot b',$$

for any $a, a', b, b' \in S$.

Consider the following families of relations:

$$\mathcal{F} = \Big\{[T,u,x][S,v,x] = [T+S, u\circ v, x] : (T,u,x),(S,v,x) \in \mathcal{P}\Big\},$$
$$\mathcal{G} = \Big\{[T,u,x][S,v,y] = [S,v,y][T,u,x] : (T,u,x),(S,v,y) \in \mathcal{P}\Big\}.$$

Notice that the elements of $\mathcal{F}$ are well defined. In fact, by the definition of '$\sim$' one has that

$$(T,u,x) \sim (\bar{T},\bar{u},\bar{x}) \text{ and } (S,v,x) \sim (\bar{S},\bar{v},\bar{x})$$

imply

$$(T+S, u\circ v, x) \sim (\bar{T}+\bar{S}, \bar{u}\circ\bar{v}, \bar{x}).$$

The union of the families $\mathcal{F}$ and $\mathcal{G}$ clearly can be seen as a relation on $\mathcal{Q}$, i.e., as a subset of $\mathcal{Q}\times\mathcal{Q}$. Now, since the intersection of congruences is again a congruence, it makes sense to consider the congruence $(\mathcal{F}\cup\mathcal{G})^{\#}$ generated by the set $\mathcal{F}\cup\mathcal{G}$, namely the intersection of all the congruences containing $\mathcal{F}\cup\mathcal{G}$ (see e.g. [7]).

Finally, we define the *dynamic index* $\mathcal{I}(D)$ of D as the quotient

$$\mathcal{I}(D) := \mathcal{Q}(D)/(\mathcal{F}\cup\mathcal{G})^{\#} \ .$$

Notice that $\mathcal{I}(D)$ is a commutative semigroup. Next we consider two easy examples.

Example 1 (Linear Systems). Consider the following linear control system with restricted control range

$$\dot{x}(t) = Ax(t) + Bu(t) \quad \text{in } \mathbb{R}^d,\ u \in \mathcal{U},$$

where $U \subset \mathbb{R}^m$ is convex and compact with $0 \in$ int U and A and B are constant matrices of dimensions $d\times d$ and $d\times m$, respectively. We assume that the pair (A,B) is controllable, i.e., that rank $[B, AB, ...A^{d-1}B] = d$. Then the index $\mathcal{I}(D)$ of the unique control set D reduces to the unity. In fact: For a T-periodic strong inner pair (u,x) in the interior of D, define a homotopy to the origin via

$$H(\alpha) := (T, \alpha u, \alpha x),\ \alpha \in [0,1].$$

Linearity implies that $\varphi(T,\alpha x,\alpha u) = \alpha x$ for all $\alpha \in [0,1]$. Hence this is a periodic solution, and for $\alpha = 0$ one obtains the equilibrium. If $0 \notin U$ then the same result holds provided that A is hyperbolic (cp. [3]).

Example 2 (Small perturbations of linear systems). Consider a control process of the form:

$$\dot{x}(t) = Ax(t) + Bu(t) + \sum_{i=1}^{m} u_i(t)F_i(x(t)),\ u \in \mathcal{U}, \tag{4}$$

with $U \subset \mathbb{R}^m$ compact and convex with non empty interior. Assume that (A, B) in (4) is controllable and A is hyperbolic, and let F_i be C^1.

It follows from the proof of the uniqueness for perturbations in [4] that there exists $M > 0$ such that, if

$$\|D_1 F_i(x,u)\| \le M, \quad \text{and} \quad \|D_2 F_i(x,u)\| \le M$$

for all (x, u) and $i = 1, ..., m$, then the control process (4) admits exactly one control set D. Furthermore, the dynamic index $\mathcal{I}(D)$ reduces to its unity. In fact, in [4], we constructed a homotopy between any two given periodic triples (possibly with different periods).

5 The index of a control set near a periodic orbit

This section is devoted to the computation of the index of the control set for (1) which arise for a small control range around an isolated attracting periodic orbit $\gamma = \varphi([0,T], x_0, 0)$, with (minimal) period $T > 0$, of the uncontrolled system, assuming that the linearized system along γ is controllable. Recall that a periodic orbit (of an autonomous differential equation) is called attracting, if the eigenvalues of the linearized Poincaré map are strictly smaller than one in modulus; compare [10].

Proposition 3. *Let γ be a attracting orbit of the uncontrolled system, and let A be a neighborhood of γ. Assume that the controllability rank condition (3) holds. Then there exist ρ_0 such that for any $0 < \rho \le \rho_0$ there exists a unique control set D^ρ with $\gamma \subset D^\rho \subset A$.*

Proof. The controllability rank condition implies by Proposition 1 that all pairs $(x, 0) \in \gamma \times \mathcal{U}^\rho$ are strong inner pairs, hence inner pairs. Then Corollary 4.7.6 in [1] implies the assertion.

We shall prove that, when ρ is small enough, the index of the control set D^ρ containing the T-periodic orbit γ is isomorphic to $\mathbb{N}$. To prove this result we need to show that when $(T_1, u_1, x_1) \in \mathcal{P}(D^\rho)$ is such that $\varphi([0, T_1], x_1, u_1)$ goes n times around γ, then $(T_1, u_1, x_1) \sim (nT, 0, x_0)$ and therefore $[T_1, u_1, x_1] = [T, 0, x_0]^n$. To make this precise we shall introduce Definition 5 below.

However, it is first necessary to establish some preliminaries on the Poincaré map for control systems. We will use some notions and results from Colonius/Sieveking [2].

Definition 4. Let $x_0 \in \mathbb{R}^n$, $L : \mathbb{R}^d \to \mathbb{R}$ linear and $\alpha > 0$. If $Lf(x, u) > \alpha$ for all x in a neighborhood W of x_0 and all $u \in U$ then the connected component of $W \cap L^{-1}(Lx_0)$ containing x_0 is called a *local transversal section* through x_0.

Obviously, trajectories "can cross a local transversal section only from one side".

Definition 5. Let Ω be a neighborhood of γ. We say that a closed orbit $\gamma_1 = \varphi([0, T_1], x_1, u_1) \subset \Omega$ goes n times around γ (relatively to Ω) if there exists a linear map L as in Definition 4 such that

1. $S := \Omega \cap L^{-1}(Lx_0)$ is a local transversal section to γ,
2. $\gamma \cap S = \{x_0\}$,
3. $x_1 \in S$, and
4. there exist exactly n times $t_i \in (0, T_1]$, $i = 1, ..., n$, such that $\varphi(t_i, x_1, u_1) \in S$.

It is a consequence of the Hahn-Banach Theorem that x_0 admits a local transversal section if $0 \notin f(x_0, U)$, since the set $f(x_0, U)$ is convex and compact. Therefore, if x_0 is not an equilibrium of the uncontrolled system, i.e., if $0 \neq f(x_0, 0)$, then x_0 admits a local transversal section for the system with control range ρU with ρ small enough.

Definition 6. Let S be a local transversal section through x_0, and let $V_1 \subset V_0$ be neighborhoods of x_0. The triple (V_0, V_1, S) is a *flow box* around x_0 if it has the following property:

If $\varphi(\cdot, x_0, u)$ satisfies

$$\varphi(t_0, x_0, u) \notin V_0, \quad \varphi(t_1, x_0, u) \in V_1, \quad \varphi(t_2, x_0, u) \notin V_0$$

for some $0 \leq t_0 < t_1 < t_2$, then there exists $t \in (t_0, t_2)$ such that $\varphi(t, x_0, u) \in S$ and $\varphi(s, x_0, u) \in V_0$ for all s between t and t_1.

The following result establishes the existence of flow boxes; it is a slight modification of Theorem 2.16 in [2].

Lemma 1. *Let S be a local transversal section through x_0. Then for every neighborhood W of S there are neighborhoods V_0 and V_1 of x_0 contained in W such that (V_0, V_1, S) is a flow box around x_0.*

Proof. There exist a linear map $L : \mathbb{R}^d \to \mathbb{R}$, a constant $\alpha > 0$, and a neighborhood $W_1 \subset W$ of x_0 with $S \supset W_1 \cap L^{-1}(Lx_0)$ and

$$Lf(y, v) > \alpha \text{ for all } y \in W_1,\ v \in U.$$

Choose a ball $V_0 = B(r_0, x_0)$ around x_0 with radius $r_0 > 0$ such that $V_0 \subset W_1$ and set $c := \sup\{|f(y, u)|,\ y \in V \text{ and } v \in U\}$. Then choose $r_1 \in (0, r_0)$ so small that

$$Lz - \alpha/2c(r_0 - r_1) \leq Ly \leq Lz + \alpha/2c(r_0 - r_1) \tag{5}$$

for all $z,\, y \in V_1 = B(r_1, x_0)$. We have for $t > t' \geq 0$:

$$\varphi(t,x,u) = \varphi(t',x,u) + \int_{t'}^{t} f(\varphi(s,x,u),u(s))\,ds$$

and hence

$$\begin{aligned} L\varphi(t,x,u) &= L\varphi(t',x,u) + \int_{t'}^{t} Lf(\varphi(s,x,u),u(s))\,ds \\ &\geq L\varphi(t',x,u) + \alpha(t-t'), \end{aligned}$$

provided that $\varphi(s,x,u) \in W_1,\ t' \leq s \leq t$. Without loss of generality, we may assume

$$\varphi(s,x,u) \in V_0 \text{ for all } t_0 \leq s \leq t_2$$

replacing, if necessary, t_0 by the last time before t_1 at which $\varphi(t,x,u)$ is in the complement of V_0 and t_2 by the first time after t_1 at which $\varphi(t,x,u)$ leaves V_0. We have

$$\begin{aligned} r_0 - r_1 &\leq |\varphi(t_1,x,u) - \varphi(t_0,x,u)| \leq c(t_1 - t_0), \\ r_0 - r_1 &\leq |\varphi(t_2,x,u) - \varphi(t_1,x,u)| \leq c(t_2 - t_1). \end{aligned}$$

If $L\varphi(t_0,x,u) \leq Lx_0 \leq L\varphi(t_2,x,u)$, or $L\varphi(t_2,x,u) \leq Lx_0 \leq L\varphi(t_1,x,u)$, the assertion follows by continuity of $t \mapsto Lf(t,x,u)$. Hence we only have to consider the following two cases:

1. $Lx_0 < \min\{L\varphi(t_0,x,u), L\varphi(t_2,x,u)\}$. Here $L\varphi(t_1,x,u) \geq L\varphi(t_0,x,u) + \alpha(t_1 - t_0) > Lx_0 + \alpha/c(r_0 - r_1)$, contradicting (5) for $y = L\varphi(t_1,x,u)$.
2. $Lx_0 > \max\{L\varphi(t_0,x,u), L\varphi(t_2,x,u)\}$. Here $L\varphi(t_2,x,u) \geq L\varphi(t_1,x,u) + \alpha(t_2 - t_1) > L\varphi(t_1,x,u) + \alpha/c(r_0 - r_1)$, again contradicting (5).

We now turn to the Poincaré map. Note that for $\rho = 0$ a local transversal section as in Definition 4 coincides with the usual notion of a local transversal section for autonomous differential equations. Then it defines for $\rho > 0$, small enough, also a local transversal section for the system with control range ρU.

Proposition 4. *Let S be a local transversal section through $x_0 \in \gamma$. If ρ is small enough, there exists a neighborhood V of x_0 in S such that the Poincaré first return map $P : V \times \mathcal{U}^\rho \to S$ is well-defined and continuous. Moreover, the map that takes (x,u) into the 'first return time' $\tau(x,u)$ is continuous.*

Proof. Let us first show that P is well-defined. Notice that the orbits can cross S only from one side; therefore it is sufficient to show that there exists a neighborhood $V \subset S$ of x_0 such that the orbits return to S after a finite time.

Let W be a neighborhood of x_0 in $\mathbb{R}^d$ and (V_0, V_1, S) be a flow box around x_0 with $\operatorname{cl} V_0 \subset W$. Taking if necessary a smaller W, we can assume that there are times t_0 and t_1, with $0 < t_0 < T < t_1$, for which $\varphi(t_0, x_0, 0)$ and $\varphi(t_1, x_0, 0)$ are in $W \setminus \mathrm{cl}\, V_0$.

By continuous dependence on initial data there exist a neighborhood $V \subset V_1$ of x_0 in S and $\rho_0 > 0$ such that, for $0 < \rho < \rho_0$ and for every $(x, u) \in V \times \mathcal{U}^\rho$

$$\varphi(t_0, x, u) \in W \setminus \mathrm{cl}\, V_0,\ \varphi(t_1, x, u) \in W \setminus \mathrm{cl}\, V_0,\ \varphi(T, x, u) \in V_1.$$

Since (V_0, V_1, S) is a flow box, for each $(x, u) \in V \times \mathcal{U}^\rho$ there exists a time $\tau(x, u)$, with $t_0 < \tau(x, u) < t_1$ such that $\varphi(\tau(x, u), x, u) \in S$. For W small enough this time is unique proving that $P(x, u) := \varphi(\tau(x, u), x, u)$ is well-defined.

We shall now prove continuity of the map $(x, u) \mapsto P(x, u)$. Consider a sequence $\{(\xi_n, u_n)\}$ in $S \times \mathcal{U}^\rho$ converging to (ξ_0, u_0). Fix a neighborhood W of $P(\xi_0, u_0)$ in S and let $\hat{W}$ be a neighborhood of $P(\xi_0, u_0)$ in $\mathbb{R}^d$ such that $W = \hat{W} \cap S$. Let (V_0, V_1, S) be a flow box around $P(\xi_0, u_0)$ with $\mathrm{cl}\, V_0 \subset \hat{W}$.

Let $\tau = \tau(\xi_0, u_0)$. As in the first part of the proof, taking W smaller if necessary, one can find times $0 < \tau_0 < \tau < \tau_1$ such that

$$\varphi(\tau_0, x_0, u_0), \varphi(\tau_1, x_0, u_0) \in W \setminus \mathrm{cl}\, V_1.$$

From [1, Lemma 4.3.2] one has

$$\begin{aligned} &\lim_{n\to\infty} \varphi(\tau, \xi_n, u_n) = \varphi(\tau, \xi_0, u_0) = P(\xi_0, u_0), \\ &\lim_{n\to\infty} \varphi(\tau_0, \xi_n, u_n) = \varphi(\tau_0, \xi_0, u_0), \\ &\lim_{n\to\infty} \varphi(\tau_1, \xi_n, u_n) = \varphi(\tau_1, \xi_0, u_0). \end{aligned}$$

Therefore, for n large enough,

$$\varphi(\tau_0, x_n, u_n), \varphi(\tau_1, x_n, u_n) \notin V_0 \quad \text{and} \quad \varphi(\tau, x_n, u_n) \in V_1.$$

Since (V_0, V_1, S) is a flow box there exists $\tau_n \in (\tau_0, \tau_1)$ such that $P(x_n, u_n) = \varphi(\tau_n, x_n, u_n) \in S \cap W$. This proves that, for n large, $P(x_n, u_n) \in W$ and continuity follows. Notice also that, in the construction above, $\tau_n = \tau(\xi_n, u_n)$ satisfies

$$\tau_1 - \tau_0 > |\tau - \tau_n|;$$

by shrinking W, we can make the differences $\tau_1 - \tau_0$ as small as we please, therefore proving the continuity of the map $(x, u) \mapsto \tau(x, u)$.

The (technically involved) proof of the following proposition is given in [5]. Here for a control function u and a time $T > 0$ the function $u \mid [0, T]$ extended periodically to $\mathbb{R}$ is denoted by $[u]_T$.

Proposition 5. *Let $\lambda \mapsto T_\lambda : [0, 1] \to \mathbb{R}$ be continuous. Then, for a (fixed) control function $u \in \mathcal{U}$, the map $\lambda \mapsto u_\lambda := \lambda\, [u]_{T_\lambda} : [0, 1] \to \mathcal{U}$ is continuous.*

The next fact is crucial for the construction of the homotopy between the orbits that wind n times around γ and $[T, 0, x]^n$.

Lemma 2. *Assume that the T-periodic orbit $\gamma = \varphi([0,T], x_0, 0)$ is attracting, and let S be a local transversal section for the orbit γ through x_0. Then there exists $\rho > 0$ and a neighborhood V of x_0 such that $P(\cdot, u)$ is a contraction, uniformly for $u \in \mathcal{U}^\rho$.*

Proof. Without loss of generality we can assume that S lies on the hyperplane $X := \{x^d = 0\} \subset \mathbb{R}^d$. Here and along all this proof the exponent d denotes the d-th component in $\mathbb{R}^d$. Take

$$\mathcal{F} := \mathcal{U}^\rho \cap C^1(\mathbb{R}, \mathbb{R}^m)$$

with the L_∞ topology, and define for $u \in \mathcal{F}$ the C^1 function $\Psi_u : \mathbb{R} \times X \to \mathbb{R}$ as

$$\Psi_u(t, y) := \varphi^d(t + \tau_u, y, u) - y^d,$$

where $\tau_u = \tau(x_0, u)$. One has $\Psi_u(0, x_0) = 0$. Observe that Ψ_u is continuous, uniformly for $u \in \mathcal{F}$. Moreover, reducing ρ if necessary, one can find $\delta > 0$ such that

$$D_1\Psi_u(0, x_0) = f^d(\varphi(\tau_u, x_0, u), u(\tau_u)) \geq 1/\delta.$$

A parametrized version of the implicit function theorem implies the existence of a neighborhood V of x_0 in S and of a C^1 function $t_u : V \to \mathbb{R}$ such that $\Psi_u(t_u(x), x) = 0$ for every $x \in V$.

Clearly, if ρ is small enough and $x \in V$, then the time $\tau(x, u)$ for the Poincaré map coincides with $t_u(x) + \tau_u$. Thus $D_1\tau(x, u)$ is well defined and the map

$$(x, u) \mapsto D_1 P(x, u) = D_1\varphi(\tau(x, u), x, u) D_1\tau(x, u) + D_2\varphi(\tau(x, u), x, u)$$

is continuous. Since γ is attracting, the eigenvalues of $D_1 P(x_0, 0)$ are strictly smaller than one in modulus. Thus there exists a norm on S such that the operator $D_1 P(x_0, 0)$ has norm smaller than one. By continuity and restricting V and ρ if necessary, we can assume that the same is true for $D_1 P(x, u)$ for every $x \in V$ and $u \in \mathcal{U}^\rho \cap C^1(\mathbb{R}, \mathbb{R}^m)$. Whence it follows that $P(\cdot, u)$ is a contraction with constant

$$k = \sup_{(\xi,\mu) \in V \times \mathcal{U}^\rho} \|D_1 P(\xi, \mu)\| < 1.$$

Let us show that $P(\cdot, u)$ remains a k-contraction when u is a general (not necessarily continuously differentiable) element of $\mathcal{U}^\rho$. Since the C^1 functions are dense in $\mathcal{U}^\rho$ in the weak* topology, there is a sequence $\{u_n\}$ of C^1 functions in $\mathcal{U}^\rho$ converging to u_0 in the weak* topology. Take x and y in V, by

Proposition 4 we know that P is continuous when $\mathcal{U}^\rho$ is endowed with the weak* topology. Therefore, for $\varepsilon > 0$, one has

$$|P(x,u) - P(x,u_n)| + |P(y,u) - P(y,u_n)| < \varepsilon,$$

for n sufficiently large. Therefore

$$\begin{aligned} |P(x,u) - P(y,u)| &\leq |P(x,u) - P(x,u_n)| + |P(x,u_n) - P(y,u_n)| \\ &\quad + |P(y,u_n) - P(y,u)| \\ &\leq k|x-y| + \varepsilon. \end{aligned}$$

Since $\varepsilon > 0$ is arbitrary, this proves the assertion.

Proposition 6. *Assume that the T-periodic orbit $\gamma = \varphi([0,T],x_0,0)$ is attracting, and let S be a local transversal section for γ through x_0. Then there exists $\rho > 0$ and a neighborhood V of x_0 such that for every $n \in \mathbb{N}$ and every $(x,u) \in V \times \mathcal{U}^\rho$ the map $(x,u) \to P^n(x,u)$ is well defined.*

Moreover, for every $u \in \mathcal{U}^\rho$, there exists a $T_u > 0$ and a unique T_u-periodic solution $\varphi(\cdot, x_u, [u]_{T_u})$ winding n times around γ, and the functions $u \mapsto T_u$ and $u \mapsto x_u$ are continuous.

Proof. In Lemma 2 we proved that for $u \in \mathcal{U}^\rho$, with ρ sufficiently small, $P(\cdot,u)$ is a contraction on cl V. Consequently, we can assume that $P(\text{cl } V, u) \subset \text{cl } V$ for any $u \in \mathcal{U}^\rho$. In particular $P^n(\cdot,u)$ is well-defined for any $n \in \mathbb{N}$ and $u \in \mathcal{U}^\rho$. Notice also that for every $n \in \mathbb{N}$ and $u \in \mathcal{U}^\rho$ also $P^n(\cdot,u)$ is a contraction. Therefore, given n and u as above, there exist a unique fixed point x_u for $P^n(\cdot,u)$ in N which depends continuously on u. Define T_u as the time needed for $\varphi(\cdot,x_u,u)$ to reach x_u after winding n times around γ. Continuous dependence of T_u on u follows from continuity of the first return time as shown in Proposition 4.

Notice that in $P^n(x,u)$ the control u restricted to $[0,T_u]$ is applied n times. When the control function is identically zero and V is small enough, the sets of fixed points of $P(\cdot,0)$ and of $P^n(\cdot,0)$ reduce to x_0. However, this needs not be true when nontrivial controls are applied. Indeed, when the controllability condition (3) is satisfied one can for $n > 1$ construct a (small) control u which yields a periodic trajectory meeting the transversal section in n distinct points. (This can be deduced from Proposition 3.) Thus, for such u, the fixed point set of $P^n(\cdot,u)$ strictly contains that of $P(\cdot,u)$.

We are finally in a position to prove the claim we made at the beginning of this section.

Theorem 1. *Assume that the uncontrolled system has an attracting T-periodic solution $\varphi(\cdot,x_0,0)$ with $T > 0$, and that the controllability condition (3) is satisfied. Then, when ρ is small enough, the dynamic index $\mathcal{I}(D^\rho)$ of the control set D^ρ containing $\gamma := \varphi([0,T],x_0,0)$ is isomorphic to $\mathbb{N}$.*

Proof. Let $N = \mathrm{cl}\, V$ be the compact neighborhood of x_0 found in the proof of Proposition 6 above. Consider a T_1-periodic orbit $\varphi(\cdot, x, u)$ with $x \in N$, $u \in \mathcal{U}^{\rho'}$ for some $0 < \rho' < \rho$ and u piecewise constant. There exists n such that $\varphi(T_1, x, u) = P^n(x, u)$. By Proposition 6, there exist $T_\lambda > 0$ and a unique T_λ-periodic solution $\varphi(\cdot, x_\lambda, [\lambda u]_{T_\lambda})$ winding n times around γ. By Proposition 5 the map $\lambda \mapsto u_\lambda := [\lambda u]_{T_\lambda}$ is continuous. Hence, again by Proposition 6, it follows that T_λ and x_λ depend continuously on λ. In particular, $T_0 = nT$. Since by Proposition 1 $([v]_{T_\lambda}, x_\lambda)$ is a strong inner pair for each λ, this yields the desired homotopy between (T_1, u_1, x_1) and $(T_0, 0, x_0)$.

We conclude the paper with a remark showing that the dynamic index allows us to distinguish control sets around an attracting periodic orbit as above from control sets around a homoclinic orbit.

Remark 2. Suppose that the uncontrolled system has a homoclinic orbit given by

$$\varphi(t, x_1, u_1),\ t \in \mathbb{R},\ \text{with} \lim_{t \to \pm\infty} \varphi(t, x_1, u_1) = x_0,$$

where x_0 is an equilibrium of the uncontrolled system. If the controllability condition (3) holds for all points in $\gamma := \{x_0\} \cup \{\varphi(t, x_1, u_1),\ t \in \mathbb{R}\}$ and this is a chain recurrent component of the uncontrolled system, then for every $\rho > 0$ there is a control set D^ρ containing this set in its interior and

$$\bigcap_{\rho > 0} D^\rho = \gamma;$$

see Corollary 4.7.6 in [1]. For any small ρ, the index $\mathcal{I}(D^\rho)$ contains an element $[T, x_0, 0]$ which is idempotent, i.e., $[T, x_0, 0]^2 = [T, x_0, 0]$. Hence $\mathcal{I}(D^\rho)$ is not isomorphic to $\mathbb{N}$.

References

1. Colonius, F., Kliemann, W. (2000) The Dynamics of Control. Birkhäuser, Boston
2. Colonius, F., Sieveking, M. (1989) Asymptotic properties of optimal solutions in planar discounted control problems. SIAM J. Control and Optimization, **27**, 608–630
3. F. Colonius, F., Spadini, M. (1999) Uniqueness of Control Sets for Perturbations of Linear Systems. In: Stability and Stabilization of Nonlinear Systems, D. Aeyels, F. Lamnabhi-Lagarrigue and A. van der Schaft, eds., Lecture Notes in Control and Information Sciences, **246**, Springer Verlag
4. Colonius, F., Spadini, M. (2001) Local control sets. submitted
5. Colonius, F., Spadini, M. (2001) A dynamic index for control sets. submitted
6. Dunford, N. Schwartz, J. T. (1977) Linear Operators, Part I: General Theory. Wiley-Interscience

7. Howie, J. M. (1976) An Introduction to Semigroups Theory. Academic Press, London
8. Katok, A., Hasselblatt, B. (1995) Introduction to the Modern Theory of Dynamical Systems. Cambridge University Press, Cambridge
9. Krener, A. J. (1974) A generalization of Chow's theorem and Bang-Bang theorem for nonlinear control problems. SIAM J. Control Optim., **14**, 43–52
10. Robinson, C. (1995) Dynamical Systems: Stability, Symbolic Dynamics, and Chaos. CRC Press
11. San Martin, L. A., Santana, A. J. (1999) The homotopy type of Lie semigroups in semisimple Lie groups. Instituto de Matematica, Universidade Estadual de Campinas, Brasil
12. Sarychev, A. V. (1991) On topological properties of trajectory space for nonlinear control systems. In: Analysis of Nonlinear Controlled Dynamical Systems, B. Bride, B. Bonnard, G.P. Gauthier, I. Kupka, eds., Birkhäuser, 358–371
13. Sarychev, A. V. (1990) On homotopy properties of trajectories for completely nonholonomic differential systems. Doklady Akad. Nauk SSSR **314**, 1336–1340. English translation (1991) Doklady **42**, 674–678

On the Frequency Theorem for Nonperiodic Systems

Roberta Fabbri[1], Russell Johnson[1], and Carmen Núñez[2]

[1] Università degli Studi di Firenze, Dipartimento di Sistemi e Informatica Via Santa Marta 3, 50139 Firenze, Italy
[2] Universidad de Valladolid, Departamento de Matemática Aplicada a la Ingeniería, Paseo del Cauce s/n, 47011 Valladolid, Spain

Abstract. Using methods of the theory of nonautonomous linear differential systems, namely exponential dichotomies and rotation numbers, we generalize some aspects of Yakubovich's Frequency Theorem from periodic control systems to systems with bounded measurable coefficients.

1 Introduction

In papers published in 1986 and 1990, Yakubovich ([17], [18]) studied control processes with periodic coefficients of the form

$$\begin{aligned} &x' = A(t)x + B(t)u \qquad x \in \mathbb{R}^n, u \in R^m \\ &x(0) = x_0 \end{aligned} \tag{1}$$

together with a quadratic functional

$$I = \int_0^\infty \{x^* G(t)x + 2x^* g(t)u + u^* R(t)u\}\, dt,$$

where $G(t) = G^*(t)$ and $R(t) = R^*(t)$ are continuous T-periodic symmetric matrix-valued functions. The integrand is not assumed to be positive definite; one does assume that $R(t) > 0$ for all $t \in \mathbb{R}$, however, $G(t)$ may have a negative part. A condition of L^2- stabilizability is assumed; thus for each $x_0 \in \mathbb{R}^n$ there exists $u \in L^2[0, \infty)$ such that the corresponding solution of (1) lies in $L^2[0, \infty)$.

Yakubovich proves that, in the periodic case, the solvability of this problem is equivalent to several other conditions of a classical nature (existence of a Lyapunov function, existence of a stabilizing feedback control, ...). But he also proves that it is equivalent to the truth of a Frequency Condition and a Nonoscillation Condition. It would seem that, in stating these conditions, Yakubovich penetrates to the essential elements of the structure of the problem of minimizing I subject to (1).

Our goal here is to state how the Frequency Condition and the Nonoscillation Condition can be generalized to the case when the coefficients in (1) and in the integrand of I are bounded measurable functions. We will use the

F. Colonius, L. Grüne (Eds.): Dynamics, Bifurcations, and Control, LNCIS 273, pp. 233–240, 2002.

concepts of exponential dichotomy and rotation number from the theory of nonautonomous differential equations. We will then generalize Yakubovich's main results from the periodic case to the case of bounded coefficients.

The first step is to pass in a standard way to the Hamiltonian formulation of the minimization problem. Let

$$\mathcal{L}(t,x,u) = \frac{1}{2}[x^*G(t)x + 2x^*g(t)u + u^*R(t)u]$$

$$\mathcal{H}(t,x,u) = \sum_{i=1}^{n} y_i x_i' - \mathcal{L}(t,x,u).$$

Using the Pontryagin Maximal Principle one proves that a minimizing control must satisfy

$$u = R^{-1}(t)[B^*(t)y - g^*(t)x],$$

and finally arrives at the linear Hamiltonian system

$$Jz' = \begin{pmatrix} -Q & A^* - gR^{-1}B^* \\ A - BR^{-1}g^* & BR^{-1}B^* \end{pmatrix} z. \tag{2}$$

Here we have written $z = \begin{pmatrix} x \\ y \end{pmatrix}$, $J = \begin{pmatrix} 0 & -I \\ I & 0 \end{pmatrix}$, and $Q = G - gR^{-1}g^*$.

Yakubovich formulates his Frequency Condition in the following way. Let $Z(T)$ be the period matrix of (2), and let $\omega \in [0, 2\pi)$; then

$$det(Z(T) - e^{i\omega} I) \neq 0. \tag{*}$$

It is immediate that this condition is equivalent to the following assertion: equation (2) admits no nontrivial solutions $z(t)$ which are bounded on $\mathbb{R}$.

He also formulates a basic Nonoscillation Condition in the following way. First note that the condition (*) states that the period matrix of the $2n$-dimensional Hamiltonian system (2) has no eigenvalues on the unit circle. Hence there exist n linearly independent solutions $z_1(t), \ldots, z_n(t)$ of (2) which decay exponentially as $t \to \infty$. Write the $2n \times n$ matrix formed by these solutions as follows:

$$(z_1(t) \ldots z_n(t)) = \begin{pmatrix} X(t) \\ Y(t) \end{pmatrix}.$$

Then the Nonoscillation Condition states that $det X(t) \neq 0$ for all $t \in \mathbb{R}$.

Yakubovich states a theorem to the effect that, in the periodic case, the following conditions (A)-(E) are equivalent. The hypothesis of L^2-stabilizability is assumed to hold.

(A) For each initial condition $x_0 \in \mathbb{R}^n$, the minimization problem is solvable: there exists a control $u(\cdot) \in L^2([0,\infty), \mathbb{R}^m)$ for which the corresponding solution $x(\cdot)$ of (1) is square-integrable and so that I takes on its minimum value when $u(\cdot), x(\cdot)$ are substituted in $\mathcal{L}$.

(B) The Frequency Condition and the Nonoscillation Condition are valid.

(C) There is a T-periodic matrix function $\Gamma(t) = \Gamma^*(t)$ so that, for some $\delta > 0$, the "Lyapunov function" $V(t) = x^*\Gamma(t)x$ satisfies

$$\begin{aligned} \frac{dV}{dt} &= x^*\frac{d\Gamma}{dt}x + 2x^*\Gamma(t)[A(t)x + B(t)u] \geq \\ &\geq -2\mathcal{L}(t,x,u) + \delta(|x|^2 + |u|^2) \end{aligned}$$

for (say) continuous functions $u : [0,\infty) \to \mathbb{R}^m$.

(D) There is a T-periodic matrix function $m(t)$, satisfying a Riccati equation, so that if

$$u = -R^{-1}(t)[B^*(t)m(t) + g^*(t)]x,$$

then the solution of (1) with $x(0) = x_0$ goes to zero exponentially fast.

(E) There exists $\delta > 0$ so that, if $\mathcal{M}_0$ is the set of processes $x(.) \in L^2[0,\infty)$, $u(.) \in L^2[0,\infty)$ satisfying (1) with $x(0) = 0$, then

$$\int_0^\infty \mathcal{L}(t,x(t),u(t))dt \geq \delta \int_0^\infty (|x(t)|^2 + |u(t)|^2)dt.$$

for all $(x,u) \in \mathcal{M}_0$.

It turns out that the theorem stating the equivalence of (A)–(E) can be formulated and proved in the context of non-autonomous control processes with bounded measurable coefficients. The main concepts used for this are those of exponential dichotomy and rotation number.

In the remaining two sections we introduce a framework in which the problem of minimizing I subject to (1) can be studied using methods of topological dynamics and ergodic theory (§2). We discuss basic facts about exponential dichotomies and the rotation number for linear Hamiltonian systems. Then in §3 we generalize some of Yakubovich's results.

2 Nonautonomous Hamiltonian systems

Suppose that A, B, G, g, R are bounded measurable matrix-valued functions defined on $\mathbb{R}$ and let $\gamma = \max\{\|A\|_\infty, \dots, \|R\|_\infty\}$. Let M_{kl} denote the set of $k \times l$ real matrices for positive integers k, l. Let B_γ be the closed ball of radius γ in $L^\infty(\mathbb{R}, M_{nn} \times M_{nm} \times M_{nn} \times M_{nm} \times M_{mm})$; here if $\omega = (A, B, G, g, R)$ is an element of this space, we put $\|\omega\| = \max\{\|A\|_\infty, \dots, \|R\|_\infty\}$. Then B_γ is compact in the $w - *$ topology, and it is invariant with respect to the translation flow $\{T_t \mid t \in \mathbb{R}\}$ given by $(T_t\omega)(s) = \omega(t+s)$. This flow was introduced and studied by Bebutov [3].

Now let $\omega_0 \in B_\gamma$ be the point defined by the data in (1) and in I. Let $\Omega = cls\{T_t(\omega_0) \mid t \in \mathbb{R}\} \subset B_\gamma$. Then (Ω, T_t) is a continuous real flow. If we let

ω vary over Ω, we obtain a family of Hamiltonian systems $(2)_\omega$, parametrized by $\omega \in \Omega$.

Somewhat more precisely, for $\omega \in \Omega$ we can write the coefficient matrix in (2) as

$$H_\omega(t) = \begin{pmatrix} -Q_\omega & A_\omega^* - g_\omega R_\omega^{-1} B_\omega^* \\ A_\omega - B_\omega R_\omega^{-1} g_\omega^* & B_\omega R_\omega^{-1} B_\omega^* \end{pmatrix} (t),$$

where it is understood that $A_\omega(.), B_\omega(.), G_\omega(.), g_\omega(.)$, and $R_\omega(.)$ are the bounded measurable functions determined by the components of ω. Also $Q_\omega = G_\omega - g_\omega R_\omega^{-1} g_\omega^*$. Thus the family is

$$Jz' = H_\omega(t)z, \quad (\omega \in \Omega) \tag{2_ω}$$

Definition 1. The equations $(2)_\omega$ are said to have an exponential dichotomy over Ω if there are constants $k > 0, \delta > 0$ and a continuous function $\omega \to P_\omega$ from Ω to the set $\mathcal{P}$ of linear projections on $\mathbb{R}^{2n}$ so that

$$\|\Phi_\omega(t) P_\omega \Phi_\omega(s)^{-1}\| \le ke^{-\delta(t-s)} \quad (t \ge s)$$
$$\|\Phi_\omega(t)(I - P_\omega \Phi_\omega)(s)^{-1}\| \le ke^{\delta(t-s)} \quad (t \le s).$$

Here $\Phi_\omega(t)$ is the principal matrix solution of $(2)_\omega$. We will sometimes use the abbreviation ED to refer to exponential dichotomy.

Theorem 1. *Suppose that all equations $(2)_\omega$ have the property that the only solution $z(t)$ of $(2)_\omega$ which is bounded on all of $\mathbb{R}$ is the identically zero solution. Then equations $(2)_\omega$ have an ED over Ω.*

We indicate the main steps in the proof of Theorem (1). Let $M \subset \Omega$ be a minimal subset; that is, for each $\omega \in M$, the trajectory $\{T_t(\omega) \mid t \in \mathbb{R}\}$ is contained in M and is dense in M. By the main Theorem of the paper [12] of Sacker and Sell, the hypothesis implies that equations $(2)_\omega$ have an ED over M for each minimal subset $M \subset \Omega$. See also Selgrade [14]. Now, one can show that for Hamiltonian linear systems of dimension $2n$, the existence of an exponential dichotomy implies that each projection P_ω must have rank n; see, e.g., ([7], Lemma 3.3) or ([11], Lemma 2.3.) The constancy of the rank now implies, via another theorem of Sacker and Sell ([13],Theorem 6) that ED must hold over all of Ω.

Remark 1. We see from Theorem 1 that a natural generalization of Yakubovich's Frequency Condition is the requirement that, for each $\omega \in \Omega$, the only bounded solution of equation $(2)_\omega$ is the trivial solution.

We next discuss the concept of rotation number for equation $(2)_\omega$. This will eventually allow us to connect the generalized version of nonoscillation with standard controllability conditions.

There are three ways to define the rotation number. One makes use of the argument functions on the symplectic group $Sp(n,\mathbb{R})$ of Yakubovich ([15],[16]). This approach has been developed by Novo-Núñez-Obaya [11]. Another one uses facts about the topological structure of the set Λ of Lagrange planes in $\mathbb{R}^{2n}$. Still another, related to the previous one, uses an argument function on Λ introduced by V.Arnold ([1]).

One of Yakubovich's argument functions on $Sp(n,\mathbb{R})$ is the following ([15],[16]). If $V \in Sp(n,\mathbb{R})$, write $\begin{pmatrix} V_1 & V_3 \\ V_2 & V_4 \end{pmatrix}$ and let $ArgV = arg\, det(V_1 + iV_3)$ where we represent by arg the usual argument of a complex number. If $\omega \in \Omega$, then the principal matrix $\Phi_\omega(t)$ of $(2)_\omega$ is in $Sp(n,\mathbb{R})$; define the rotation number α by

$$\alpha = -\lim_{t\to\infty} \frac{1}{t} Arg\Phi_\omega(t). \tag{3}$$

This definition is the starting point of Novo-Núñez-Obaya.

A second definition takes as its point of departure the set $\Lambda = \{l\}$ of Lagrange planes in $\mathbb{R}^{2n}$. Say that a vector subspace $l \subset \mathbb{R}^{2n}$ is a Lagrange plane if $dim\, l = n$ and if, whenever $z_1, z_2 \in l$, there holds $\langle z_1, Jz_2\rangle = 0$. The set Λ is a smooth manifold of dimension $\frac{n(n+1)}{2}$. It contains a $\mathbb{Z}_2$-cycle C of codimension 1 called the (vertical) Maslov cycle. This cycle C is defined as follows. The subspace $l_0 = \left\{ \begin{pmatrix} 0 \\ y \end{pmatrix} \mid y \in \mathbb{R}^n \right\} \subset \mathbb{R}^{2n}$ is a Lagrange plane; define $C = \{l \mid dim(l \cap l_0) \geq 1\}$. The complement $\Lambda \setminus C$ is homeomorphic to the Euclidean space $\mathbb{R}^{\frac{n(n+1)}{2}}$.

Fix $l \in \Lambda \setminus C$, and note that $c_T(t) = \Phi_\omega(t)l$ $(0 \leq t \leq T)$ defines a curve in Λ for fixed $\omega \in \Omega$. One can close c_T by first moving its terminal point $c_T(T)$ off C if necessary, then sliding it through $\Lambda \setminus C$ until it meets $c_T(0)$. Now, the cycle C is two-sided, and one can speak of the oriented number $N(c)$ of points of intersection of a closed curve c in Λ with C. Define

$$\alpha = -\lim_{T\to\infty} \frac{1}{T} N(c_T). \tag{4}$$

This definition is discussed in [7]. A third definition uses facts about Λ presented in [1]; we forego the discussion of this and refer again to [7].

It is not clear that the limits in (3) and (4) exist, nor that they are equal. To discuss the sense in which they exist (and are then equal), we need to fix an ergodic measure μ on Ω.

Definition 2. A regular Borel probability measure μ on Ω is called $\{T_t\}$-invariant if for each Borel subset $B \subset \Omega$ one has $\mu(T_t(B)) = \mu(B)$ $(t \in \mathbb{R})$. An invariant measure μ is called ergodic if in addition, whenever $\mu(T_t(B)\Delta B) = 0$ for all $t \in \mathbb{R}$, the either $\mu(B) = 0$ or $\mu(B) = 1$. Here Δ indicates the symmetric difference of sets.

The second condition above is a type of indecomposibility criterion. A theorem of Krylov-Bogoliubov [10] states that the flow $(\Omega, \{T_t\})$ supports at least one ergodic measure μ.

Theorem 2. *Let μ be an ergodic measure on Ω. Then for μ-a.a. $\omega \in \Omega$ the limits in* (3) *and* (4) *exist, are equal, and do not depend on the choices made in the definition* (4).

Núñez understood how to unify the various definitions of the rotation number [5].

There is a relation between the rotation number and the exponential dichotomy concept. To explain it, let K be a continuous function on Ω with values in the set of $2n \times 2n$ real symmetric matrices. Suppose that $K(\omega) \geq 0$ for each $\omega \in \Omega$, and that

$$\int_{-\infty}^{\infty} \Phi_\omega(t)^* K(T_t)(\omega)\Phi_\omega(t)dt > 0$$

for each $\omega \in \Omega$. This is a type of Atkinson condition [2]. Introduce an Atkinson-type spectral problem of the form

$$Jz' = [H_\omega(t) + \lambda K(T_t(\omega))]z.$$

If μ is a fixed ergodic measure on Ω, then the rotation number becomes a function $\alpha(\lambda)$ of the real parameter λ. The following result is proved in [9]. The proof depends on a trace formula proved in ([7], [6]).

Theorem 3. *Suppose that Ω is the topological support of an ergodic measure μ. Suppose that $\alpha(\lambda) = const$ in an open interval $(\lambda_1, \lambda_2) \subset \mathbb{R}$. Then the Atkinson equation admits an exponential dichotomy for each $\lambda_1 < \lambda < \lambda_2$.*

3 Generalization of Yakubovich's theorem

Let us return to the condition (A)-(E), proved to be equivalent by Yakubovich when the functions A, B, G, g and R are T-periodic. Let us first discuss an appropriate formulation of Condition (B) in the non-autonomous case.

First, it is natural to redefine the Frequency Condition to state that no equation $(2)_\omega$ admits a non-trivial bounded solution. Then by Theorem 1, equations $(2)_\omega$ admit an ED over Ω.

Second, as regards the Nonoscillation Condition, a natural generalization is obtained as follows. For each $\omega \in \Omega$ let P_ω be the projection of the definition of ED. Let $l_\omega \subset \mathbb{R}^{2n}$ be the image of P_ω; then l_ω is a Lagrange plane in $\mathbb{R}^{2n}$. We reexpress the Nonoscillation Condition as

$$l_\omega \notin C, \text{ the vertical Maslov cycle } (\omega \in \Omega).$$

We have now reformulated Condition (B). We assume from now on that, for each $\omega \in \Omega$, the equation $(1)_\omega$ defined by $A_\omega(t)$ and $B_\omega(t)$ is L^2-stabilizable. Then it is straightforward, using Yakubovich's arguments, to prove the following

Fact 4 *One has (A) $\Leftrightarrow$ (B) $\Leftrightarrow$ (C) $\Leftrightarrow$ (D) $\Leftrightarrow$ (E) just as in the periodic case, with natural modifications. In particular, for each $\omega \in \Omega$, the matrix functions $\Gamma(t)$ and $m(t)$ are obtained as restrictions of continuous matrix-valued functions defined on Ω to the trajectory $\{T_t(\omega) \mid t \in \mathbb{R}\}$. This is a "conservation of recurrence" phenomenon which reduces to T-periodicity of Γ and m if A, B, G, g and R are T-periodic.*

The fact just stated makes no reference to controllability conditions. One would like to relate this result to others which involve controllability conditions. We state two such results; the proofs will be given elsewhere [4].

Theorem 5. *Assume that the Frequency Condition holds and that, for each ergodic measure μ on Ω, the rotation number satisfies $\alpha = 0$. Assume further that (1) is null controllable for at least one point in each minimal subset $M \subset \Omega$. Then the condition (B) holds. As before, one has the equivalences (A) $\Leftrightarrow$ (B) $\Leftrightarrow$ (C) $\Leftrightarrow$ (D) $\Leftrightarrow$ (E).*

Yakubovich makes an observation equivalent to the fact that, in the constant coefficient case, the condition $\alpha = 0$ follows from the Frequency Condition, i.e., from the non-existence of non-trivial bounded solution of (2). He also points out that this is no longer true in the periodic case because of the existence of infinitely many stability zones for linear periodic Hamiltonian systems. However, it is sometimes true again in the non-periodic case that the Frequency Condition implies that $\alpha = 0$. A sufficient condition for this is that Ω be the topological support of an ergodic measure μ and that the first integral Čech cohomology group $\check{H}^1(\Omega, \mathbb{Z}) = \{0\}$. See ([7],[5]).

Finally we give a sufficient condition for the simultaneous validity of the Frequency Condition and of the condition $\alpha = 0$. For this, let $\{\hat{Q}_\omega(t) \mid \omega \in \Omega\}$ be a family of functions with values in the set of $n \times n$ symmetric matrices such that $\hat{Q}_\omega$ is positive definite $(\omega \in \Omega, t \in \mathbb{R})$.

Theorem 6. *Suppose that* (1) *is null controllable on minimal sets $M \subset \Omega$, and observe further that*

$$y' = -A^*_\omega(t)y + \hat{Q}_\omega(t)u$$

is null controllable on minimal sets. Let η be a positive real number, and consider the Hamiltonian equations obtained from $(2)_\omega$ by replacing Q_ω by $Q_\omega + \eta\hat{Q}_\omega$ and $B_\omega {R_\omega}^{-1} {B_\omega}^*$ by $(1+\eta)B_\omega {R_\omega}^{-1} {B_\omega}^*$. Call the new equations $(2)_{\omega,\eta}$. If for some $\eta > 0$, the rotation number α is 0 for all ergodic measures μ on Ω, then the Frequency Condition and the Nonoscillation Condition hold for equation $(2)_\omega$. Hence Condition (B) is verified; also one has the equivalences (A) $\Leftrightarrow$ (B) $\Leftrightarrow$ (C) $\Leftrightarrow$ (D) $\Leftrightarrow$ (E).

The proof of this result uses Theorem 3 and certain arguments in [9]. If Q_ω is negative semi-definite for all $\omega \in \Omega$ (the "worst case"), then it turns out that one can take $\hat{Q}_\omega = -Q_\omega$ in Theorem 6

References

1. Arnold, V.I. (1969) On a characteristic class entering in a quantum condition. Funk. Anal. Appl, **1**, 1–14.
2. Atkinson, F. (1964) Discrete and Continuous Boundary Value Problems. Academic Press, New York.
3. Bebutov, M. (1941) On dynamical systems in the space of continuous functions. Boll. Moskov. Univ. Matematica, 1–52.
4. Fabbri, R., Johnson, R., Núñez, C. On the Yakubovich frequency theorem for non-autonomous control processes, in preparation.
5. Fabbri, R., Johnson, R., Núñez, C. The rotation number for non-autonomous linear Hamiltonian systems I: basic properties, submitted.
6. Fabbri, R., Johnson, R., Núñez, C. The rotation number for non-autonomous linear Hamiltonian systems II: the Floquet coefficient, submitted.
7. Johnson, R.(1987) m-functions and Floquet exponents for linear differential systems. Ann. Mat. Pura Appl., **147**, 211–248.
8. Johnson, R., Nerurkar, M. (1994) Exponential dichotomy and rotation number for linear Hamiltonian systems. Jour. Diff. Eqns., **108**, 201–216.
9. Johnson, R., Nerurkar, M. (1996) Stabilization and random linear regulator problem for random linear control processes. Jour. Math. Anal. Appl., **197**, 608–629.
10. Nemytskii, V., Stepanov, V. (1960) Qualitative Theory of Differential Equations. Princeton Univ. Press, Princeton.
11. Novo, S., Núñez, C., Obaya, R. (1998) Ergodic properties and rotation number for linear Hamiltonian systems. Jour. Diff. Eqns, **148**, 148–185.
12. Sacker, R.J., Sell, G.R.(1974) Dichotomies and invariant splittings for linear differential equations I. Jour. Diff. Eqns., **15**, 429–458.
13. Sacker, R.J., Sell, G.R. (1978) A spectral theory for linear differential systems. Jour. Diff. Eqns., **27**, 320–358.
14. Selgrade, J. (1975) Isolated invariant sets for flows on vector bundles. Trans. Amer. Math. Soc., **203**, 359–390.
15. Yakubovich, V. (1961) Arguments on the group of symplectic matrices. Math. Sb., **55**, 255–280 (Russian).
16. Yakubovich, V. (1964) Oscillatory properties of the solutions of canonical equations. Amer. Math. Soc. Transl. Ser. 2, **42**, 247–288.
17. Yakubovich, V. (1986) Linear-quadratic optimization problem and the frequency theorem for periodic systems I. Siber. Math. Jour., **27**, 614–630.
18. Yakubovich, V. (1990) Linear-quadratic optimization problem and the frequency theorem for periodic systems II. Siber. Math. Jour. **31**, 1027–1039.

Longtime Dynamics in Adaptive Gain Control Systems

Gennady A. Leonov[1] and Klaus R. Schneider[2]

[1] St. Petersburg State University, Department of Mathematics and Mechanics, Petrodvoretz, Bibliotechnaya pl. 2, 198904 St. Petersburg, Russia
[2] Weierstrass Institute for Applied Analysis and Stochastics, Mohrenstraße 39, D–10117 Berlin, Germany

Abstract. We study the longtime dynamics of a nonlinear adaptive control system introduced by Mareels et al. [10] to control the behavior of a plant which can be described by a finite dimensional SISO linear time invariant system stabilizable by a high gain output feedback. We apply frequency domain methods to derive conditions for global stability, to approximate the region containing the global attractor and to estimate its Hausdorff dimension.

1 Introduction

Adaptive output gain control has been considered by I. Mareels [8], A.S. Morse [11], C.I. Byrnes and J.C. Willems [1], A. Ilchmann [4], H. Kaufman, I. Bar–Kana and K. Sobel [5] and I. Mareels et al. [10] to name but a few. The goal of this paper is to study the longtime dynamics of a class of adaptive gain control systems considered in [10].

We assume that the plant to be controlled can be described by a finite dimensional single input single output linear time invariant system that can be stabilized by a high gain output feedback. Such systems have a transfer function with stable zeroes and relative degree one. As has been proved in [9,10], the class of systems under consideration can be transformed into the form

$$\begin{aligned}\frac{dx}{dt} &= Ax + by,\\ \frac{dy}{dt} &= -c^T x - dy + u,\end{aligned} \tag{1}$$

where u is the input, y the output, $(x, y) \in \mathrm{R}^n \times \mathrm{R}$ is the state of the system, A is an $n \times n$-matrix, i.e. $A \in L(\mathrm{R}^n, \mathrm{R}^n)$, $b, c \in \mathrm{R}^n$, $d \in \mathrm{R}$. In [10] the adaptive feedback law

$$\begin{aligned}u &= -zy + e,\\ \frac{dz}{dt} &= -\sigma z + y^2, \quad z(0) > 0\end{aligned} \tag{2}$$

F. Colonius, L. Grüne (Eds.): Dynamics, Bifurcations, and Control, LNCIS 273, pp. 241–254, 2002.

has been applied to (1). Here, σ is a positive constant representing the so-called sigma-modification, and e characterizes the control offset error. Substituting (2) into (1) we obtain after some rescaling

$$\begin{aligned} \frac{dx}{dt} &= Ax + by, \\ \frac{dy}{dt} &= -c^T x - dy - zy + e, \\ \frac{dz}{dt} &= -\sigma z + y^2, \quad z(0) > 0. \end{aligned} \tag{3}$$

Under the assumptions that (A, b) is controllable, A is Hurwitz, and $\sigma > 0$, it has been proved in [10] that system (3) is dissipative in the sense of Levinson, that is, every trajectory enters finally a uniformly bounded region G of the phase space, moreover an estimate of G and conditions for global stability has been derived. An essential aim of [10] was to show by a bifurcation analysis and by numerical investigations that, for $n = 1$, the longtime dynamics of system (3) can be very rich, including chaotic behavior. Therefore, from the point of control theory it is desirable to find conditions for (3) to be globally stable or to minimize the region G containing the global attractor.

The goal of this paper is to study the longtime dynamics of (3) by frequency methods. We derive estimates for the global attractor and give conditions for global asymptotic stability which improve corresponding results in [10] at least for the case $n = 1$, furthermore, we derive an upper bound for the Hausdorff dimension of the global attractor.

2 Assumptions and preliminaries

Throughout this paper we assume

(A_1). The matrix A is Hurwitz, that is, all eigenvalues of A are located in the left half plane.

(A_2). The pair (A, b) is controllable.

Since we are using frequency methods we have to introduce some transfer functions. First we introduce the function $\chi : \mathrm{C} \to \mathrm{C}$ by

$$\chi(s) := c^T (sI - A)^{-1} b \tag{4}$$

which is the transfer function of the input y to the output v of the system

$$\begin{aligned} \frac{dx}{dt} &= Ax + by, \\ v &= c^T x. \end{aligned}$$

By $\psi : \mathrm{C} \to \mathrm{C}$ we denote the transfer function of system (1) which can be represented in the form

$$\psi(s) := \frac{1}{s + d + \chi(s)}. \tag{5}$$

Using the notation

$$\eta(s) := det(sI - A), \quad p(s) := det\begin{pmatrix} sI - A & -b \\ c & s \end{pmatrix} \tag{6}$$

$\psi(s)$ can be represented also in the form

$$\psi(s) = \frac{\eta(s)}{p(s) + d\,\eta(s)}. \tag{7}$$

The investigation of the longtime behavior of system (3) will be based on the construction of appropriate Lyapunov functions. An essential part of these functions is some quadratic form defined by means of a symmetric positive definite matrix H. For the existence and also for the construction of H we use frequency domain methods, in particular, we will apply results of V.A. Yakubovich, R.E. Kalman and V.M. Popov. For convenience of the reader we recall these results, also a theorem due to A. Douady and J. Oesterlé that will be used to estimate the Hausdorff dimension of the global attractor.

The following result represents a version of the Yakubovich - Kalman frequency domain theorem (see Theorem 1.10.1 in [7]).

Theorem 1. *Let $P \in L(\mathrm{R}^n, \mathrm{R}^n)$ be Hurwitz, let $q, \delta \in \mathrm{R}^n$, let $g \in \mathrm{R}$. We assume (P, q) to be controllable, and (P, δ) to be observable. Let $\mathcal{G}(\xi, \eta)$ be the Hermitian form defined by*

$$\mathcal{G}(\xi, \eta) := 2Re\, \xi^* \delta\eta + g\,|\eta|^2, \quad \xi \in \mathrm{C}^n,\ \eta \in \mathrm{C}. \tag{8}$$

Then there is a positive definite symmetric matrix $H \in L(\mathrm{R}^n, \mathrm{R}^n)$ satisfying

$$2Re\ \xi^* H(P\xi + q\eta) + \mathcal{G}(\xi, \eta) \le 0 \quad \forall \xi \in \mathrm{C}^n, \quad \forall \eta \in \mathrm{C}$$

if and only if

$$Re\, \mathcal{G}\Big((i\omega I - P)^{-1} q\eta, \eta\Big) \le 0 \quad \forall \eta \in \mathrm{C}, \forall\, \omega \in \mathrm{R}.$$

The following result is basically an application of Theorem 1 (see Theorem 1.12.1 in [7]).

Theorem 2. *Let $P \in L(\mathrm{R}^n, \mathrm{R}^n)$ be Hurwitz, let $q, r \in \mathrm{R}^n$. We assume the pair (P, q) to be controllable, and the pair (P, r) to be observable. Let $\kappa : \mathrm{C} \to \mathrm{C}$ be the transfer function defined by*

$$\kappa(s) := r^T (P^T - sI) q. \tag{9}$$

Then there exists a positive definite symmetric matrix $H \in L(\mathrm{R}^n, \mathrm{R}^n)$ satisfying the relations

$$HP + P^T H \le 0 \quad and \quad Hq + r = 0$$

if and only if

$$Re\left[\kappa(i\omega)\right] > 0 \quad \forall \omega \in R.$$

The next result represents a special form of the criterion of Popov and coincides essentially with the circle criterion (see Theorem 1.14.1 in [7]).

Theorem 3. *Let the matrix $P \in L(\mathrm{R}^n, \mathrm{R}^n)$ be Hurwitz, let $r, q \in \mathrm{R}^n$, let the pair (P, q) be controllable. Suppose that for a certain number $\mu > 0$ the following inequality holds*

$$\mu^{-1} + Re\big[\zeta(i\omega)\big] > 0 \quad \forall \omega \in \mathrm{R},$$

where $\zeta(s) := r^T(P - sI)^{-1}q$. Then the system

$$\begin{aligned} \frac{dx}{dt} &= Px + qy, \quad \sigma = r^T x, \\ y &= \varphi(t, \sigma), \end{aligned} \tag{10}$$

where $\varphi : \mathrm{R} \times \mathrm{R} \to \mathrm{R}$ is continuous and such that

$$0 \le \frac{\varphi(t,\sigma)}{\sigma} \le \mu \quad \forall\, t, \sigma \in \mathrm{R},$$

is globally asymptotically stable.

It is well-known [3] that a dissipative autonomous system

$$\frac{dx}{dt} = f(x) \tag{11}$$

with $f \in C^1(\mathrm{R}^n, \mathrm{R}^n)$ has a global attractor K. Let $J(x)$ be the Jacobian of $f(x)$. The following theorem due to A. Douady and J. Oesterlé [2] aims to estimate the Hausdorff dimension of K by means of the eigenvalues $\lambda_1(x) \ge \cdots \ge \lambda_n(x)$ of the symmetric matrix

$$M(x) := \frac{1}{2}\Bigg[J(x) + J(x)^T\Bigg]. \tag{12}$$

It follows from a more general result (see Theorem 5.5.1 in [7]).

Theorem 4. *Assume $f \in C^1(\mathrm{R}^n, \mathrm{R}^n)$ and (11) to be dissipative. Let $\lambda_1(x) \ge \ldots \ge \lambda_n(x)$ be the eigenvalues of the symmetric matrix $M(x)$ defined in (12). Furthermore, we suppose that for $x \in G$, where G is an open bounded region in R^n, and for some $s \in [0, 1]$ and some j, $1 \le j < n$, the following inequality holds*

$$\lambda_1(x) + \cdots + \lambda_j(x) + s\lambda_{j+1}(x) < 0. \tag{13}$$

Then the Hausdorff dimension $\dim_H K$ of the global attractor K of system (11) can be estimated by

$$\dim_H K \le j + s.$$

Under some additional conditions, Theorem 4 yields a criterion for global stability (see Theorem 3.1.1 in [6]).

Theorem 5. *Suppose $f \in C^1(\mathrm{R}^n, \mathrm{R}^n)$ and that there exists a bounded region G with smooth boundary ∂G such that the trajectories of* (11) *transversally enter G for increasing t. Furthermore, we assume that G contains only a finite set of equilibria of* (11) *and that for all $x \in G$ the following inequality holds*

$$\lambda_1(x) + \lambda_2(x) < 0.$$

Then any solution $x(t, x_0)$ of system (11) *with initial data $x_0 \in G$ tends to some equilibrium as t tends to $+\infty$.*

In the next section we derive some estimates for the region where the global attractor K of (3) is located.

3 Localization of the global attractor

First we note that from the last equation in (3) we get

$$z(t) = e^{-\sigma t} z(0) + \int_0^t e^{-\sigma(t-\tau)} y^2(\tau) d\tau \geq e^{-\sigma t} z(0).$$

Thus, if the z-component of a solution of system (3) satisfies $z(0) > 0$ then $z(t) > 0$ holds for all $t \geq 0$. This implies

$$\liminf_{t \to \infty} \; z(t) \geq 0 \quad \text{for} \quad z(0) \geq 0. \tag{14}$$

Theorem 6. *Suppose the hypotheses* (A_1), (A_2) *and $\sigma > 0$ to be valid. Moreover, we assume the pair (A, c) to be observable and that for some numbers γ, ν, λ satisfying $\nu \geq 0, \;\; \lambda \in (0, \sigma], \gamma \in R$ the following relations hold*

(i) *all eigenvalues of the matrix $A + \lambda I$ have negative real parts.*
(ii) $\nu + Re\left[\chi(i\omega - \lambda)\right] > 0 \quad \forall\, \omega \in R.$
(iii)

$$(2\lambda - \sigma)\gamma \leq 0, \tag{15}$$

$$2(d - \nu - \lambda) - \gamma > 0. \tag{16}$$

Then there exists a positive definite symmetric matrix H such that the global attractor of system (3) *is contained in*

$$\Omega := \left\{ (x, y, z) \in \mathrm{R}^{n+2} : x^T H x + y^2 + z^2 + \gamma z \leq \frac{e^2}{2\lambda(2(d - \nu - \lambda) - \gamma)} \right\}.$$

Theorem 7. *Suppose the hypotheses of Theorem 6 are valid except condition* (15). *Then there exists a positive definite matrix H such that the global attractor of system* (3) *is contained in the set*

$$\Phi := \left\{ (x,y,z) \in \mathrm{R}^{n+2} : x^T H x + y^2 + z^2 + \gamma z \right.$$
$$\left. \leq \frac{1}{2\lambda}\left[\frac{e^2}{2(d-\nu-\lambda)-\gamma} + \frac{(2\lambda-\sigma)^2\gamma^2}{8(\sigma-\lambda)}\right] \right\}.$$

Theorem 8. *Suppose the hypotheses* (A_1),(A_2) *and $\sigma > 0$ to be valid. Additionally, we assume $e = 0$ and that for $\lambda = 0$ and for some numbers $\nu \geq 0$, $\gamma \in R$ the relations* (i) - (iii) *of Theorem* 6 *are valid. Then, any solution of* (3) *tends to the origin as t tends to $+\infty$.*

Proofs of Theorem 6 - Theorem 8.

First we prove Theorem 6. To this end we construct a Lyapunov function in the form

$$V(x,y,z) := x^T H x + y^2 + z^2 + \gamma z \tag{17}$$

where H is a real positive definite symmetric matrix with some special property. We will apply Theorem 1 to establish its existence. To this end we set $P = A + \lambda I, q = b, \delta = -c,\ g = -2\nu$. From (8) we get

$$\begin{aligned}\mathcal{G}((i\omega I - P)^{-1} q\eta, \eta) &= \mathcal{G}((i\omega - \lambda)I - A)^{-1} b\eta, \eta) \\ &= -2\Big(c^T\left((i\omega - \lambda)I - A\right)^{-1} b + \nu\Big)|\eta|^2.\end{aligned}$$

Taking into account the definition of the transfer function χ in (4) we obtain

$$-Re\left[\mathcal{G}((i\omega I - P)^{-1} q\eta, \eta)\right] = 2(Re\left[\chi(i\omega - \lambda)\right] + \nu)|\eta|^2.$$

Applying Theorem 1 we get that under the conditions (i) and (ii) of Theorem 6 there exists a positive symmetric matrix H satisfying

$$2x^T H[(A + \lambda I)x + by] - 2c^T xy - 2\nu y^2 \leq 0 \quad \forall\, x \in \mathrm{R}^n, \quad \forall\, y \in \mathrm{R}. \tag{18}$$

An algorithm to construct the matrix H satisfying (18) can be found in [7].

Using the inequality (18) we get from (17) and (3)

$$\begin{aligned}\frac{dV}{dt} + 2\lambda V = {} & 2x^T H[(A + \lambda I)x + by] - \\ & 2c^T xy - 2\nu y^2 + 2(\nu - d)y^2 + 2ey - 2\sigma z^2 \\ & -\sigma\gamma z + \gamma y^2 + 2\lambda(y^2 + z^2 + \gamma z) \\ & \leq -[2(d - \nu - \lambda) - \gamma]y^2 + 2ey \\ & -2(\sigma - \lambda)z^2 + (2\lambda - \sigma)\gamma z.\end{aligned}$$

From the validity of the relations (15) and (16) and taking into account (14) we obtain

$$\frac{dV}{dt} + 2\lambda V \leq \frac{e^2}{2(d - \nu - \lambda) - \gamma}. \tag{19}$$

Therefore, dV/dt is negative outside Ω, and the global attractor K is located in Ω. This proves Theorem 6.
In case that only the inequality (16) holds we have

$$\frac{dV}{dt} + 2\lambda V \leq \frac{e^2}{2(d - \nu - \lambda) - \gamma} + \frac{(2\lambda - \sigma)^2\gamma^2}{8(\sigma - \lambda)}. \tag{20}$$

This inequality implies the validity of Theorem 7.
In case $\lambda = 0,\ e = 0$ we have the inequality

$$\frac{dV}{dt} \leq -\Big[2(d - \nu) - \gamma\Big]y^2 - 2\sigma z^2.$$

From this inequality and from the relation

$$V(x, y, x) \longrightarrow +\infty \quad \text{as} \quad |x| + |y| + |z| \longrightarrow \infty$$

we get that any solution $(x(t), y(t), z(t))$ of system (3) is uniformly bounded for $t \geq 0$. Obviously all conditions of the theorem of LaSalle (see [7]) are satisfied. Hence, the ω-limit set of any trajectory of system (3) is contained in the subspace $\{y = 0,\ z = 0\}$. From the invariance of the ω-limit set and from the first differential equation in (3) we get that for the ω-limit set the relation $x = 0$ is valid. Therefore, the ω-limit set of any trajectory of system (3) consists of the equilibrium point $x = y = z = 0$. This completes the proof of Theorem 8.
Remark 1. *It is easy to see that the conditions (i) and (ii) of Theorem 6 can be satisfied if we choose λ sufficiently small and ν sufficiently large. Then, for negative γ and for sufficiently large $|\gamma|$ condition (16) can be fulfilled. By this way, we can always find parameters λ, ν, γ such that the hypotheses of Theorem 7 are satisfied. Thus, system (3) is dissipative. From this point of view, Theorem 6 yields an improvement of the region of dissipativity compared with Theorem 7.*

Theorem 6 is of special interest in case $e = 0$. Here, we can draw the following conclusion.

Corollary 1. *Let the hypotheses of Theorem 6 be valid. Additionally we assume $e = 0$. Then, on the global attractor of system (3) we have*

$$0 \leq z \leq |\gamma|. \tag{21}$$

4 Longtime behavior and estimates of the Hausdorff dimension of the global attractor

In this section we estimate the Hausdorff dimension of the global attractor K of system (3) by means of Theorem 4. At the same time we derive conditions for global stability.

To be able to apply Theorem 4 to system (3) we first derive conditions for the existence of a coordinate transformation such that the Jacobian $J(x)$ of the transformed system has the property that $J(x) + J(x)^T$ possesses a block-diagonal structure.

Let S be an invertible $n \times n$-matrix. By means of the coordinate transformation

$$x \to Sx, \; z \to \sqrt{2}\, z, \; y \to y \tag{22}$$

we obtain from (3)

$$\begin{aligned} \frac{dx}{dt} &= S^{-1}ASx + S^{-1}by, \\ \frac{dy}{dt} &= -c^T Sx - dy - \sqrt{2}zy + e, \\ \frac{dz}{dt} &= -\sigma z + \frac{\sqrt{2}}{2}y^2. \end{aligned} \tag{23}$$

The Jacobian of (23) reads

$$J(x) := \begin{pmatrix} S^{-1}AS & S^{-1}b & 0 \\ -c^T S & -d-\sqrt{2}z & -\sqrt{2}y \\ 0 & \sqrt{2}y & -\sigma \end{pmatrix}.$$

If we assume

$$S^{-1}b = (c^T S)^T = S^T c$$

which is equivalent to

$$b = SS^T c \tag{24}$$

then $J(x) + J(x)^T$ has the block diagonal structure

$$J(x) + J(x)^T = \begin{pmatrix} S^{-1}AS + (S^{-1}AS)^T & 0 & 0 \\ 0 & -2(d+\sqrt{2}z) & 0 \\ 0 & 0 & -2\sigma \end{pmatrix}. \tag{25}$$

Our goal is to guarantee the existence of a positive definite symmetric matrix H such hat

$$b = Hc, \quad (A+\mu I)H + H(A+\mu I)^T \le 0. \tag{26}$$

It is clear that the existence of a symmetric positive definite matrix H satisfying (26) implies the existence of a regular matrix S ($H = SS^T$) satisfying (24).

The proof of the existence of the matrix H is based on the application of Theorem 2. To this end we set in Theorem 2 $P = (A + \mu I)^T$, $q = c$, $r = -b$ and assume

(H_1). There is a positive number μ such that

(i). $A + \mu I$ is Hurwitz.
(ii). $(A + \mu I, c)$ is controllable
(iii). $(A + \mu I, b)$ is observable.

(H_2).

$$Re\left[\kappa(i\omega - \mu)\right] < 0 \quad \forall\, \omega \in \mathrm{R}, \tag{27}$$

where $\kappa(s)$ is defined according to (9) by

$$\kappa(s) := b^T(A^T - sI)^{-1}c. \tag{28}$$

Under the assumptions (H_1) and (H_2), it follows from Theorem 2 that there exists a positive definite matrix H satisfying (26). Thus, the following lemma is valid.

Lemma 1. *Assume the hypotheses (H_1) and (H_2) hold. Then there exists a regular matrix S such that by means of the coordinate transformation (22) system (3) can be mapped into system (23) whose Jacobian $J(x)$ satisfies the relation (25), moreover the inequality*

$$S^{-1}AS + (S^{-1}AS)^T \leq -2\mu I \tag{29}$$

is valid

We note that (29) is equivalent to

$$ASS^T + SS^TA^T \leq -2\mu SS^T \tag{30}$$

which follows from (26) by setting $H = SS^T$.

Now we are able to apply Theorem 4 to system (23) in order to estimate the Hausdorff dimension of the global attractor K.

Theorem 9. *Suppose the hypotheses of Lemma 1 hold. Then, under the additional condition*

$$\min\,(\mu, \sigma) + d > 0 \tag{31}$$

any solution of system (3) tends to a stationary solution for $t \to +\infty$. Under the condition

$$\min\,(\mu, \sigma) + d \leq 0,\ d + \sigma + \mu \geq 0$$

the Hausdorff dimension $\dim_H K$ *of the global attractor* K *satisfies*

$$\dim_H K \le 2 - \frac{\min(\mu,\sigma)+d}{\max(\mu,\sigma)}. \tag{32}$$

The same estimate holds for $\sigma < \mu, \sigma + d \le 0, \sigma + n\mu + d > 0$.
In case $\sigma \ge \mu, d + n\mu > 0$ *we have*

$$\dim_H K \le 1 - \frac{d}{\mu}.$$

For $\sigma \ge \mu, d + n\mu \le 0, d + n\mu + \sigma > 0$ *it holds*

$$\dim_H K \le n + 1 - \frac{d+n\mu}{\sigma}.$$

Proof. Under our assumptions, we get from (25) and (30)

$$\begin{aligned} J(x) + J(x)^T &:= \begin{pmatrix} S^{-1}AS + (S^{-1}AS)^T & 0 & 0 \\ 0 & -2(d+\sqrt{2}\ z) & 0 \\ 0 & 0 & -2\sigma \end{pmatrix} \\ &\le \begin{pmatrix} -2\mu I & 0 & 0 \\ 0 & -2d & 0 \\ 0 & 0 & -2\sigma \end{pmatrix}. \end{aligned} \tag{33}$$

We consider condition (31) and assume $\min(\mu,\sigma) = \sigma$. Then we obtain from (31) and (33)

$$\lambda_1(x) + \lambda_2(x) \le -2(d+\sigma) < 0.$$

Thus, according to Theorem 5, any solution of (3) tends to an equilibrium point as t tends to $+\infty$. The case $min(\mu,\sigma) = \mu$ is treated analogously.
Let $\min(\mu,\sigma) + d \le 0, d + \sigma + \mu \ge 0$ and $\min(\mu,\sigma) = \mu$. In that case we have for $s > -(d+\mu)/\sigma$

$$\lambda_1(x) + \lambda_2(x) + s\lambda_3(x) \le -2(d + \mu + s\sigma) < 0.$$

This proves the estimate (4.9). The other cases can be treated similarly. This completes the proof of the theorem.

For $n = 1,\ A = -a < 0,\ b = 1$ we obtain from (28)

$$\kappa(s) = \frac{-c}{s+a}.$$

In that case it is easy to see that the relations (27) holds for $c > 0$ and $\mu \in (0,a)$. Thus, we have

Corollary 2. *Assume* $n = 1,\ a > 0,\ c > 0,\ \sigma > 0$ *and*

$$\min\,(a, \sigma) + d > 0.$$

Then any solution of system (3) *tends to an equilibrium for* $t \to +\infty$.

In case

$$\min\,(a, \sigma) + d < 0$$

the Hausdorff dimension of the global attractor K *can be estimated by*

$$\dim_H K \le 2 - \frac{\min(a, \sigma) + d}{\max(a, \sigma)}.$$

We note that Mareels et al. [10] in case $a = 1,\ \sigma = 0.1,\ d = -\alpha,\ c = 3/4 - \alpha/4,\ \alpha \in [0, 1]$ have got numerically for $e = 0$ that the origin is globally stable for $\alpha \in (0, 0.6)$. In this case we obtain from Corollary 2 that for any e the origin is globally stable for $\alpha \in (0, 0.1)$. For $\alpha > 0.1$ we obtain the following estimate of the Hausdorff dimension of the global attractor K

$$\dim_H K \le \alpha + 1.9.$$

We wish to underline that this result holds true for any e.

In what follows we consider system (1.3) in case $e = 0$, and under the condition $\gamma < 0$. Our goal is to derive a frequency criterion for the global asymptotical stability of the origin which extends a corresponding result in [10].

For this purpose we study the system

$$\begin{aligned} \dot{x} &= Ax + by, \\ \dot{y} &= -c^T x - dy - z(t)y, \end{aligned} \tag{34}$$

where we assume

$$0 \le z(t) \le -\gamma \quad \text{for} \quad t \in R. \tag{35}$$

We will apply Theorem 3 to system (34) in order to get a criterion guaranteeing the global asymptotic stability of the origin.

First we note that the transfer function of system (34) with the input $z(t)y$ and the output $-y$ coincides with the function $\psi(s)$ defined in (5).

To satisfy the assumptions of Theorem 3 we have to assume

($\tilde{A}_3$). The matrix

$$\tilde{A} = \begin{pmatrix} A & b \\ -c^T & -d \end{pmatrix}$$

is Hurwitz.

Under the assumption (A_1) we have due to Schur's lemma and taking into account the notation introduced in (6) and the relation (7)

$$\det(s\tilde{I} - \tilde{A}) = \det(sI - A)\det(s + d + c^T(A - pI)^{-1}b) = p(s) + d\eta(s) = \frac{\eta(s)}{\psi(s)}.$$

Thus, if assumption (A_1) holds, then hypothesis ($\tilde{A}_3$) is equivalent to the following hypothesis

(A_3). $\psi(s)$ has only poles with negative real parts.

If we additionally assume

(A_4).

$$-\gamma^{-1} + \text{Re}\,\psi(i\omega) > 0 \qquad \forall\,\omega \in R, \tag{36}$$

then Theorem 3 can be applied to system (34) and we get that the origin of system (34) is asymptotically stable, that is, any solution of system (34) satisfies

$$\lim_{t\to+\infty} x(t) = 0, \quad \lim_{t\to+\infty} y(t) = 0. \tag{37}$$

Under the assumptions of Theorem 6, the z-component of system (3) satisfies by Corollary 1 the condition (35). Thus, from (37) and from the last equation in (3) we get that in case $\sigma > 0$ the relation

$$\lim_{t\to+\infty} z(t) = 0$$

holds and the origin of (3) is also asymptotically stable.

Theorem 10. *Let all hypotheses of Theorem 6 be satisfied. Additionally we suppose $e = 0$ and that the assumptions (A_3) and (A_4) are valid. Then the origin of system (1.3) is globally asymptotically stable.*

Remark 2. *We note that Theorem 10 improves Theorem 3.3 in [10] at least in the case $n = 1$ where instead of (36) the condition $\text{Re}\,\psi(i\omega) > 0$ is used.*

Now we apply Theorem 10 to system (1.3) in the case $n = 1,\ b = 1,\ e = 0, c > 0, A = -a$.

By (2.1) and (2.2) the corresponding transfer function reads

$$\psi(s) = \frac{s+a}{(s+a)(s+d)+c}. \tag{38}$$

With $\nu = 0$, condition (ii) of Theorem 6 reads

$$\text{Re}\,\frac{c}{i\omega + a - \lambda} > 0 \quad \forall\,\omega \in R.$$

This relation is valid for any $\lambda \in (0, a)$. For the same λ also condition (i) of Theorem 6 holds.

If we assume

$$a > \sigma/2 > d,$$

then for

$$\lambda = \sigma/2, \ -\gamma > \sigma - 2d, \ \nu = 0$$

all conditions of Theorem 6 are satisfied. Taking into account the explicit form of the transfer function $\psi(s)$ defined in (38) we get the result:

Corollary 3. *Consider the case* $n = 1$, $b = 1$, $e = 0$, $a > 0$, $\sigma > 0$, $c > 0$, $a > \sigma/2 > d$. *We assume that the polynomial*

$$(s+d)(s+a)+c \tag{39}$$

has only zeros with negative real parts and that the frequency inequality

$$\frac{1}{\sigma - 2d} + Re\left[\frac{i\omega + a}{(i\omega + a)(i\omega + d) + c}\right] > 0 \quad \forall\, \omega \in R \tag{40}$$

holds true. Then system (3) *is globally stable.*

It can be easily verified that all zeros of the polynomial (39) are located in the left half plane if we have

$$a + d > 0, \;\; ad + c > 0.$$

Condition (40) can be written in the form

$$\frac{1}{\sigma - 2d} + \frac{d\omega^2 + a(ad+c)}{(ad+c-\omega^2)^2 + (a+d)^2\omega^2} > 0.$$

Note that in [10] it has been shown numerically that in the case

$$e = 0, \;\; a = 1, \;\; \sigma = 0.1, \;\; c = \frac{-\alpha}{4} + \frac{3}{4}, \;\; d = -\alpha, \;\; \alpha \in (0,1)$$

the origin is globally stable for $\alpha < 0.6$. From Corollary 3 we get that the origin is globally stable for $\alpha < 0.5463$.

We note that Theorem 3.3 in [10] is not applicable since from $d < 0$ it follows that the inequality $\psi(i\omega) > 0$ cannot be satisfied for sufficiently large ω.

Acknowledgment

The authors acknowledge stimulating discussions with Achim Ilchmann and a careful reading of the manuscript by the referee.

References

1. Byrnes, C.I., Willems, J.C. (1984) Global adaptive stabilization in the absence of information on the sign of the high frequency gain. Part I: Analysis and optimization of systems, In: Proc. 6th Intern. Conf. on Anal. and Optimiz. of Syst., Eds. A. Bensoussan and J.L. Lions, Springer Verlag New York, 49–57
2. Douady, A., Oesterle, J. (1980) Dimension de Hausdorff des attracteurs. C. R. Acad. Sci., Paris, Ser. A **290**, 1135-1138

3. Hale, J.K. (1988) Asymptotic behavior of dissipative systems. Mathematical Surveys and Monographs, **25**, AMS, Providence
4. Ilchmann, A. (1993) Non-identifier-based high-gain adaptive control. Springer-Verlag, London
5. Kaufman, H., Bar-Kana, I., Sobel, K. (1994) Direct adaptive control algorithms: Theory and Applications. Springer-Verlag, New York
6. Leonov,G.A., Burkin, I.M., Shepeljavyi, A.I. (1996) Frequency methods in oscillation theory. Mathematics and its Applications **357** , Kluwer Academic Publishers, Dordrecht
7. Leonov, G.A., Ponomarenko, D.V., Smirnova, V.B. (1996) Frequency-domain methods for nonlinear analysis. Theory and applications. World Scientific Series on Nonlinear Science, Series A, Vol. **9**
8. Mareels, I. (1984) A simple selftuning controller for stably invertible systems. Systems & Control Letters, **4**, 5–16
9. Mareels, I., Polderman, J.W. (1996) Adaptive Systems. An Introduction, Birkhäuser-Verlag
10. Mareels, I., Van Gils, S., Polderman, J.W., Ilchmann, A. (1999) Asymptotic dynamics in adaptive gain control. In: Advances in control: Highlights of ECC '99, Ed. P.M. Frank, 29–63 , Springer - Verlag, 29–63
11. Morse, A.S. (1983) Recent problems in parameter adaptive control, In: Outils et Modèles Mathématiques pour l'Automatique, l'Analyse de Systèmes et le Traitment du Signal. Ed. I.D.Landau, Editions du CNRS **3** Paris

Model Reduction for Systems with Low-Dimensional Chaos

Carlo Piccardi and Sergio Rinaldi

Dipartimento di Elettronica e Informazione, Politecnico di Milano,
Piazza Leonardo da Vinci 32, 20133 Milano, Italy
piccardi@elet.polimi.it, rinaldi@elet.polimi.it

Abstract. A method for deriving a reduced model of a continuous-time dynamical system with low-dimensional chaos is discussed. The method relies on the identification of peak-to-peak dynamics, i.e. the possibility of approximately (but accurately) predicting the next peak amplitude of an output variable from the knowledge of at most the two previous peaks. The reduced model is a simple one-dimensional map or, in the most complex case, a set of one-dimensional maps. Its use in control system design is discussed by means of some examples.

1 Introduction

Model reduction is certainly one of the most thoroughly studied topics in control system design, as witnessed by hundreds of papers published in the past decades, mostly in the area of linear systems [1]. In this paper, the issue of model reduction is considered in the context of controlled chaotic systems.

As a matter of fact, the control of chaotic systems has received wide attention in the last decade (see e.g. the survey books [2],[3]). Whereas many contributions have reported successful applications of well-known classic or modern techniques (such as adaptive-, optimal-, H_∞-, fuzzy-, or sliding mode control), a few contributions have exploited the complex nature of chaos in a genuinely original fashion. Among them, it is worthwhile to recall the OGY approach, the harmonic balance (or distortion control) method, and the bifurcation control method [2].

In this paper, we discuss a control technique for continuous-time systems that, as those mentioned above, is strictly specific to chaotic systems. The key point is the existence of peak-to-peak dynamics [4], i.e. the possibility of approximately (but accurately) predicting the next peak of a scalar output variable (i.e. the amplitude of the next relative maximum) from the knowledge of the last peak value or, in the most complex case, of the last two peaks. If peak-to-peak dynamics exist, the system can be described, regardless of its dimension, by a reduced model that consists of one or a few one-dimensional maps. Moreover, the reduced model can be fruitfully used for efficiently designing a control system.

The existence of peak-to-peak dynamics is crucially related to the low-dimensionality of the chaotic attractor [4]. Although many examples have

F. Colonius, L. Grüne (Eds.): Dynamics, Bifurcations, and Control, LNCIS 273, pp. 255–268, 2002.

been pointed out in all fields of science and engineering (see e.g. [5] and [6] for an early and a recent contribution), peak-to-peak dynamics have generally been considered as a curious hallmark of chaos or, at most, as a descriptive tool. Only recently the study of peak-to-peak dynamics has been proposed as a systematic tool for deriving reduced models of systems with low-dimensional chaos and for designing a control system.

This paper is a comprehensive review of several contributions in the field [4],[7],[8],[9]. The first part (Sects. 2 to 4) is devoted to finite-dimensional systems. The conditions giving rise to peak-to-peak dynamics are discussed in detail, and the structure of the reduced model is analyzed. Then a control problem is formulated and solved taking advantage of the reduced model and some examples are presented. The second part (Sect. 5) considers one of the possible extensions, namely the application of the approach to delay-differential systems, a special class of infinite-dimensional systems. It is shown that the theoretical framework is actually unchanged, so that the model reduction technique that has been developed for finite-dimensional systems can also be applied in this context. The concluding remarks (Sect. 6) briefly touch upon some other extensions.

2 Peak-to-peak dynamics

We begin by considering an autonomous continuous-time finite-dimensional system of the form

$$\dot{x}(t) = f(x(t)), \tag{1}$$

where $t \in R_+ = \{t \geq 0\}$ is time, $x \in R^n$ is the state vector, and $f : R^n \to R^n$ is a smooth function. Assume that (1) has a single attractor $X \subset R^n$ which is a *chaotic attractor*, i.e. an invariant, attractive set exhibiting sensitive dependence on initial conditions (e.g. [10],[11],[12]). Thus $x(t)$ is aperiodic but bounded. The system is observed through a scalar output variable

$$y(t) = g(x(t)), \tag{2}$$

where $g : R^n \to R$ is a smooth function.

Now suppose that $x(0) \in X$ (i.e. neglect the transient motion toward the attractor) and denote by t_k and y_k, respectively, the time instant and the amplitude of the k-th relative maximum (*peak*) of $y(t)$ $(0 < t_1 < t_2 < \cdots)$, i.e. $y_k = y(t_k)$, $\dot{y}(t_k) = 0$, $\ddot{y}(t_k) < 0$ (provided the maximum is quadratic). Then the *peak-to-peak plot* (PPP) is defined as the set $S = \{(y_k, y_{k+1}), k = 1, 2, \cdots\}$, i.e. the set of all ordered pairs of consecutive peaks. Figure 1 shows four PPPs obtained from the simulation of four different dynamical systems: Lorenz system [5], a chemical reactor [13], Chua system [14], and Rössler hyperchaotic system [15].

Although the sets S displayed in Fig. 1 are all fractal sets [12], i.e. they have non integer dimension (e.g. capacity dimension), in the examples (a),

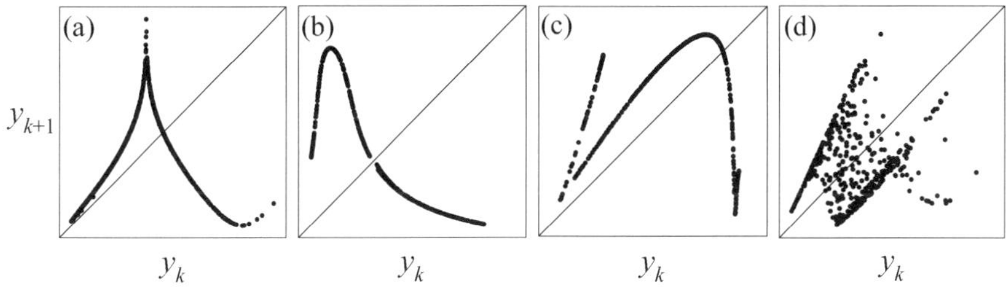

Fig. 1. Four peak-to-peak plots. (**a**) Lorenz system [5]. (**b**) Chemical reactor [13]. (**c**) Chua system [14]. (**d**) Rössler hyperchaotic system [15]

(b) and (c) they can be accurately approximated by a suitable set of smooth curves. By contrast, this cannot be done in example (d). In the first three cases, we say that system (1),(2) has *peak-to-peak dynamics* (PPD) [4]. Once a fitting criterion has been specified, the PPP of a system with PPD defines a (possibly multi-valued) function

$$y_{k+1} = \Psi(y_k)\,. \tag{3}$$

The PPD are said to be *complex* when Ψ is actually a multi-valued function (i.e. multiple values of y_{k+1} are associated to some y_k), and *simple* otherwise. Thus Fig. 1 shows two examples of simple PPD ((a) and (b)) and one of complex PPD ((c)).

It is relatively easy to understand that the condition under which system (1),(2) has PPD is that $\dim(X) \approx 2$, where $\dim(X)$ denotes the (fractal) dimension of the set X. Indeed, denote by $x_k = x(t_k)$ the state corresponding to the k-th peak, i.e. $y_k = g(x_k)$. Note that $x_k \in \Sigma$, where Σ is the Poincaré section defined by $\dot{y} = 0$, i.e. the manifold

$$\sum_{i=1}^{n} \frac{\partial g(x)}{\partial x_i} f_i(x) = 0\,. \tag{4}$$

Let us denote by $P \subset \Sigma$ the set of such states, i.e. $P = \{x_k, k = 1, 2, \cdots\}$ (Fig. 2). If $\dim(X) \approx 2$, then $\dim(P) = \dim(X) - 1 \approx 1$, so that P can be effectively approximated by a curve segment (or a few curve segments) $P^* \in \Sigma$. This amounts to replace P, which has non-integer dimension, with the one-dimensional set P^* or, equivalently, to assume that the relationship $x_k \in P^*\ \forall k$ holds approximately.

Without loss of generality, a coordinate $p : P^* \to [0,1]$ can be introduced on P^*. Denoting by $p_k = p(x_k)$ the value associated to x_k, a map $\varphi : [0,1] \to [0,1]$ can be defined such that $p_{k+1} = \varphi(p_k)$, since given x_k the state x_{k+1} is univocally identified by the Poincaré map $\Pi_\Sigma : \Sigma \to \Sigma$, i.e. $x_{k+1} = \Pi_\Sigma(x_k)$. On the other hand, denoting by $\rho : [0,1] \to R$ the restriction of g to P^*, i.e. $y_k = \rho(p_k) = g(x_k)$, we have

$$y_{k+1} = \rho(p_{k+1}) = \rho(\varphi(p_k))\,. \tag{5}$$

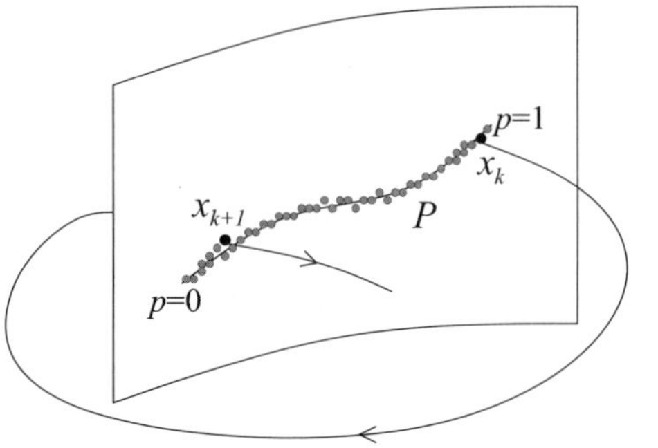

Fig. 2. The Poincaré section Σ, with some points of the set P and a segment of the trajectory $x(t)$. The output $y(t)$ has a peak when $x(t) = x_k$ and when $x(t) = x_{k+1}$

Now, if ρ is invertible then $y_{k+1} = \rho(\varphi(p_k)) = \rho(\varphi(\rho^{-1}(y_k)))$, which is of the form

$$y_{k+1} = Y(y_k) \,. \tag{6}$$

This is the case of simple PPD (see Figs. 1(a)–(b), where the knowledge of y_k is sufficient to accurately estimate y_{k+1}). The one-dimensional map (6) is a reduced model of system (1),(2) which, regardless of its order n, captures the essential behavior of the system in its chaotic regime. From a practical point of view, the map $Y : \Omega \to \Omega$, where $\Omega = [\inf_k y_k, \sup_k y_k]$ is the domain of the peak values y_k, can easily be identified from the PPP by standard fitting methods.

If, on the contrary, the map ρ is not invertible, the equation $\rho(p_k) = y_k$ has $m > 1$ solutions, that can be indexed by $\alpha_k \in \{1, 2, \cdots, m\} = M$. Then the pair (y_k, α_k) uniquely identifies p_k and hence $p_{k+1} = \varphi(p_k)$ which, in turn, corresponds to a new pair (y_{k+1}, α_{k+1}). Therefore, the PPD are described, in abstract terms, by a hybrid system of the form

$$\begin{aligned} y_{k+1} &= Y(y_k, \alpha_k) \,, \\ \alpha_{k+1} &= A(y_k, \alpha_k) \,, \end{aligned} \tag{7}$$

whose state is the pair $(y_k, \alpha_k) \in R \times M$. Such a system can be regarded as the feedback connection of a one-dimensional map with a finite (m) state automaton, as portrayed in Fig. 3. It represents a reduced model in the case of complex PPD, such as those of Fig. 1(c).

Let us further analyze the properties of complex PPD. Equation (7) points out that y_{k+1} can be predicted if one knows y_k and α_k. But it can be shown that the latter information depends only on the predecessor pair (y_{k-1}, y_k), i.e. $\alpha_k = \upsilon(y_{k-1}, y_k)$, so that, in conclusion, y_{k+1} is a function of the two previous peaks, i.e.

$$y_{k+1} = Y(y_k, \upsilon(y_{k-1}, y_k)) \,. \tag{8}$$

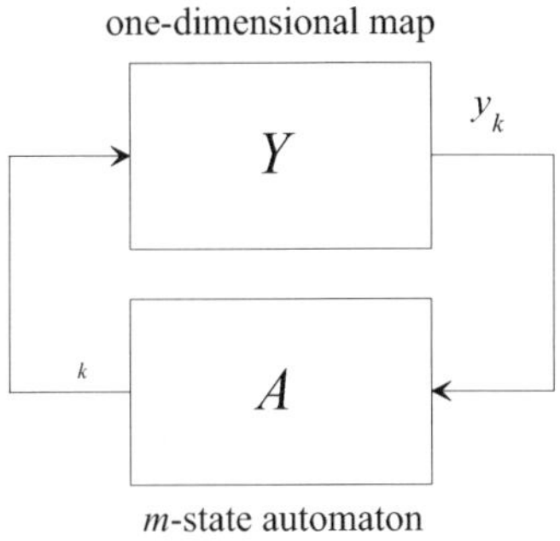

Fig. 3. When peak-to-peak dynamics are complex, the reduced model is a hybrid system composed of a one-dimensional map and an m-state automaton

To clarify this point, consider the PPP of Fig. 1(c) (Chua system). As shown in Fig. 4(a), the set $S = \{(y_k, y_{k+1})\}$ can be partitioned into $m = 2$ subsets S_1, S_2, each one defining a (single-value) function $Y(y, \alpha)$, $\alpha = 1, 2$. But this partition induces a corresponding partition S_1^-, S_2^- on the set $S^- = \{(y_{k-1}, y_k)\}$ of the predecessor pairs, namely $(y_{k-1}, y_k) \in S_\alpha^-$ iff $(y_k, y_{k+1}) \in S_\alpha$. Therefore, given y_k, the next peak y_{k+1} will be given by $y_{k+1} = Y(y_k, \alpha)$ if $(y_{k-1}, y_k) \in S_\alpha^-$. To clarify further, Fig. 4(c) shows the *cobweb* corresponding to 5 steps of the time evolution of (7) starting from an arbitrary pair (y_0, y_1) (the small square in the figure). Each iteration $(y_{k-1}, y_k) \to (y_k, y_{k+1})$ consists of two segments (one horizontal and one vertical). If $(y_{k-1}, y_k) \in S_1^-$ then the segments are solid and terminate on $(y_k, y_{k+1}) \in S_1$. Vice versa, if $(y_{k-1}, y_k) \in S_2^-$ then the two segments are dashed and terminate on $(y_k, y_{k+1}) \in S_2$. By this procedure one can also easily verify that (7) has an unstable fixed-point.

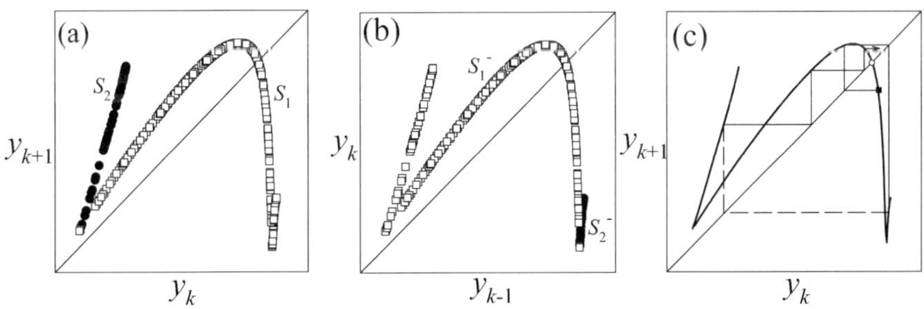

Fig. 4. Chua system. A partition of the set $S = \{(y_k, y_{k+1})\}$ into two subsets S_1, S_2 (**a**) induces a partition of the set $S^- = \{(y_{k-1}, y_k)\}$ of the predecessor pairs into two corresponding subsets S_1^-, S_2^- (**b**). In (**c**) 5 steps of the time evolution of (7) starting from an arbitrary pair (y_0, y_1) are shown

To summarize, the identification of the reduced model (7) from a time series amounts to derive the PPP and the m maps $y_{k+1} = Y(y_k, \alpha)$ $(\alpha \in M)$

through some standard fitting method, and to identify the relationship $\alpha_k = v(y_{k-1}, y_k)$ as discussed above by means of the example. As shown in the next sections, such a reduced model can effectively be exploited for control purposes by suitably extending it to the case of time-varying control.

3 The control problem

If the system is not autonomous but has a control input, system (1),(2) must be modified into

$$\begin{aligned} \dot{x}(t) &= f(x(t), u(t))\,, \\ y(t) &= g(x(t))\,, \end{aligned} \tag{9}$$

where $u : R_+ \to R$ is a piecewise-continuous function. For a nominal (constant) input $u(t) = \bar{u}_{\mathrm{nom}}$, system (9) reduces to (1),(2). Moreover, we assume that, for all $t \in R_+$, the control $u(t)$ takes values in a prescribed sufficiently small interval $U = [u_{\min}, u_{\max}] \subset R$, and that, for any constant control $\bar{u} \in U$, system (9) has a chaotic attractor $X(\bar{u}) \in R^n$ with PPD. Typically, the PPP is (roughly speaking) smoothly deformed as $\bar{u}$ is changed. Then a family of models of the form (7), parameterized in $\bar{u} \in U$, can be defined

$$\begin{aligned} y_{k+1} &= Y(y_k, \alpha_k, \bar{u})\,, \\ \alpha_{k+1} &= A(y_k, \alpha_k, \bar{u})\,, \end{aligned} \tag{10}$$

where $Y : \Omega \times M \times U \to \Omega$, $A : \Omega \times M \times U \to M$, and $\Omega = \bigcup_{\bar{u}} \Omega(\bar{u})$ is the domain of the peak values y_k.

In order to exploit (10) to effectively design a controller for system (9), the control u is now allowed to vary in time in a piecewise-constant fashion, i.e. $u(t) = u_k$ $\forall t \in (t_k, t_{k+1}]$. In other words, $u(t)$ is kept constant between two subsequent peaks of $y(t)$. Then, by replacing $\bar{u}$ by u_k in (10) we obtain

$$\begin{aligned} y_{k+1} &= Y(y_k, \alpha_k, u_k)\,, \\ \alpha_{k+1} &= A(y_k, \alpha_k, u_k)\,, \end{aligned} \tag{11}$$

which is a reduced model of system (9) with piecewise-constant control.

Conceptually, (11) is justified only if the transient time from the chaotic attractor $X(u_{k-1})$ to $X(u_k)$ is much shorter than the average time between consecutive peaks. Although it is difficult to rigorously assess whether this property is owned by a system, a rough *a priori* evaluation can be based on the knowledge of the negative Lyapunov exponents which, in the neighborhood of the attractor, correspond to the rate of convergence toward the attractor itself. For brevity, we do not discuss this issue in this work; see [8] for details.

A control law for the reduced model (11) has the form

$$u_k = q(y_k, \alpha_k)\,, \tag{12}$$

where $q : \Omega \times M \to U$, and must be designed in accordance with a prescribed objective. In this respect, several control problems can be formulated and solved (see [7] for a few examples). In this section we consider only the most popular goal in chaos control, namely the suppression of chaos in favor of a regular, cyclic regime. It is the typical goal when the aim is to extend the region (in parameter space) of cyclic behavior of the system (e.g. [2],[3]).

If system (9) has PPD, such a goal can be reformulated as the stabilization of a fixed-point of the reduced model. Indeed, if we fix a control value $\bar{u} \in U$, a fixed-point of (11), namely a pair $(\bar{y}, \bar{\alpha})$ satisfying

$$\begin{aligned} \bar{y} &= Y(\bar{y}, \bar{\alpha}, \bar{u}) \,, \\ \bar{\alpha} &= A(\bar{y}, \bar{\alpha}, \bar{u}) \,, \end{aligned} \tag{13}$$

corresponds to a periodic orbit of system $\dot{x} = f(x, \bar{u})$ with output peaks $y_k = \bar{y}$ for all k. A stabilizing control law can be found by solving a quadratic optimal control problem, i.e. by finding the control law q that minimizes the cost functional

$$J(y_0, \alpha_0) = \lim_{N \to \infty} \sum_{k=0}^{N-1} (y_k - \bar{y})^2 \,, \tag{14}$$

where $J : \Omega \times M \to R$, subject to (11),(12),(13). It is a problem extensively treated in many texbooks (e.g. [16]). If the optimal cost $J_{\text{opt}}(y_0, \alpha_0)$ is bounded for all (y_0, α_0), then $y_k \to y$ for all initial conditions. Therefore, if we initially apply a constant control value $\bar{u} \in U$ and system (9) is in its chaotic regime, then switching on the optimal control law will steer the system to the prescribed periodic orbit. Notice that the control law q is not constrained to belong to any *a priori* class of functions. Consequently, the optimal control law will be, in general, a nonlinear function.

The optimal control problem (14) can be numerically solved by standard dynamic programming (e.g. [16] for details), i.e. by computing a sequence of functions $L_i : \Omega \times M \to R$, for $i = 0, 1, \cdots$, by

$$L_{i+1}(y, \alpha) = \inf_{u \in U} \left((y - \bar{y})^2 + L_i(Y(y, \alpha, u), A(y, \alpha, u)) \right) , \tag{15}$$

starting from $L_0(y, \alpha) = 0 \; \forall (y, \alpha) \in \Omega \times M$. The optimal cost $J_{\text{opt}}(y, \alpha)$ is the limit of the sequence $L_i(y, \alpha)$, and is obtained by stopping the algorithm (15) at some $i = \bar{i}$ with a suitable convergence criterion. Then $J_{\text{opt}}(y, \alpha) \approx L_{\bar{i}}(y, \alpha)$ $\forall (y, \alpha)$, and the optimal control law is finally obtained by

$$q(y, \alpha) = \arg \inf_{u \in U} \left((y - \bar{y})^2 + J_{\text{opt}}(Y(y, \alpha, u), A(y, \alpha, u)) \right) . \tag{16}$$

4 Examples of application

4.1 Lorenz system

The Lorenz system [5] is the most popular example of a continuous-time chaotic system and, as such, it has become a benchmark for testing chaos control techniques. A recent application of the Lorenz system is related to modeling a thermal convection loop ([17],[18]), where the equations take the form

$$\begin{aligned} \dot{x}_1 &= \sigma(x_2 - x_1)\,, \\ \dot{x}_2 &= -x_1 x_3 - x_2\,, \\ \dot{x}_3 &= x_1 x_2 - b x_3 - b u\,. \end{aligned} \tag{17}$$

From (17) the traditional Lorenz model is easily recovered (provided u is constant) by the transformation $\tilde{x}_1 = x_1$, $\tilde{x}_2 = x_2$, $\tilde{x}_3 = x_3 + u$. We set $\sigma = 4$, $b = 1$ and $y = x_3$.

As already pointed out with reference to Fig. 1(a), the Lorenz system has simple PPD. The three PPPs presented in Fig. 5 are related to the extreme and central values of the control interval $U = [23, 33]$. Linear interpolation was used, in this example, to derive the equation

$$y_{k+1} = Y(y_k, \bar{u}) \tag{18}$$

from a set of 21 PPPs obtained for equally spaced values of $\bar{u} \in U$. At $\bar{u} = 28$ the model (18) has an unstable fixed-point at $\bar{y} = 8.200$. By designing a control law aimed at stabilizing the corresponding periodic orbit, as described in the previous section, one obtains the remarkably nonlinear function of Fig. 5(b). The application of this control law to system (9) leads to the performance shown in Fig. 6(a). After the control law is switched on at $t = 30$, the chaotic oscillations is very quickly replaced by regular (periodic) oscillations. The output $y(t)$ and the control $u(t)$ are affected by small residual oscillations which are due to the numerical approximations introduced in deriving the reduced model, in computing its fixed-point, and in deriving the optimal control law.

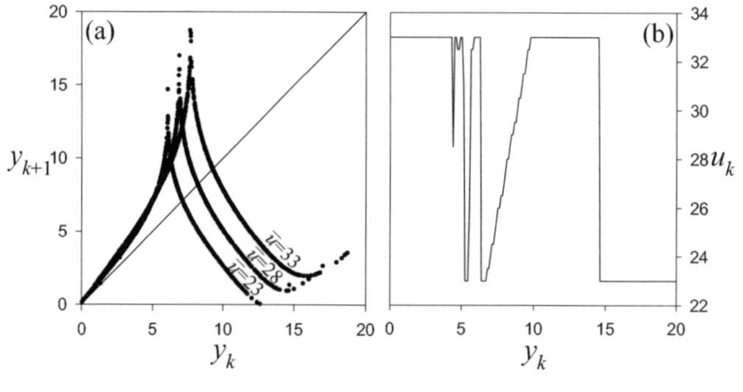

Fig. 5. Lorenz system. **(a)** Three PPPs for three different constant control values. **(b)** The optimal control law

4.2 Chua system

As well as Lorenz system, Chua system is a continuous-time nonlinear system that has been recently studied in great depth. A thorough comparative analysis of these two popular systems can be found in [14], where the following version of Chua system (with $u = 0$) is analyzed in detail

$$\begin{aligned} \dot{x}_1 &= a(x_2 - bx_1^3 - cx_1)\,, \\ \dot{x}_2 &= x_1 - x_2 + x_3 + u\,, \\ \dot{x}_3 &= -dx_2 - ex_3\,. \end{aligned} \tag{19}$$

Figures 1(c) and 4 have been obtained letting $a = -80$, $b = -1$, $c = 0.2$, $d = 31.25$, $e = -3.125$, $u(t) = \bar{u} = 0$, and $y = x_1$. As already noticed, for these parameter values the system displays complex PPD with $m = 2$. For the control set $U = [-0.005, 0.005]$, the reduced model (11), with $\alpha_k \in \{1, 2\}$, has been identified as described in the previous section. As illustrated in Fig. 4, $\alpha_k = i$ means that $(y_{k-1}, y_k) \in S_i^-$. Then, it can easily be ascertained that, for $u(t) = \bar{u} = 0$, the reduced model has an unstable fixed point at $(\bar{y}, \bar{\alpha}) = (0.656, 1)$. A control law $u_k = q(y_k, \alpha_k)$ has been derived to stabilize such a fixed-point. Figure 6(b) shows that also in this case the desired periodic regime is rapidly reached after the control law is switched on, and a small residual oscillation affects $u(t)$.

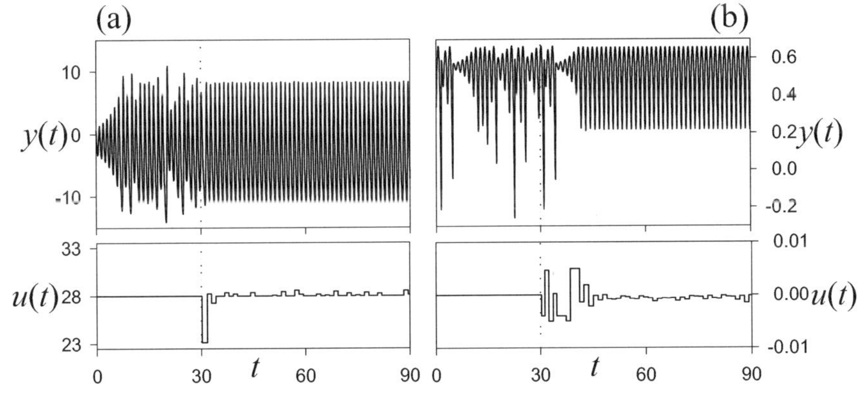

Fig. 6. The output and input time series: the control law is activated at $t = 30$. **(a)** Lorenz system. **(b)** Chua system

5 Delay-differential systems

The approach to model reduction and control design described in this paper is essentially a model-free method. In fact, one has only to be able to perform experiments with different (constant) control values: if the resulting PPPs reveal the existence of PPD, then a reduced model can be derived and a control law can be designed as described in the previous sections. As a consequence, it is natural to wonder whether PPD can be found in other classes of dynamical systems.

Indeed, PPD can also be found in *delay-differential systems* (DDSs) (e.g. [19],[20]), a special class of infinite-dimensional dynamical systems frequently used in science and engineering. These systems are known since long to display both low- and high-dimensional chaotic behavior (e.g. [21],[22]). Consistently, we can expect that our method can be successfully applied in the case of low-dimensional chaos, which is typically related to a *small delay*. In this respect, the method is somehow complementary to the approach followed in [22],[23], which derives a first-order map that approximately describes the DDS behavior in the presence of a *large delay*.

We consider systems described, for simplicity, by a single differential equation with delay, i.e.

$$\dot{x}(t) = f(x(t), x(t-\tau)) \, , \tag{20}$$

where $t \in R_+ = \{t \geq 0\}$, $x(t) \in R$, and τ is a given delay. The function $f : R \times R \to R$ is such to guarantee that $x(t)$ is univocally defined, for all $t \in R_+$, by the initial condition $x(\cdot) : [-\tau, 0] \to R$. For brevity, in the following we will write x for $x(t)$ and x_ξ for $x(t-\xi)$.

Assume that (20) has a chaotic attractor (i.e. $\lambda_1 > 0$, where $\lambda_1 \geq \lambda_2 \geq \cdots$ is the Lyapunov spectrum, which is now formed by infinitely many exponents, e.g. [11],[21]). It is known that the dimension of such an attractor is finite, although the system is infinite-dimensional. This implies that, for a generic $T > 0$, we can find a finite integer q such that the q-dimensional space E of the delay-coordinate vectors $e(t) = (x, x_T, \cdots, x_{(q-1)T})$ is an embedding space for the attractor. We will denote by $X \subset E$ the attractor in the embedding space and by $\dim(X) < q$ its dimension.

In this context, it is convenient to introduce a scalar output variable as a function of r equally-spaced samples of x, i.e.

$$y(t) = g(x, x_T, \cdots, x_{(r-1)T}) \, , \tag{21}$$

where $g : R^r \to R$ is a smooth function. Then, when $y(t)$ has a peak the delay-coordinate vector $e(t)$ crosses the manifold defined by $\dot{y}(t) = 0$, i.e.

$$\sum_{i=0}^{r-1} \frac{\partial g}{\partial x_{iT}} f(x_{iT}, x_{iT+\tau}) = 0 \, . \tag{22}$$

It is not restrictive to choose T as an integer divisor of τ (i.e. $\tau = lT$, l integer), since T is arbitrary, and to select an embedding space E of dimension $q \geq r+l$. Then (22) is actually of the form

$$\Theta(x, x_T, \cdots, x_{(q-1)T}) = 0 \, , \tag{23}$$

i.e. it defines a manifold $\Sigma \subset E$ that can be regarded as a Poincaré section. Therefore, we are exactly in the framework of finite-dimensional systems: if $\dim(X) \approx 2$ then the PPP (which is defined in exactly the same way) can be accurately fitted by one or a few curves. If so, the system has simple or complex PPD, and the dynamics on the attractor are captured by an equation of the form (6) or (7), respectively. Finally, a control variable $u(t)$, taking values in a (sufficiently small) prescribed set U, can be introduced in the DDS and a reduced model with piecewise-constant control can be obtained. It has the general form (11), and can be used to effectively formulate and solve a control problem as in the case of finite-dimensional systems.

For an application we consider one of the first examples of chaotic DDS, namely the Mackey-Glass equation [24] (see also [21] for a detailed analysis). It has the form

$$\dot{x} = u\frac{ax_\tau}{(1+x_\tau)^c} - bx\,. \tag{24}$$

As in [21], we set $a = 0.2$, $b = 0.1$, $c = 10$, and $\tau = 17$. Moreover, we let $U = [0.975, 1.025]$. Figure 7, obtained with $u(t) = \bar{u} = 1$, shows the output time series and the corresponding PPP for

$$y(t) = x(t)\,, \tag{25}$$

and for

$$y(t) = \frac{x + x_T + \cdots + x_{(r-1)T}}{r}\,, \tag{26}$$

i.e. for the moving average of r equally-spaced samples of x. Whereas the existence of PPD is evident in both situations (recall that the existence of PPD depends only on the dimension of the chaotic attractor), the use of a moving average output yields a definitely simpler PPP, which can be accurately fitted by a single curve (simple PPD). The analysis of the output patterns of Fig. 7 shows that the moving average output smooths out small amplitude, high-frequency oscillations without destroying the information contained in the large amplitude, low-frequency ones. This feature has been observed in several examples of DDS [9].

At $u(t) = \bar{u} = 1$ the reduced model has an unstable fixed point at $\bar{y} = -0.616$. A control law $u_k = q(y_k)$ has been derived to stabilize such a fixed-point, using the same procedure adopted for finite-dimensional systems. Its application yields the performance shown in Fig. 8: after the control law is activated, the chaotic oscillations are very quickly replaced by periodic oscillations, and the control $u(t)$ settles down to a practically constant value.

6 Concluding remarks

The model reduction and control design method discussed in this paper proves to be an effective alternative to other chaos control techniques in

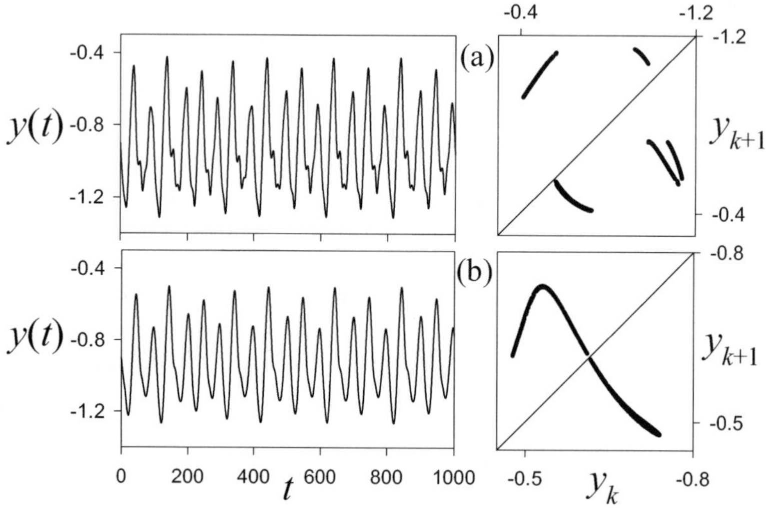

Fig. 7. Mackey-Glass equation [24]. The output time series and the corresponding PPP: **(a)** $y = x$. **(b)** $y = (x + x_T + \cdots + x_{(r-1)T})/r$, $r = 110$, $T = 0.1$

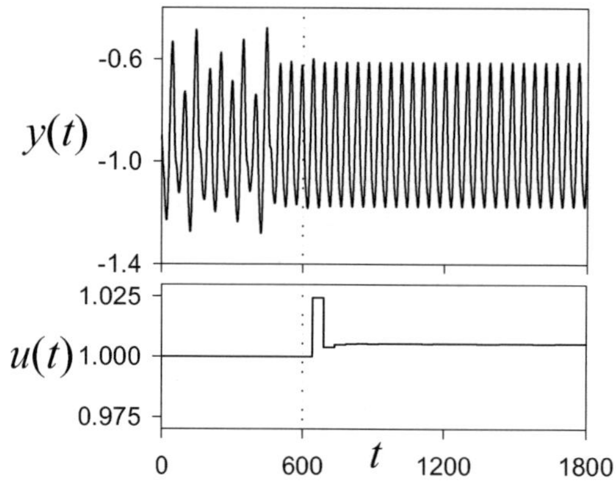

Fig. 8. Mackey-Glass equation. The output and input time series: the control law is activated at $t = 600$

the case of systems with low-dimensional chaos. It is essentially a numerical method, as it requires to perform simulations (or experiments) on the continuous-time system, and then to identify the reduced model by means of a suitably organized fitting procedure.

For brevity, the exposition could not touch upon some other interesting aspects of the method. For example, it can easily be ascertained that, when peak-to-peak dynamics exist, not only the amplitude of the next peak but also its time of occurrence can be accurately predicted [4]. This extra-information allows one to enrich the scope of applicability of the method by formulating control problems with different goals, e.g. min-max problems involving return times or mixed amplitude-return times indicators [7].

References

1. Obinata G., Anderson B.D.O. (2001) Model Reduction for Control System Design. Springer, New York
2. Chen G., Dong X. (1998) From Chaos to Order, Methodologies, Perspectives and Applications. World Scientific, Singapore
3. Chen G. (Ed.) (2000) Controlling Chaos and Bifurcations in Engineering Systems. CRC, Boca Raton, FL
4. Candaten M., Rinaldi S. (2000) Peak-to-Peak Dynamics: a Critical Survey. Int. J. Bif. Chaos 10:1805–1819
5. Lorenz E.N. (1963) Deterministic Nonperiodic Flow. J. Atm. Sci. 20:130–141
6. Abarbanel H.D.I., Korzinov L., Mees A.I., Rulkov N.F. (1997) Small Force Control of Nonlinear Systems to Given Orbits. IEEE Trans. Circ. Sys. I 44:1018–1023
7. Piccardi C., Rinaldi S. (2000) Optimal Control of Chaotic Systems via Peak-to-Peak Maps. Physica D 144:298–308
8. Piccardi C., Rinaldi S. (2001) Control of Complex Peak-to-Peak Dynamics. Int. J. Bif. Chaos, to appear
9. Piccardi C. (2001) Controlling Chaotic Oscillations in Delay-Differential Systems via Peak-to-Peak Maps. IEEE Trans. Circ. Sys. I, to appear
10. Guckenheimer J., Holmes P. (1983) Nonlinear Oscillations, Dynamical Systems, and Bifurcations of Vector Fields. Springer, New York
11. Eckmann J.P., Ruelle D. (1985) Ergodic Theory of Chaos and Strange Attractors. Rev. Mod. Phys. 57:617–656
12. Alligood K.T., Sauer T.D., Yorke J.A. (1996) Chaos, An Introduction to Dynamical Systems. Springer, New York
13. Peng B., Scott S.K., Showalter K. (1990) Period Doubling and Chaos in a Three Variable Autocatalator. J. Phys. Chem. 94:5243-5246
14. Pivka L., Wu C.W., Huang A. (1996) Lorenz Equation and Chua's Equation. Int. J. Bif. Chaos 6:2443–2489
15. Rössler O.E. (1979) An Equation for Hyperchaos. Phys. Lett. A 71:155-157
16. Bertsekas D.P. (1995) Dynamic Programming and Optimal Control. Athena, Belmont, Mass
17. Singer J., Wang Y.Z., Bau H.H. (1991) Controlling a Chaotic System. Phys. Rev. Lett. 66:1123–1125
18. Wang H.O., Abed E.H. (1995) Bifurcation Control of a Chaotic System. Automatica 31:1213–1226
19. Kuang J. (1993) Delay Differential Equations with Applications in Population Dynamics. Academic, Boston, Mass
20. Diekmann O., van Gils S.A., Verduyn Lunel S.M., Walther H.O. (1995) Delay Equations: Functional-, Complex-, and Nonlinear Analysis. Springer, New York
21. Farmer J.D. (1982) Chaotic Attractors of an Infinite-Dimensional Dynamical System. Physica D 4:366–393
22. Ikeda K., Matsumoto K. (1987) High-Dimensional Chaotic Behavior in Systems with Time-Delayed Feedback. Physica D 29:223–235

23. Celka P. (1997) Delay-Differential Equation versus 1D-Map: Application to Chaos Control. Physica D 104:127–147
24. Mackey M.C., Glass L. (1977) Oscillation and Chaos in Physiological Control Systems. Science 197:287–289

Feedback Equivalence to Feedforward Forms for Nonlinear Single-Input Systems

Issa Amadou Tall* and Witold Respondek

Laboratoire de Mathématiques
INSA de Rouen
76131 Mont Saint Aignan, France
{tall,wresp}@lmi.insa-rouen.fr

Abstract. We study the problem of transforming a single-input nonlinear control system to feedforward form via a static state feedback. We provide checkable necessary and sufficient conditions to bring the homogeneous terms of any fixed degree of the system into homogeneous feedforward form. If those conditions are satisfied, this leads to a constructive procedure which transforms the system, step by step, into feedforward form. We also formulate a similar result to bring a system to nice feedforward form. We illustrate our solution by analyzing the three- and four-dimensional cases. In particular, we calculate the codimension of four-dimensional systems that are feedback equivalent to feedforward form.

1 Introduction

Consider a single-input nonlinear control system of the form

$$\Pi \ : \ \dot{\xi} = F(\xi, u),$$

where $\xi(\cdot) \in \mathbb{R}^n$ and $u(\cdot) \in \mathbb{R}$. We say that the system Π is in *feedforward form* (resp. in *strict feedforward form*) if we have

$$\begin{aligned} \dot{\xi}_1 &= F_1(\xi_1, \ldots, \xi_n, u) \\ \dot{\xi}_2 &= F_2(\xi_2, \ldots, \xi_n, u) \\ &\vdots \\ \dot{\xi}_{n-1} &= F_{n-1}(\xi_{n-1}, \xi_n, u) \\ \dot{\xi}_n &= F_n(\xi_n, u). \end{aligned} \qquad \left(\text{resp.} \quad \begin{aligned} \dot{\xi}_1 &= F_1(\xi_2, \ldots, \xi_n, u) \\ \dot{\xi}_2 &= F_2(\xi_3, \ldots, \xi_n, u) \\ &\vdots \\ \dot{\xi}_{n-1} &= F_{n-1}(\xi_n, u) \\ \dot{\xi}_n &= F_n(u) \end{aligned} \right).$$

One of the most appealing features of systems in (strict) feedforward form is that we can construct for them a stabilizing feedback. This important result goes back to Teel [21] and has been followed by a growing literature on stabilization and tracking for systems in (strict) feedforward form (see e.g. [6], [12], [14], [22], [3], [13]).

* on leave from the Departement de Mathématiques et Informatique, Université de Dakar, Senegal

F. Colonius, L. Grüne (Eds.): Dynamics, Bifurcations, and Control, LNCIS 273, pp. 269–286, 2002.

Feedforward systems can be viewed as duals of feedback linearizable systems. To see this, recall that in the single-input case, the class of feedback linearizable systems coincides with that of flat systems. Single-input flat systems are defined as systems for which we can find a function of the state that, together with its *derivatives*, gives all the states and the control of the system [4]. In a dual way, for systems in strict feedforward form, we can find all states via a successive *integration* starting from a function of the control. Indeed, knowing $u(t)$ we integrate $F_n(u(t))$ to get $\xi_n(t)$, then we integrate $F_{n-1}(\xi_n(t), u(t))$ to get $\xi_{n-1}(t)$, we keep doing that, and finally we integrate $F_1(\xi_2(t), \ldots, \xi_n(t), u(t))$ to get $\xi_1(t)$. For systems in *nice feedforward form*, that is for feedforward systems such that $F_j(\xi_j, \ldots, \xi_n)$ is affine with respect to ξ_j, we can also find all solutions via integrations, the only difference being that we need two quadratures for each component. For feedforward systems, solutions can be found by solving scalar differential equations: for each component we have to solve one scalar differential equation.

It is therefore natural to ask which systems are equivalent to one of feedforward forms defined above. In [11], the problem of transforming a system, linear with respect to controls, into (strict) feedforward form via a diffeomorphism, i.e., via a nonlinear change of coordinates, was studied. Recently, a geometric description of systems transformable into feedforward form, either via a diffeomorphism or via feedback, has been given in [2]. The conditions of [2], although being intrinsic, are not checkable. In [18] and [19] we proposed a constructive procedure which allows to verify, step by step, whether a given system is feedback equivalent to strict feedforward form and to bring it to that form whenever it is possible. Our solution was inspired by a fruitful method to study nonlinear systems proposed by Kang and Krener [9] and then followed by Kang [8]. Their idea, which is based on classical Poincaré's technique for linearization of dynamical systems (see e.g. [1]), is to consider the action of homogeneous transformations of any fixed degree on the homogeneous part, of the corresponding degree, of the system and then to analyze the action of the feedback group on the system step by step. In the present paper we use a similar approach in order to characterize systems that can be transformed via feedback to feedforward form or, under some additional assumption, to nice feedforward form.

The paper is organized as follows. In Section 2 we give basic definitions. In Section 3 we introduce a feedforward normal form and give a result stating that any system that is feedback equivalent to feedforward form can be transformed to that normal form. We recall m-invariants of homogeneous transformations in Section 4. Our main results are given in Section 5. The first result states that if a system is feedback equivalent to feedforward form, then the first nonlinearizable term must be feedforward when brought to Kang normal form. The second result gives necessary and sufficient conditions for feedback equivalence to feedforward form of the homogeneous part (of any fixed degree higher than the degree of the first nonlinearizable term) of the system. We compare our solution with that obtained earlier for strict feedforward form. We also state a result on feedback equivalence to nice feedforward form. In Section 6 we illustrate our results by simple examples on $\mathbb{R}^3$ and $\mathbb{R}^4$. In Section 7, we compute the codimension of the space of homogeneous systems that are feedback equivalent to feedforward form. We send the reader to [20] for proofs and a complete analysis of various feedforward forms.

2 Definitions and notations

Consider a nonlinear control-affine system

$$\Sigma \ : \ \dot{\xi} = f(\xi) + g(\xi)u,$$

where $\xi(\cdot) \in \mathbb{R}^n$ and $u(\cdot) \in \mathbb{R}$. Throughout the paper we assume that $f(0) = 0$ and $g(0) \neq 0$. We say that Σ is in *feedforward form*, or that it is a *feedforward system*, if we have

$$f(\xi) = \begin{pmatrix} f_1(\xi_1, \cdots, \xi_n) \\ f_2(\xi_2, \cdots, \xi_n) \\ \vdots \\ f_{n-1}(\xi_{n-1}, \xi_n) \\ f_n(\xi_n) \end{pmatrix} \quad \text{and} \quad g(\xi) = \begin{pmatrix} g_1(\xi_1, \cdots, \xi_n) \\ g_2(\xi_2, \cdots, \xi_n) \\ \vdots \\ g_{n-1}(\xi_{n-1}, \xi_n) \\ g_n(\xi_n) \end{pmatrix}.$$

Apply to the system Σ a feedback transformation of the form

$$\Gamma : \begin{array}{l} x = \phi(\xi) \\ u = \alpha(\xi) + \beta(\xi)v. \end{array}$$

The transformation Γ brings Σ to the system

$$\tilde{\Sigma} \ : \ \dot{x} = \tilde{f}(x) + \tilde{g}(x)v,$$

whose dynamics are given by

$$\begin{array}{l} \tilde{f} = \phi_*(f + g\alpha) \\ \tilde{g} = \phi_*(g\beta), \end{array}$$

where for any vector field f and any diffeomorphism ϕ we denote

$$(\phi_* f)(x) = d\phi(\phi^{-1}(x)) \cdot f(\phi^{-1}(x)).$$

In the paper we will be interested in the following question.

Problem: *When is the system Σ feedback equivalent to feedforward form?*

All objects, i.e., functions, maps, vector fields, control systems, etc., are considered in a neighborhood of $0 \in \mathbb{R}^n$ and assumed to be C^∞-smooth. Let h be a smooth $\mathbb{R}$-valued function. By

$$h(x) = h^{[0]}(x) + h^{[1]}(x) + h^{[2]}(x) + \cdots = \sum_{m=0}^{\infty} h^{[m]}(x)$$

we denote its Taylor series expansion around zero, where $h^{[m]}(x)$ stands for a homogeneous polynomial of degree m.

Similarly, for a map ϕ of an open subset of $\mathbb{R}^n$ to $\mathbb{R}^n$ (resp. for a vector field f on an open subset of $\mathbb{R}^n$) we will denote by $\phi^{[m]}$ (resp. by $f^{[m]}$) the term of degree m of its Taylor series expansion at zero, i.e., each component $\phi_j^{[m]}$ of $\phi^{[m]}$ (resp. $f_j^{[m]}$ of $f^{[m]}$) is a homogeneous polynomial of degree m in x. Denote also $\bar{x}_i = (x_1, \ldots, x_i)$.

Together with the system

$$\Sigma \ : \ \dot{\xi} = f(\xi) + g(\xi)u,$$

we will consider its Taylor series expansion

$$\Sigma^\infty \ : \ \dot{\xi} = F\xi + Gu + \sum_{m=2}^{\infty} (f^{[m]}(\xi) + g^{[m-1]}(\xi)u), \tag{1}$$

where $F = \frac{\partial f}{\partial \xi}(0)$ and $G = g(0)$.

Consider the Taylor series expansion Γ^∞ of the feedback transformation Γ given by

$$\Gamma^\infty \ : \ \begin{array}{l} x = \phi(\xi) = T\xi + \sum_{m=2}^{\infty} \phi^{[m]}(\xi) \\ u = \alpha(\xi) + \beta(\xi)v = K\xi + Lv + \sum_{m=2}^{\infty} (\alpha^{[m]}(\xi) + \beta^{[m-1]}(\xi)v), \end{array} \tag{2}$$

where T is an invertible matrix and $L \neq 0$. Let us analyze the action of Γ^∞ on the system Σ^∞ step by step.

To start with, consider the linear system

$$\dot{\xi} = F\xi + Gu.$$

Throughout the paper we will assume that it is controllable. It can be thus transformed by a linear feedback transformation of the form

$$\Gamma^1 : \begin{array}{l} x = T\xi \\ u = K\xi + Lv \end{array}$$

to the Brunovský canonical form (A, B), see e.g [7]. Assuming that the linear part (F, G), of the system Σ^∞ given by (1), has been transformed to the Brunovský canonical form (A, B), we follow an idea of Kang and Krener [9], [8] (see also [16], [18]) and apply successively a series of transformations

$$\Gamma^m \; : \; \begin{array}{l} x = \xi + \phi^{[m]}(\xi) \\ u = v + \alpha^{[m]}(\xi) + \beta^{[m-1]}(\xi)v, \end{array}$$

for $m = 2, 3, \ldots$. A feedback transformation defined as a series of successive compositions of Γ^m, $m = 1, 2, \ldots$ will also be denoted by Γ^∞ because, as a formal power series, it is of the form (2). We will not address the problem of convergence and we will call such a series of successive compositions a *formal feedback transformation.*

Notations. Because of various normal forms and various transformations which are used throughout the paper, we will keep the following notation. We will denote, respectively, by $\Sigma^{[m]}$, $\Sigma^{[\leq m]}$, and Σ^∞ the following systems

$$\begin{aligned} \Sigma^{[m]} \; &: \dot{\xi} = A\xi + Bu + f^{[m]}(\xi) + g^{[m-1]}(\xi)u + O(\xi, u)^{m+1}, \\ \Sigma^{[\leq m]} \; &: \dot{\xi} = A\xi + Bu + f^{[\leq m]}(\xi) + g^{[\leq m-1]}(\xi)u + O(\xi, u)^{m+1} \\ &\quad = A\xi + Bu + \sum_{k=2}^{m} (f^{[k]}(\xi) + g^{[k-1]}(\xi)u) + O(\xi, u)^{m+1}, \\ \Sigma^\infty \; &: \dot{\xi} = A\xi + Bu + \sum_{m=2}^{\infty} (f^{[m]}(\xi) + g^{[m-1]}(\xi)u). \end{aligned}$$

The symbols $\Sigma^{[m]}$, $\Sigma^{[\leq m]}$, and Σ^∞ will stand for the systems under consideration. Their state vector will be denoted by ξ and their control by u. The system $\Sigma^{[m]}$ (resp. $\Sigma^{[\leq m]}$ and Σ^∞) transformed via a feedback transformation Γ^m (resp. $\Gamma^{\leq m}$ and Γ^∞) will be denoted by $\tilde{\Sigma}^{[m]}$ (resp. $\tilde{\Sigma}^{[\leq m]}$ and $\tilde{\Sigma}^\infty$). Its state vector will be denoted by x, its control by v, and the vector fields, defining its dynamics, by $\tilde{f}^{[m]}$ and $\tilde{g}^{[m-1]}$. Feedback equivalence of systems $\Sigma^{[m]}$ and $\tilde{\Sigma}^{[m]}$ and of systems $\Sigma^{[\leq m]}$ and $\tilde{\Sigma}^{[\leq m]}$ will be established via a smooth feedback. To be more precise, via a homogeneous feedback Γ^m in the former case and via a polynomial feedback $\Gamma^{\leq m}$ in the latter. On the other hand, feedback equivalence of systems Σ^∞ and $\tilde{\Sigma}^\infty$ will be established via a formal feedback Γ^∞.

We will use two kinds of normal forms: Kang normal forms and feedforward normal forms. The symbol "bar" will correspond to the vector field $\bar{f}^{[m]}$ defining Kang normal forms $\Sigma_{NF}^{[m]}$, $\Sigma_{NF}^{[\leq m]}$, and Σ_{NF}^∞. The symbol "hat" will correspond to the vector field $\hat{f}^{[m]}$ defining feedforward normal forms $\Sigma_{FNF}^{[m]}$, $\Sigma_{FNF}^{[\leq m]}$, and Σ_{FNF}^∞. Analogously, the m-invariants of the system $\Sigma^{[m]}$ will be denoted by $a^{[m]j,i+2}$, the m-invariants of the system $\Sigma_{NF}^{[m]}$ by $\bar{a}^{[m]j,i+2}$, and the m-invariants of the system $\Sigma_{FNF}^{[m]}$ by $\hat{a}^{[m]j,i+2}$.

3 Feedforward normal form

Since the linear part (F, G) of the system Σ^∞, given by (1), is controllable, we can assume, without loss of generality, that the system is in the form

$$\Sigma^\infty : \quad \dot{\xi} = A\xi + Bu + \sum_{m=2}^{\infty} (f^{[m]}(\xi) + g^{[m-1]}(\xi)u), \tag{3}$$

where (A, B) is in Brunovský canonical form.

Recall that, as proved by Kang [8], see also [16] and [17], any nonlinear system of the form (3) can be put, via a formal feedback transformation Γ^∞, to the following normal form, which we will call *Kang normal form*,

$$\Sigma^\infty_{NF} \; : \dot{x} = Ax + Bv + \sum_{m=2}^{\infty} \bar{f}^{[m]}(x),$$

where for any $m \geq 2$,

$$\bar{f}_j^{[m]}(x) = \begin{cases} \sum\limits_{i=j+2}^{n} x_i^2 P_{j,i}^{[m-2]}(x_1, \cdots, x_i) & \text{if } 1 \leq j \leq n-2, \\ 0 & \text{if } n-1 \leq j \leq n. \end{cases} \tag{4}$$

It is natural to ask whether it is possible to bring a system, that is feedback equivalent to feedforward form, to Kang normal form (4) which would be simultaneously feedforward, that is, which would satisfy

$$P_{j,i}^{[m-2]}(x) = P_{j,i}^{[m-2]}(x_j, \cdots, x_i).$$

Although, this is always possible for the first nonlinearizable term, see Theorem 2 below, in general the answer to the above question is negative. For this reason we will introduce the following notion.

Definition 1. *Feedforward normal form* is the system

$$\Sigma^\infty_{FNF} \; : \dot{x} = Ax + Bv + \sum_{m=2}^{\infty} \hat{f}^{[m]}(x),$$

such that for any $m \geq 2$,

$$\hat{f}_j^{[m]}(x) = \begin{cases} x_j k_j^{[m-1]}(x_j, x_{j+1}) \\ \quad + \sum\limits_{i=j+2}^{n} x_i^2 P_{j,i}^{[m-2]}(x_j, \cdots, x_i), & 1 \leq j \leq n-2, \\ 0 & n-1 \leq j \leq n, \end{cases}$$

where $k_j^{[m-1]}$ and $P_{j,i}^{[m-2]}$ are homogeneous polynomials, of degree $m-1$ and $m-2$, respectively, depending on the indicated variables.

Usefulness of feedforward normal form is justified by the following result.

Theorem 1. *The system Σ^∞, given by (1), is feedback equivalent to feedforward form if and only if it is feedback equivalent to feedforward normal form.*

4 m-invariants

Consider the following homogeneous feedback transformation

$$\Gamma^m : \begin{array}{l} x = \xi + \phi^{[m]}(\xi) \\ u = v + \alpha^{[m]}(\xi) + \beta^{[m-1]}(\xi)v. \end{array}$$

Observe that each transformation Γ^m, for $m \geq 2$, leaves invariant all homogeneous terms of degree smaller than m of the system Σ^∞. We will call Γ^m a *homogeneous feedback transformation of degree m.*

We will recall invariants, found by Kang [8], of the action of Γ^m on the homogeneous system

$$\Sigma^{[m]} : \quad \dot{\xi} = A\xi + Bu + f^{[m]}(\xi) + g^{[m-1]}(\xi)u + O(\xi, u)^{m+1}. \tag{5}$$

Let us define

$$X_i^{m-1}(\xi) = (-1)^i ad^i_{A\xi + f^{[m]}(\xi)}(B + g^{[m-1]}(\xi)).$$

Following Kang [8], we denote by $a^{[m]j,i+2}(\xi)$ the homogeneous part of degree $m-2$ of

$$C_j \left[X_i^{m-1}, X_{i+1}^{m-1}\right]_{|W_{n-i}},$$

where $C_j = (0, \ldots, 0, 1, 0, \cdots, 0)$, with j-th component being equal to 1, and the submanifolds W_i are defined as follows:

$$W_i = \{\xi \in \mathbb{R}^n \ : \ \xi_{i+1} = \cdots = \xi_n = 0\}.$$

Denote $\bar{\xi}_i = (\xi_1, \cdots, \xi_i)$ and set

$$\Delta = \{(j,i) \in \mathbb{N} \times \mathbb{N} \ : \ 1 \leq j \leq n-2, \ 0 \leq i \leq n-j-2\}.$$

The functions $a^{[m]j,i+2}(\xi)$, for $(j,i) \in \Delta$, will be called *m-invariants* of $\Sigma^{[m]}$.

The following result of Kang [8] asserts that m-invariants $a^{[m]j,i+2}(\xi)$ are complete invariants of homogeneous feedback and, moreover, illustrates their meaning for the normal form $\Sigma^{[m]}_{NF}$. Consider two systems: $\Sigma^{[m]}$, given by (5), and $\tilde{\Sigma}^{[m]}$, given by

$$\tilde{\Sigma}^{[m]} : \quad \dot{x} = Ax + Bv + \tilde{f}^{[m]}(x) + \tilde{g}^{[m-1]}(x)v + O(x, v)^{m+1}.$$

Let

$$\{ a^{[m]j,i+2} \ : \ (j,i) \in \Delta\} \ \text{ and } \ \{ \tilde{a}^{[m]j,i+2} \ : \ (j,i) \in \Delta\}$$

denote, respectively, their m-invariants. The following result was proved by Kang [8].

Proposition 1. *The m-invariants have the following properties:*

(i) *The systems $\Sigma^{[m]}$ and $\tilde{\Sigma}^{[m]}$ are equivalent via a homogeneous feedback transformation Γ^m, modulo higher order terms, if and only if*

$$a^{[m]j,i+2} = \tilde{a}^{[m]j,i+2},$$

for any $(j,i) \in \Delta$.

(ii) *The m-invariants $\bar{a}^{[m]j,i+2}$ of the normal form*

$$\Sigma_{NF}^{[m]} \ : \ \dot{x} = Ax + Bv + \bar{f}^{[m]}(x) + O(x,v)^{m+1},$$

where $\bar{f}^{[m]}(x)$ is of the form (4), are given by

$$\bar{a}^{[m]j,i+2} = \frac{\partial^2}{\partial x_{n-i}^2} x_{n-i}^2 P_{j,n-i}^{[m-2]}(\bar{x}_{n-i}),$$

for any $(j,i) \in \Delta$.

5 Main results

Consider the system Σ^∞, given by (3), and let m_0 be the largest integer such that all distributions

$$\mathcal{D}^k = \text{span}\left\{g, ad_f g, \cdots, ad_f^{k-1} g\right\},$$

for $1 \leq k \leq n-1$, are involutive modulo terms of order $m_0 - 2$. As proved by Krener [10], this means that the system Σ^∞, given by (3), is feedback linearizable up to order $m_0 - 1$. Since a system linearizable up to order m is also feedback equivalent to a feedforward form up to the same order, we can assume without loss of generality that the system Σ^∞ is in feedforward normal form up to order $m_0 - 1$ (see Theorem 1), that is, it takes the form

$$\begin{aligned} \Sigma^{[\leq m_0]} \ : \ \dot{\xi} = A\xi + Bu + &\sum_{m=2}^{m_0-1} h^{[m]}(\xi) \\ &+ f^{[m_0]}(\xi) + g^{[m_0-1]}(\xi)u + O(\xi,u)^{m_0+1}, \end{aligned} \tag{6}$$

where, for any $2 \leq m \leq m_0 - 1$, we have

$$h_j^{[m]}(\xi) = \begin{cases} \xi_j k_j^{[m-1]}(\xi_j, \xi_{j+1}) & \text{if } \ 1 \leq j \leq n-2, \\ 0 & \text{if } \ n-1 \leq j \leq n. \end{cases} \tag{7}$$

5.1 Feedforward form: first nonlinearizable term

Let us denote by $a^{[m_0]j,i+2}$ the m_0-invariants associated to the homogeneous part of degree m_0 of the system (6)-(7). We get the following theorem.

Theorem 2. *There exists a transformation $\Gamma^{\leq m_0}$ bringing the system* (6) *into feedforward form, up to order m_0, if and only if for any $(j,i) \in \Delta$ and any $1 \leq q \leq j-1$ we have*

$$L_{A^{n-q}B}a^{[m_0]j,i+2} = 0. \tag{8}$$

Corollary 1. *If there exists a transformation Γ^{∞} bringing the system*

$$\Sigma^{\infty} \ : \ \dot{\xi} = A\xi + Bu + \sum_{m=m_0}^{\infty} (f^{[m]}(\xi) + g^{[m-1]}(\xi)u)$$

to feedforward form, then the condition (8) *is satisfied for any $(j,i) \in \Delta$ and any $1 \leq q \leq j-1$.*

In other words, the above result says that if a system is feedback equivalent to feedforward form, then, after having linearized lower order terms, the first nonlinearizable term must be feedforward when transformed to Kang normal form. This is the case if and only if the condition (8) is satisfied. If $m_0 = 2$, then the m_0-invariants $a^{[2]j,i+2}$ are constant and the condition (8) is satisfied which implies that any system is equivalent to feedforward form up to order 2. Actually, in this case, Kang-Krener normal form [9] is strict feedforward and can serve as feedforward normal form (we do not have to add the vector field $h^{[2]}$).

Corollary 2. (Kang-Krener) *If $m_0 = 2$ then the system* (6) *is always equivalent to feedforward form up to order 2.*

5.2 Feedforward form: the general step

According to Theorem 2, Kang normal form of the first nonlinearizable term of a system, which is feedback equivalent to feedforward form, must be feedforward. We will see in the sequel, that the situation gets different when we proceed to higher order terms. Let us assume that the system Σ^{∞}, given by (3), is in feedforward normal form up to order $m_0 + l - 1$, that is Σ^{∞} takes the form

$$\begin{aligned} \Sigma^{\infty} : \ \ \dot{\xi} = A\xi + Bu + \sum_{m=2}^{m_0+l-1} h^{[m]}(\xi) + \sum_{m=m_0}^{m_0+l-1} \bar{f}^{[m]}(\xi) \\ + f^{[m_0+l]}(\xi) + g^{[m_0+l-1]}(\xi)u + O(\xi,u)^{m_0+l+1}, \end{aligned} \tag{9}$$

where for any $m_0 \le m \le m_0 + l - 1$,

$$\bar{f}_j^{[m]}(\xi) = \begin{cases} \sum\limits_{i=j+2}^{n} \xi_i^2 P_{j,i}^{[m-2]}(\xi_j, \cdots, \xi_i) & \text{if } 1 \le j \le n-2, \\ 0 & \text{if } n-1 \le j \le n, \end{cases} \tag{10}$$

and

$$h_j^{[m]}(\xi) = \begin{cases} \xi_j k_j^{[m-1]}(\xi_j, \xi_{j+1}) & \text{if } 1 \le j \le n-2, \\ 0 & \text{if } n-1 \le j \le n. \end{cases} \tag{11}$$

By the definition of m_0, it follows that there exists $1 \le j \le n-2$ such that $\bar{f}_j^{[m_0]} \ne 0$. Throughout the paper we will assume that $\bar{f}_{n-2}^{[m_0]} \ne 0$, which simplifies the exposition. The analysis of the general case, although more technical, follows the same line and will be given in [20].

For any $1 \le k \le n-3$ and any $0 \le s \le m$, consider the homogeneous vector fields

$$Y_{k,s}^{[m]} = \xi_k^s \xi_{k+1}^{m-s} \frac{\partial}{\partial \xi_{k+1}} + L_{A\xi}(\xi_k^s \xi_{k+1}^{m-s}) \frac{\partial}{\partial \xi_{k+2}} + \cdots + L_{A\xi}^{n-k-1}(\xi_k^s \xi_{k+1}^{m-s}) \frac{\partial}{\partial \xi_n},$$

which are crucial objets in our study. Denote by $a^{[m_0+l]j,i+2}$ the (m_0+l)-invariants of the homogeneous system $\Sigma^{[m_0+l]}$, defined by (9)-(11), and by $a_{k,s}^{[m_0+l]j,i+2}$ the (m_0+l)-invariants of the homogeneous system

$$\Sigma_{k,s}^{[m_0+l]} \ : \ \dot{\xi} = A\xi + Bu + \left[\bar{f}^{[m_0]}, Y_{k,s}^{[l+1]}\right](\xi) + O(\xi, u)^{m_0+l+1}.$$

Recall that

$$\Delta = \{(j,i) \in \mathbb{N} \times \mathbb{N} \ : \ 1 \le j \le n-2, \ 0 \le i \le n-j-2\}.$$

Our main result can be stated as follows.

Theorem 3. *The system Σ^∞, defined by (9)-(11), is feedback equivalent, up to order $m_0 + l$, to feedforward form if and only if there exist real constants $\sigma_{k,s}$ for $1 \le k \le n-3$ and $1 \le s \le l+1$ such that for any $(j,i) \in \Delta$ and any $1 \le q \le j-1$*

$$L_{A^{n-q}B} \left(a^{[m_0+l]j,i+2} - \sum_{k=1}^{n-3} \sum_{s=1}^{l+1} \sigma_{k,s} a_{k,s}^{[m_0+l]j,i+2} \right) = 0. \tag{12}$$

Notice that (12) is an invariant way of expressing the fact that

$$a^{[m_0+l]j,i+2} - \sum_{k=1}^{n-3} \sum_{s=1}^{l+1} \sigma_{k,s} a_{k,s}^{[m_0+l]j,i+2} = Q_{j,i}^{[m_0+l-2]},$$

where $Q_{j,i}^{[m_0+l-2]} = Q_{j,i}^{[m_0+l-2]}(\xi_j, \cdots, \xi_{n-i})$ are homogeneous polynomials of degree $m_0 + l - 2$ depending on the indicated variables only.

Observe that checking the conditions (8) and (12) involves only differentiation of polynomials and algebraic operations. Therefore Theorem 2 followed by a successive application of Theorem 3 yields to a constructive procedure that allows us to check whether a given system can be transformed into feedforward form. Moreover, for any system satisfying the conditions (8) and (12), we can calculate, step by step, explicit feedback transformations bringing it into feedforward form using transformations constructed in [16] and [17].

5.3 Strict feedforward form

We will say that the system

$$\Sigma \ : \ \dot{\xi} = f(\xi) + g(\xi)u$$

is *strict feedforward* or that it is in *strict feedforward form* if for every $1 \le j \le n-1$ we have $f_j(\xi_{j+1}, \ldots, \xi_n)$ and $g_j(\xi_{j+1}, \ldots, \xi_n)$. and, moreover, $f_n = 0$, and $g_n = c$, where $c \in \mathbb{R}$, $c \neq 0$. In [18] and [19] we gave necessary and sufficient conditions, as well as a constructive procedure, to bring any system, for which it is possible, into strict feedforward form. A necessary condition is, compare Theorem 2 and Corollary 1, that the first nonlinearizable term must be strict feedforward when brought to Kang normal form. Assume that the system

$$\begin{aligned} \Sigma^\infty \ : \ \dot{\xi} = A\xi + Bu + \sum_{m=2}^{m_0+l-1} h^{[m]}(\xi) + \sum_{m=m_0}^{m_0+l-1} \bar{f}^{[m]}(\xi) \\ + f^{[m_0+l]}(\xi) + g^{[m_0+l-1]}(\xi)u + O(\xi, u)^{m_0+l+1}, \end{aligned}$$

is in strict feedforward form up to order $m_0 + l - 1$ and, additionally, that $\bar{f}_{n-2}^{[m_0]} \neq 0$. In this case, the main result of [19] can be stated as follows.

Theorem 4. *The system Σ^∞ is feedback equivalent to strict feedforward form, up to order $m_0 + l$, if and only if there exist real constants $\sigma_{0,0}$, $\sigma_{1,0}$, $\cdots$, $\sigma_{n-3,0}$ such that for any $(j,i) \in \Delta$ and any $1 \le q \le j$*

$$L_{A^{n-q}B} \left(a^{[m_0+l]j,i+2} - \sum_{k=0}^{n-3} \sigma_{k,0} a_{k,0}^{[m_0+l]j,i+2} \right) = 0. \tag{13}$$

Above, $a_{k,0}^{[m_0+l]j,i+2}$ stand for homogeneous invariants associated to the homogeneous system

$$\Sigma_{k,0}^{[m_0+l]} \ : \ \dot{\xi} = A\xi + Bu + \left[\bar{f}^{[m_0]}, Y_k^{[l+1]} \right](\xi) + O(\xi, u)^{m_0+l+1},$$

where, for any $0 \le k \le n-3$,

$$Y_k^{[l+1]} = Y_{k,0}^{[l+1]} = \xi_{k+1}^{l+1} \frac{\partial}{\partial \xi_{k+1}} + L_{A\xi} \xi_{k+1}^{l+1} \frac{\partial}{\partial \xi_{k+2}} + \cdots + L_{A\xi}^{n-k-1} \xi_{k+1}^{l+1} \frac{\partial}{\partial \xi_n}.$$

Observe that the condition (12) can be seen as a natural generalization of (13). Indeed, in the second sum of (12) we can start the summation with $s=0$. It is so, because the action of $Y_{k,0}^{[l+1]}$ on all components, starting form the $(k+2)^{nd}$-component, can be compensated by that of $Y_{k,s}^{[l+1]}$, for $s\geq 1$, and the action of $Y_{k,0}^{[l+1]}$ on the $(k+1)^{st}$-component is irrelevant since, for a feedforward system, this component can depend on all variables $\xi_{k+1},\cdots,\xi_n$.

5.4 Nice feedforward form

Now we will consider an intermediate case between feedforward and strict feedforward forms. A system

$$\Sigma\ :\ \dot{\xi}=f(\xi)+g(\xi)u$$

is called *nice feedforward* if it is feedforward and if additionally the components $f_j(\xi_j,\ldots,\xi_n)$ and $g_j(\xi_j,\ldots,\xi_n)$ are affine with respect to the variable ξ_j. We can prove, see [20], that a system Σ with controllable linear approximation is feedback equivalent to a nice feedforward form if and only if it is feedback equivalent to the following *nice feedforward normal form*

$$\Sigma^{\infty}_{NFNF}\ :\ \dot{x}=Ax+Bv+\sum_{m=2}^{\infty}\hat{f}^{[m]}(x),$$

such that for any $m\geq 2$,

$$\hat{f}_j^{[m]}(x)=\begin{cases} h_j^{[m]}(x)+\sum\limits_{i=j+2}^{n}x_i^2P_{j,i}^{[m-2]}(x_j,\cdots,x_i) & \text{if } 1\leq j\leq n-2,\\ \\ 0 & \text{if } n-1\leq j\leq n,\end{cases}$$

where $h_j^{[m]}(x)=c_{m,j}x_{j+1}^m+d_{m,j}x_jx_{j+1}^{m-1}$, with $c_{m,j},d_{m,j}\in\mathbb{R}$, and $P_{j,i}^{[m-2]}$ are homogeneous polynomials affine with respect to the variable x_j.

Suppose that the system

$$\begin{aligned}\Sigma^{\infty}\ :\ \dot{\xi}=A\xi+Bu+\sum_{m=2}^{m_0+l-1}h^{[m]}(\xi)+\sum_{m=m_0}^{m_0+l-1}\bar{f}^{[m]}(\xi)\\ +f^{[m_0+l]}(\xi)+g^{[m_0+l-1]}(\xi)u+O(\xi,u)^{m_0+l+1}\end{aligned}$$

is in nice feedforward form up to order m_0+l-1. Let us assume that $\frac{\partial\bar{f}_{n-2}^{[m_0]}}{\partial x_{n-2}}\neq 0$. In this case we have (see [20]) the following result.

Theorem 5. *The system Σ^{∞} is feedback equivalent to nice feedforward form, up to order m_0+l, if and only if there exist real constants $\sigma_{0,s}$, $\sigma_{1,s}$, $\cdots$,*

$\sigma_{n-3,s}$, *for* $s = 0, 1$, *such that for any* $(j,i) \in \Delta$ *and any* $1 \leq q \leq j-1$

$$\begin{aligned} L_{A^{n-q}B}\left(a^{[m_0+l]j,i+2} - \sum_{k=0}^{n-3}\left(\sigma_{k,0}a_{k,0}^{[m_0+l]j,i+2} + \sigma_{k,1}a_{k,1}^{[m_0+l]j,i+2}\right)\right) &= 0 \\ L^2_{A^{n-j}B}\left(a^{[m_0+l]j,i+2} - \sum_{k=0}^{n-3}\left(\sigma_{k,0}a_{k,0}^{[m_0+l]j,i+2} + \sigma_{k,1}a_{k,1}^{[m_0+l]j,i+2}\right)\right) &= 0. \end{aligned} \tag{14}$$

In the general case, that is if we drop the assumption $\frac{\partial \bar{f}^{[m_0]}_{n-2}}{\partial x_{n-2}} \neq 0$, the conditions (14) involve not only linear but also quadratic relations between invariants (see [20]).

6 Examples

In this section we will give three examples illustrating results developed in the paper.

Example 1 Consider the following system

$$\Sigma \; : \; \dot{\xi} = f(\xi) + g(\xi)u, \quad \xi(\cdot) \in \mathbb{R}^3, \quad u(\cdot) \in \mathbb{R},$$

whose linear approximation is controllable. The system Σ is always feedback equivalent to feedforward form. Indeed, it is always possible to normalize the vector field g and to linearize the two last components of the vector field f in order to get

$$\begin{aligned} \dot{x}_1 &= x_2 + f_1(x_1, x_2, x_3) \\ \dot{x}_2 &= x_3 \\ \dot{x}_3 &= v, \end{aligned}$$

which obviously is feedforward. In particular, we can transform f_1 to its Kang normal form $f_1 = x_3^2 P(x_1, x_2, x_3)$, which gives feedforward normal form of the system.

Example 2 We consider the well known ball-and-beam example [5], whose Lagrange equations are given by

$$\begin{aligned} (\tfrac{J_b}{R^2} + 1)\ddot{r} + G\sin\theta + \beta\dot{r} - r\dot{\theta}^2 &= 0 \\ (r^2 + J_b)\ddot{\theta} + 2r\dot{r}\dot{\theta} + Gr\cos\theta &= \tau, \end{aligned}$$

where we take the mass of the ball equal to one and the momentum of the beam equal to zero. Above, J_b denotes the momentum of the ball, r its position, τ the torque applied to the beam, θ its angle with respect to the horizontal and $\beta \geq 0$ the viscous friction constant, and g the gravity constant. We set $\tau = 2r\dot{r}\dot{\theta} + Gr\cos\theta + (r^2 + J)u$, where u denotes the control variable.

In the coordinates system $(\xi_1, \xi_2, \xi_3, \xi_4) = (r, \dot{r}, \theta, \dot{\theta})$, we obtain the following equations

$$\begin{aligned}\dot{\xi}_1 &= \xi_2 \\ \dot{\xi}_2 &= -\beta\xi_2 - G\sin\xi_3 + \xi_1\xi_4^2 \\ \dot{\xi}_3 &= \xi_4 \\ \dot{\xi}_4 &= u.\end{aligned} \tag{15}$$

Applying a suitable feedback transformation (see [15] for details), we show that the ball-and-beam system is feedback equivalent to the following normal form

$$\begin{aligned}\dot{x}_1 &= x_2 + \beta x_1 x_3^2 + O(x)^4 \\ \dot{x}_2 &= x_3 + x_1 x_4^2 + O(x)^4 \\ \dot{x}_3 &= x_4 \\ \dot{x}_4 &= v.\end{aligned}$$

The ball-and-beam system is thus feedback linearizable up to order 2 and we have $m_0 = 3$. Applying Theorem 2, with $m_0 = 3$, we conclude that the vector field $\bar{f}^{[3]}(x) = \beta x_1 x_3^2 \frac{\partial}{\partial x_1} + x_1 x_4^2 \frac{\partial}{\partial x_1}$, which is in Kang normal form, should be feedforward. It is not the case and thus, independently of the value of the friction constant β, the ball-and-beam system is not feedback equivalent, up to order 3, to a feedforward form, which completes an observation, which we made in [19], that the ball-and-beam system is not feedback equivalent, up to order 3, to a strict feedforward form.

In the next example, we will consider third order terms of systems on $\mathbb{R}^4$.

Example 3 Consider the system Σ defined on $\mathbb{R}^4$ by

$$\begin{aligned}\dot{\xi}_1 &= \xi_2 + a\xi_3^2 + b\xi_4^2 \\ \dot{\xi}_2 &= \xi_3 + c\xi_4^2 + d\xi_1\xi_4^2 \\ \dot{\xi}_3 &= \xi_4 \\ \dot{\xi}_4 &= u.\end{aligned} \tag{16}$$

This system is in Kang normal form. Notice that we do not specify terms of order 3 of the first component because they are always feedforward. We would like to express, in terms of the coefficients a, b, c, d, conditions for the system to be equivalent to a feedforward form. If $d = 0$, then the system is already in feedforward form so in what follows we will assume that $d \neq 0$.

We have $\bar{f}^{[2]}(\xi) = (a\xi_3^2 + b\xi_4^2)\frac{\partial}{\partial \xi_1} + c\xi_4^2 \frac{\partial}{\partial \xi_2}$. Moreover,

$$Y_{1,1}^{[2]} = \xi_1\xi_2 \frac{\partial}{\partial \xi_2} + (\xi_1\xi_3 + \xi_2^2)\frac{\partial}{\partial \xi_3} + (\xi_1\xi_4 + 3\xi_2\xi_3)\frac{\partial}{\partial \xi_4}$$

and

$$Y_{1,2}^{[2]} = \xi_1^2 \frac{\partial}{\partial \xi_2} + 2\xi_1\xi_2 \frac{\partial}{\partial \xi_3} + 2(\xi_1\xi_3 + \xi_4^2)\frac{\partial}{\partial \xi_4}.$$

For $s = 1, 2$, the 3-invariants $a_{1,s}^{[3]j,i+2}$ associated to the homogeneous system

$$\dot{\xi} = A\xi + Bu + [\bar{f}^{[2]}, Y_{1,s}^{[2]}](\xi) + O(\xi, u)^4$$

are given, for any $(j,i) \in \Delta = \{(1,0),(1,1),(2,0)\}$, by (see Section 4)

$$a_{1,s}^{[3]j,i+2} = C_j(-ad_{A^{i+1}B}X_i^{[2]} + ad_{A^iB}X_{i+1}^{[2]})_{|W_{n-i}},$$

where

$$X_i^{[2]} = (-1)^i ad^i_{A\xi+[\bar{f}^{[2]},Y_{1,s}^{[2]}]}(B).$$

Therefore we obtain

$$a_{1,1}^{[3]2,2}(\xi) = 2b\xi_2 - 2c\xi_1$$

and

$$a_{1,2}^{[3]2,2}(\xi) = 4b\xi_1.$$

We also have $\bar{f}^{[3]}(\xi) = d\xi_1\xi_4^2\frac{\partial}{\partial\xi_2}$. By Theorem 3, we conclude that the system (16) is feedback equivalent to feedforward form up to order 3 if and only if there exist real constants $\sigma_{1,1}$ and $\sigma_{1,2}$ such that for $(j,i) = (2,0) \in \Delta$, we have

$$L_{A^3B}(a^{[3]2,2} - \sigma_{1,1}a_{1,1}^{[3]2,2} - \sigma_{1,2}a_{1,2}^{[3]2,2}) = 0.$$

This implies the following relation

$$d + c\sigma_{1,1} - 2b\sigma_{1,2} = 0.$$

Then a necessary (resp. a necessary and sufficient) condition for the system to be feedback equivalent (resp. feedback equivalent up to order 3) to feedforward form is given by $(b,c) \neq (0,0)$.

7 Feedforward systems in $\mathbb{R}^4$

In this section we will analyze, for systems on $\mathbb{R}^4$, the general step of our procedure that allows to verify whether the system is feedback equivalent to feedforward form. Consider a system on $\mathbb{R}^4$ and assume that it is feedback equivalent to feedforward form up to terms of degree m. Then, according to Theorem 1, we can assume, without loss of generality, that the system takes the form

$$\begin{aligned}
\dot{\xi}_1 &= \xi_2 + \sum_{k=2}^{m} \hat{f}_1^{[k]}(\xi) + \bar{f}_1^{[m+1]}(\xi) \\
\dot{\xi}_2 &= \xi_3 + \sum_{k=2}^{m} \hat{f}_2^{[k]}(\xi) + \bar{f}_2^{[m+1]}(\xi) \\
\dot{\xi}_3 &= \xi_4 \\
\dot{\xi}_4 &= u,
\end{aligned} \tag{17}$$

where the homogeneous vector fields $\hat{f}_1^{[k]}\frac{\partial}{\partial \xi_1} + \hat{f}_2^{[k]}\frac{\partial}{\partial \xi_2}$, for $2 \le k \le m$, are in feedforward normal form (see Definition 1) and

$$
\begin{aligned}
\bar{f}^{[m+1]} &= \bar{f}_1^{[m+1]}\frac{\partial}{\partial \xi_1} + \bar{f}_2^{[m+1]}\frac{\partial}{\partial \xi_2} \\
&= (\xi_3^2 P_1^{[m-1]}(\xi_1,\xi_2,\xi_3) + \xi_4^2 Q_1^{[m-1]}(\xi_1,\xi_2,\xi_3,\xi_4))\frac{\partial}{\partial \xi_1} \\
&\quad + (\xi_4^2 Q_2^{[m-1]}(\xi_1,\xi_2,\xi_3,\xi_4))\frac{\partial}{\partial \xi_2}
\end{aligned}
$$

is in Kang normal form. Let us consider, for any $1 \le s \le m$, the following homogeneous vector field

$$
Y_{1,s}^{[m]} = \xi_1^s \xi_2^{m-s}\frac{\partial}{\partial \xi_2} + L_{A\xi}(\xi_1^s \xi_2^{m-s})\frac{\partial}{\partial \xi_3} + L_{A\xi}^2(\xi_1^s \xi_2^{m-s})\frac{\partial}{\partial \xi_4}
$$

and construct the corresponding homogeneous system

$$
\Sigma_{1,s}^{[m+1]} \;:\; \dot{\xi} = A\xi + Bu + [\bar{f}^{[2]}, Y_{1,s}^{[m]}](\xi) + O(\xi,u)^{m+2},
$$

where Kang-Krener quadratic normal form $\bar{f}^{[2]}$ is given by

$$
\bar{f}^{[2]} = (a\xi_3^2 + b\xi_4^2)\frac{\partial}{\partial \xi_1} + c\xi_4^2\frac{\partial}{\partial \xi_2}.
$$

Let us denote by $a_{1,s}^{[m+1]j,i+2}$ and $a^{[m+1]j,i+2}$, for $(j,i) \in \Delta$, the homogeneous invariants associated, respectively, to the homogeneous systems $\Sigma_{1,s}^{[m+1]}$ and to the normal form

$$
\Sigma_{NF}^{[m+1]} \;:\; \dot{\xi} = A\xi + Bu + \bar{f}^{[m+1]}(\xi) + O(\xi,u)^{m+2}.
$$

Here (A,B) is the Brunovský canonical form of dimension 4. Denote $C_2 = (0,1,0,0)$. We have $\Delta = \{(1,0);(1,1);(2,0)\}$. Since the only term that is not feedforward is present if $Q_2^{[m-1]}$ depends on ξ_1, we will focus our attention only on invariants $a_{1,s}^{[m+1]2,2}$ and $a^{[m+1]2,2}$ given, respectively, by

$$
a_{1,s}^{[m+1]2,2} = C_2 ad_B^2 [\bar{f}^{[2]}, Y_{1,s}^{[m]}]
$$

and

$$
a^{[m+1]2,2} = C_2 ad_B^2 \bar{f}^{[m+1]} = \frac{\partial^2 \xi_4^2 Q_2^{[m-1]}}{\partial \xi_4^2}.
$$

A direct computation gives

$$
a_{1,s}^{[m+1]2,2} = 2bs\xi_1^{s-1}\xi_2^{m-s} - 2c(m-s)\xi_1^s\xi_2^{m-s-1}.
$$

Then the system (17) is feedback equivalent to feedforward form up to order $m+1$ if and only if there exist real constants $\sigma_{1,s}$, for $1 \le s \le m-1$, such that

$$a^{[m+1]2,2} - \sum_{s=1}^{m-1} \sigma_{1,s} a_{1,s}^{[m+1]2,2} = Q^{[m-1]}(\xi_2, \xi_3, \xi_4),$$

for some homogeneous polynomial $Q^{[m-1]}(\xi_2, \xi_3, \xi_4)$, which is equivalent to the condition that

$$Q_2^{[m-1]}(\xi) = S_1^{[m-1]}(\xi_1, \xi_2) + S_2^{[m-1]}(\xi_2, \xi_3, \xi_4).$$

The codimension c_m of the space of systems which are feedback equivalent to feedforward form, up to order $m+1$, is equal to the dimension of the space of all homogeneous polynomials of degree $m-1$ of the form

$$Q_2^{[m-1]}(\xi) = \xi_1 \xi_3 R_1^{[m-3]}(\xi_1, \xi_2, \xi_3) + \xi_1 \xi_4 R_2^{[m-3]}(\xi_1, \xi_2, \xi_3, \xi_4).$$

We thus get

$$c_m = \frac{(m-1)(m-2)}{2} + \frac{m(m-1)(m-2)}{6} = \frac{(m+3)(m-1)(m-2)}{6}.$$

References

1. V.I. Arnold, *Geometrical Methods in the Theory of Ordinary Differential Equations*, Second Edition, Springer-Verlag, 1988.
2. A. Astolfi and F. Mazenc, A geometric characterization of feedforward forms, in *Proc. MTNS'2000*, Perpignan, France, 2000.
3. S. Battilotti, Semiglobal stabilization of uncertain block-feedforward forms via measurement feedback, in *Proc.of NOLCOS'98*, Enschede, the Netherlands, (1998), pp. 342-347.
4. M. Fliess, J. Lévine, P. Martin, and P. Rouchon, Flatness and defect of non-linear systems: Introductory theory and examples, *International Journal of Control*, 6 (1995), pp. 1327-1361.
5. J. Hauser, S. Sastry, P. Kokotović, Nonlinear control via approximate input-output liearization, *IEEE Trans. Aut. Control*, 37 (1992), pp. 392-398.
6. M. Jankovic, R. Sepulchre, and P. Kokotovic, Constructive Lyapunov stabilization of nonlinear cascade systems, *IEEE Trans. Automat. Control*, 41 (1996), pp. 1723-1735.
7. T. Kailath, *Linear Systems*, Prentice-Hall, Englewood Cliffs, NJ, 1980.
8. W. Kang , Extended controller form and invariants of nonlinear control systems with single input, *J. of Mathem. Systems, Estimat. and Control*, 4 (1994), pp. 253-256.
9. W. Kang and A.J. Krener, Extended quadratic controller normal form and dynamic feedback linearization of nonlinear systems, *SIAM J. Control and Optim.*, 30 (1992), pp.1319-1337.

10. A.J. Krener, Approximate linearization by state feedback and coordinate change, *Systems Control Letters*, 5 (1984), pp. 181-185.
11. A. Marigo, Constructive necessary and sufficient conditions for strict triangularizability of driftless nonholonomic systems,in *Proc. 34th CDC*, Phoenix, Arizona, USA, (1999), pp. 2138-2143.
12. F. Mazenc and L. Praly, Adding integrations, saturated controls, and stabilization for feedforward forms, *IEEE Trans. Automat. Control*, 41 (1996), pp. 1559-1578.
13. F. Mazenc and L. Praly, Asymptotic tracking of a reference state for systems with a feedforward structure, *Automatica*, 36 (2000), pp. 179-187.
14. Sepulchre R., Janković M., and Kokotović P. *Constructive Nonlinear Control*, Springer, Berlin-Heidelberg-New York, 1996.
15. I.A. Tall, Classification par bouclage des systèmes de contrôles non linéaires mono-entrée: les formes normales, les formes canoniques, les invariants et les symétries. PhD Thesis, INSA de Rouen, 2000.
16. I.A. Tall and W. Respondek, Feedback classification of nonlinear single-input control systems with controllable linearization: normal forms, canonical forms, and invariants, to appear in *SIAM Journal on Control and Optimization.*
17. I.A. Tall and W. Respondek, Normal forms, canonical forms, and invariants of single input nonlinear systems under feedback, in *Proc. 39th IEEE Conference on Decision and Control*, Sydney, Australia, 2000, 1625-1630.
18. I.A. Tall and W. Respondek, Transforming a single-input nonlinear system to a feedforward form via feedback, in *Nonlinear Control in the Year 2000*, A. Isidori, F. Lamnabhi-Lagarrigue, and W. Respondek (eds.), Springer, 2 (2000), pp. 527-542.
19. I.A. Tall and W. Respondek, Feedback equivalence to a strict feedforward form of nonlinear single-input control systems, to appear in *International Journal of Control.*
20. I.A. Tall and W. Respondek, Feedforward forms of nonlinear single-input control systems under feedback, in preparation.
21. A. Teel, *Feedback stabilization: nonlinear solutions to inherently nonlinear problems*, Memorandum UCB/ERL M92/65.
22. A. Teel, A nonlinear small gain theorem for the analysis of control systems with saturation, *IEEE Trans Autom Control*, 41 (1996), pp. 1256-1270.

Conservation Laws in Optimal Control*

Delfim F. M. Torres

Control Theory Group
R&D Unit "Mathematics and Applications"
Department of Mathematics, University of Aveiro
3810-193 Aveiro, Portugal
delfim@mat.ua.pt

Abstract. Conservation laws, i.e. conserved quantities along Euler–Lagrange extremals, which are obtained on the basis of Noether's theorem, play a prominent role in mathematical analysis and physical applications. In this paper we present a general and constructive method to obtain conserved quantities along the Pontryagin extremals of optimal control problems, which are invariant under a family of transformations that explicitly change all (time, state, control) variables.

1 Introduction

A number of conservation laws – first integrals of Euler–Lagrange differential equations – are well known in physics (see e.g. [2], [29]). The most famous conservation law is the *energy integral*, discovered by Leonhard Euler in 1744 [10]: when the Lagrangian L corresponds to a conservative system of point masses, then

$$-L + \frac{\partial L}{\partial \dot{x}} \cdot \dot{x} \equiv \text{constant} \tag{1}$$

holds along the solutions of the Euler–Lagrange equations. In 1877, Erdmann published a generalization of the above [9]: in the *autonomous* case, i.e. when the *Lagrangian* L does not depend explicitly on time t, relation (1) is a first-order necessary optimality condition of the corresponding basic problem of the calculus of variations. Conservation law (1) is now known as the second Erdmann condition.

Emmy Amalie Noether, a distinguished german mathematician, was the one to prove, in 1918 (cf. [19]), that conservation laws in the calculus of variations are a manifestation of a very general principle: "The invariance of a system with respect to a parameter-transformation, implies the existence of a conservation law for that system". Her result comprises all theorems on first integrals known to classical mechanics. Thus, for example, the invariance relative to translation with respect to time yields the energy integral, while

* The work is part of the author's Ph.D. project which is carried out at the University of Aveiro, Portugal, under supervision of A. V. Sarychev. The research has been partially supported by the program PRODEP III 5.3/C/200.009/2000 and the European TMR Research Network ERB-4061-PL-97.

F. Colonius, L. Grüne (Eds.): Dynamics, Bifurcations, and Control, LNCIS 273, pp. 287–296, 2002.

conservation of linear and angular momenta reflect, respectively, translational and rotational invariance. Noether's theorem is applicable also in quantum mechanics, field theory, electromagnetic theory, and has deep implications in the general theory of relativity. Her result is so general and powerful and the paper [19] so deep and rich that a overwhelming number of "generalizations" of Noether's theorem are indeed particular cases (see [29]).

Typical application of conservation laws is to lower the order of the differential equations (see e.g. [3], [5], [20]). In this direction, conservation laws may also simplify the solution of the optimal control problems (see [28]). They are, however, a useful tool for many other reasons. Several important applications of conservation laws, both physical and mathematical, can be found in the literature. For example, in the calculus of variations they have been used to prove Lipschitzian regularity of the minimizers (see [8]), to construct examples with the Lavrentiev phenomenon (see [14]), and to prove existence of minimizers (see [7]). In control theory, for purposes of analyzing stability, controllability, etc. of nonlinear control systems, conservation laws have been used for the system decomposition in terms of lower dimensional subsystems (see [13] and [22]), while in optimal control they were used also to prove Lipschitzian regularity of the minimizers (see [24]). Extensions of Noether's theorem can also be found in the literature. For example, an analog of Noether's theorem for discrete systems, such as cellular automata on finite lattices, can be found in [4] or [12]. Here we are interested in formulations of Noether's theorem for *optimal control* problems.

Resolution of problems of optimal control is reduced to integration of Hamiltonian differential equations. Generalizations of E. Noether's theorem for control systems whose dynamics can be described by Hamiltonian equations of motion (*Hamiltonian control systems*) are found in the works of van der Schaft [27] (see also [18, Ch. 12] and references therein). Generalizations of the theorem of E. Noether from the calculus of variations to optimal control have been carried out by Sussmann in [25], by Jurdjevic in [16, Ch. 13] (see also [17]) and by Torres in [26]. In [25] and [16] the parameter-transformation depend on the state variables while in [26] the transformations may also depend on the independent and control variables. In [25] and [16] the parameter-transformation acts on the state variables. In [26] a second step was done, via time-reparameterization, and, besides the state variables, time-transformation is also permitted. In this work we extend previous optimal control formulations of Noether's theorem (Theorem 3 in §3.2). We deal with a parameter-transformation depending and acting simultaneously on time, state, and control variables. The proof of the result is based on a necessary and sufficient condition that is trivially obtained from the definition of conservation law and from the definition of the Pontryagin extremal (Theorem 2 in §3.1). Our approach is straightforward and, in contrast to [26], no time-reparameterization is required in the proof. The result is constructive and easily leads to extra conservation laws and new information (§4).

2 Preliminaries

Problems of optimal control can be formulated in several different ways. Well known examples are the Mayer, Lagrange, Bolza and time-optimal forms. All these are theoretically equivalent (see e.g. [6]).

2.1 The Lagrange Problem of Optimal Control

We will deal with problems of optimal control in Lagrange form.

Problem (P).

$$J\left[x(\cdot),u(\cdot)\right] = \int_a^b L\left(t,x(t),u(t)\right)\,\mathrm{d}t \longrightarrow \text{extr}\,, \tag{2a}$$

$$\dot{x}(t) = \varphi\left(t,x(t),u(t)\right) \quad \text{a.e. on } [a,b]\,. \tag{2b}$$

Here $\dot{x}(t) = \frac{\mathrm{d}x(t)}{\mathrm{d}t}$. The integral *cost functional* (2a), to be minimized or maximized, and the *dynamic control system* (2b), involve functions $L : \mathbb{R} \times \mathbb{R}^n \times \mathbb{R}^r \to \mathbb{R}$ and $\varphi : \mathbb{R} \times \mathbb{R}^n \times \mathbb{R}^r \to \mathbb{R}^n$, assumed to be continuously differentiable with respect to all variables. Problem (P) will be considered in the space

$$\mathcal{A} = W_{1,1}^n\left([a,b];\mathbb{R}^n\right) \times L_\infty^r\left([a,b];\Omega\right)\,,$$

that is, we shall consider the pairs $(x(\cdot),u(\cdot))$ where the *state trajectory* $x(\cdot)$ is an absolutely continuous n-dimensional vector function, and the *control* $u(\cdot)$ is a measurable and essentially bounded r-dimensional vector function taking values on a given arbitrary set $\Omega \subseteq \mathbb{R}^r$, called the *control constraint set*.

Definition 1. A pair $(x(\cdot),u(\cdot)) \in \mathcal{A}$ is said to be a *process* of (P) if it satisfies (2b).

All classical problems of the calculus of variations, and in particular all problems from mechanics, can be formulated in the form of (P). For example, for the *basic problem of the calculus of variations* we put $\varphi = u$ and $\Omega = \mathbb{R}^n$.

2.2 The Maximum Principle

The best known first-order necessary optimality condition for the Lagrange problem (P) is the *maximum principle*. The maximum principle is one of the central results of optimal control theory. It was conjectured by L. S. Pontryagin and proved in the middle of 1950's by him and his collaborators (see [21]).

Theorem 1 (Maximum principle for the optimal control problem (P)). *If* $(x(\cdot), u(\cdot))$ *is an* optimal process *for problem* (P), *then there exists a nonzero pair of* Hamiltonian multipliers $(\psi_0, \psi(\cdot))$, *where* $\psi_0 \leq 0$ *is a constant and* $\psi(\cdot)$ *is an absolutely continuous n-dimensional vector function with domain* $[a, b]$, *such that for almost all* $t \in [a, b]$ *the quadruple* $(x(\cdot), u(\cdot), \psi_0, \psi(\cdot))$ *satisfies:*

(i) *the* Hamiltonian system

$$\begin{cases} \dot{x}(t) = \dfrac{\partial H}{\partial \psi}(t, x(t), u(t), \psi_0, \psi(t)) \\ \dot{\psi}(t) = -\dfrac{\partial H}{\partial x}(t, x(t), u(t), \psi_0, \psi(t)) \end{cases}$$

with the Hamiltonian

$$H(t, x, u, \psi_0, \psi) = \psi_0 L(t, x, u) + \psi \cdot \varphi(t, x, u) ; \tag{3}$$

(ii) *the* maximality condition

$$H(t, x(t), u(t), \psi_0, \psi(t)) = \sup_{u \in \Omega} H(t, x(t), u, \psi_0, \psi(t)) .$$

Furthermore, the function $H(t, x(t), u(t), \psi_0, \psi(t))$ *is an absolutely continuous function of* t *and satisfies the equality*

$$\frac{\mathrm{d}H(t, x(t), u(t), \psi_0, \psi(t))}{\mathrm{d}t} = \frac{\partial H}{\partial t}(t, x(t), u(t), \psi_0, \psi(t)) . \tag{4}$$

Definition 2. A quadruple $(x(\cdot), u(\cdot), \psi_0, \psi(\cdot))$ satisfying the conditions of Theorem 1 is called a *Pontryagin extremal* (or just *extremal* for brevity) for the problem (P). An extremal is called *normal* if $\psi_0 \neq 0$ and *abnormal* if $\psi_0 = 0$.

Classical Noether's theorem is formulated for Euler–Lagrange extremals of the basic problem of the calculus of variations. We will formulate our Noether theorem for Pontryagin extremals of optimal control problems.

Definition 3. A quantity $C(t, x, u, \psi_0, \psi)$ which is constant along every extremal $(x(\cdot), u(\cdot), \psi_0, \psi(\cdot))$ of (P) will be called a *conservation law*.

Definition 4. A quantity $C(t, x, u, \psi)$ which is constant along every abnormal extremal $(x(\cdot), u(\cdot), 0, \psi(\cdot))$ of (P) will be called an *abnormal conservation law*.

3 Main results

Both classical Noether's theorem for the calculus of variations (cf., e.g., [1], [2], [11], [15], [23]) and respective versions for optimal control (cf. [16], [25], [26]) have been formulated as implications: "invariance implies a conservation law". Here we shall begin with a weak notion of *invariance* (5) and formulate a necessary and sufficient condition (§3.1). Afterwards, in order to build conservation laws constructively, we impose a stronger notion of invariance (Def. 5) and derive as corollary (§3.2) a new version of Noether's theorem where the control variable is explicitly changed by the parameter-transformation.

3.1 A Necessary and Sufficient Condition

The following proposition is a consequence of Definition 3, the Hamiltonian system (i) in the maximum principle and equality (4).

Theorem 2. *The absolutely continuous function*

$$C(t,x,u,\psi_0,\psi) = \psi_0 F(t,x,u) + \psi \cdot X(t,x,u) - H(t,x,u,\psi_0,\psi)T(t,x,u)$$

is a conservation law, with H the Hamiltonian (3) associated to the problem (P), if and only if the equality

$$\psi_0 \left(\frac{\partial L}{\partial t}T + \frac{\partial L}{\partial x}\cdot X + L\frac{\mathrm{d}}{\mathrm{d}t}T - \frac{\mathrm{d}}{\mathrm{d}t}F\right) + \psi\cdot\left(\frac{\partial \varphi}{\partial t}T + \frac{\partial \varphi}{\partial x}\cdot X + \varphi\frac{\mathrm{d}}{\mathrm{d}t}T - \frac{\mathrm{d}}{\mathrm{d}t}X\right) = 0 \quad (5)$$

holds along every extremal $(x(\cdot),u(\cdot),\psi_0,\psi(\cdot))$ of (P).

Proof. By definition, C is a conservation law if and only if $\frac{\mathrm{d}}{\mathrm{d}t}C = 0$ along every extremal. Having in mind Theorem 1, the proof follows by direct computations:

$$\begin{aligned}
0 &= \frac{\mathrm{d}}{\mathrm{d}t}\left(\psi_0 F + \psi\cdot X - HT\right)\\
&= \psi_0\frac{\mathrm{d}F}{\mathrm{d}t} + \frac{\mathrm{d}}{\mathrm{d}t}\left(\psi\cdot X\right) - \frac{\mathrm{d}}{\mathrm{d}t}\left(HT\right)\\
&= \psi_0\frac{\mathrm{d}F}{\mathrm{d}t} + \dot{\psi}\cdot X + \psi\cdot\frac{\mathrm{d}X}{\mathrm{d}t} - \frac{\mathrm{d}H}{\mathrm{d}t}T - H\frac{\mathrm{d}T}{\mathrm{d}t}\\
&= \psi_0\frac{\mathrm{d}F}{\mathrm{d}t} - \frac{\partial H}{\partial x}\cdot X + \psi\cdot\frac{\mathrm{d}X}{\mathrm{d}t} - \frac{\partial H}{\partial t}T - \psi_0 L\frac{\mathrm{d}T}{\mathrm{d}t} - \psi\cdot\varphi\frac{\mathrm{d}T}{\mathrm{d}t}\\
&= -\psi_0\left(\frac{\partial L}{\partial t}T + \frac{\partial L}{\partial x}\cdot X + L\frac{\mathrm{d}T}{\mathrm{d}t} - \frac{\mathrm{d}F}{\mathrm{d}t}\right)\\
&\quad -\psi\cdot\left(\frac{\partial \varphi}{\partial t}T + \frac{\partial \varphi}{\partial x}\cdot X + \varphi\frac{\mathrm{d}T}{\mathrm{d}t} - \frac{\mathrm{d}X}{\mathrm{d}t}\right).
\end{aligned}$$

Remark 1. The maximality condition is the only condition of Theorem 1 which is not explicitly used in the proof of Theorem 2. It appears implicitly, however, as long as equality (4) is required.

3.2 Noether Theorem for Optimal Control

The following notion of invariance, admitting an explicit one-parameter-transformation of the control variable u, generalize previous notions of invariance used in connection with the Noether theorem.

Definition 5. Let h^s be a one-parameter family of C^1 mappings,

$$\begin{gathered} h^s : [a,b] \times \mathbb{R}^n \times \Omega \to \mathbb{R} \times \mathbb{R}^n \times \mathbb{R}^r , \\ h^s(t,x,u) = (\mathcal{T}(t,x,u,s), \mathcal{X}(t,x,u,s), \mathcal{U}(t,x,u,s)) , \\ h^0(t,x,u) = (t,x,u) , \quad \forall (t,x,u) \in [a,b] \times \mathbb{R}^n \times \Omega . \end{gathered} \tag{6}$$

If for all processes $(x(\cdot), u(\cdot))$ the following two conditions hold:

(i) There exists a function $\mathcal{F}(t,x,u,s) \in C^1([a,b], \mathbb{R}^n, \Omega; \mathbb{R})$ such that

$$\begin{aligned} L(t,x(t),u(t)) &+ \frac{\mathrm{d}}{\mathrm{d}t}\mathcal{F}(t,x(t),u(t),s) \\ &= L \circ h^s(t,x(t),u(t)) \frac{\mathrm{d}}{\mathrm{d}t}\mathcal{T}(t,x(t),u(t),s) ; \end{aligned}$$

(ii) $\frac{\mathrm{d}}{\mathrm{d}t}\mathcal{X}(t,x(t),u(t),s) = \varphi \circ h^s(t,x(t),u(t)) \frac{\mathrm{d}}{\mathrm{d}t}\mathcal{T}(t,x(t),u(t),s)$;

then the problem (P) is said to be *invariant* under the transformations h^s.

Remark 2. The functions $\mathcal{T}(t,x(t),u(t),s)$ and $\mathcal{X}(t,x(t),u(t),s)$ are both assumed to be differentiable in t.

Theorem 3. *If the Lagrange optimal control problem (P) is invariant under the family of transformations (6), then*

$$\psi_0 F(t,x,u) + \psi \cdot X(t,x,u) - H(t,x,u,\psi_0,\psi) T(t,x,u)$$

is a conservation law, where

$$\begin{aligned} T(t,x,u) &= \left.\frac{\partial \mathcal{T}(t,x,u,s)}{\partial s}\right|_{s=0} , \\ X(t,x,u) &= \left.\frac{\partial \mathcal{X}(t,x,u,s)}{\partial s}\right|_{s=0} , \\ F(t,x,u) &= \left.\frac{\partial \mathcal{F}(t,x,u,s)}{\partial s}\right|_{s=0} . \end{aligned}$$

Remark 3. For the basic problem of the calculus of variations, and for $\mathcal{F} \equiv 0$, Theorem 3 coincides with the Noether's theorem in [11]. The triple

$$(T(t,x,u), X(t,x,u), U(t,x,u))$$

is called the *tangent vector field* of $\{h^s\}$ and the constructed conservation law is the value of the Cartan differential 1-form $w = \psi \, \mathrm{d}x - H \, \mathrm{d}t$ on the tangent vector field. For a survey of these questions and the role of E. Noether's results in mechanics see [29].

Proof (Theorem 3). Let $(x(\cdot), u(\cdot), \psi_0, \psi(\cdot))$ be an arbitrary extremal of (P). By virtue of Theorem 2, we only need to prove that conditions (i) and (ii) of Definition 5 imply the equality (5) along the extremal. Differentiating the relations (i) and (ii) with respect to s, at $s = 0$ we get:

$$\frac{\mathrm{d}}{\mathrm{d}t}F = \frac{\partial L}{\partial t}T + \frac{\partial L}{\partial x} \cdot X + \frac{\partial L}{\partial u} \cdot U + L\frac{\mathrm{d}}{\mathrm{d}t}T\,; \tag{7a}$$

$$\frac{\mathrm{d}}{\mathrm{d}t}X = \frac{\partial \varphi}{\partial t}T + \frac{\partial \varphi}{\partial x} \cdot X + \frac{\partial \varphi}{\partial u} \cdot U + \varphi\frac{\mathrm{d}}{\mathrm{d}t}T\,; \tag{7b}$$

where all functions are evaluated at $(t, x(t), u(t))$ and

$$U(t, x(t), u(t)) = \left.\frac{\partial \mathcal{U}(t, x(t), u(t), s)}{\partial s}\right|_{s=0}.$$

Multiplying (7a) by ψ_0, (7b) by $\psi(t)$, and summing the two equalities, one gets:

$$\psi_0\left(\frac{\partial L}{\partial t}T + \frac{\partial L}{\partial x} \cdot X + \frac{\partial L}{\partial u} \cdot U + L\frac{\mathrm{d}}{\mathrm{d}t}T - \frac{\mathrm{d}}{\mathrm{d}t}F\right) \\ + \psi(t) \cdot \left(\frac{\partial \varphi}{\partial t}T + \frac{\partial \varphi}{\partial x} \cdot X + \frac{\partial \varphi}{\partial u} \cdot U + \varphi\frac{\mathrm{d}}{\mathrm{d}t}T - \frac{\mathrm{d}}{\mathrm{d}t}X\right) = 0\,. \tag{8}$$

According to the maximality condition (ii) of Theorem 1, the function

$$M_t(s) = \psi_0 L\left(t, x(t), \mathcal{U}\left(t, x(t), u(t), s\right)\right) + \psi(t) \cdot \varphi\left(t, x(t), \mathcal{U}\left(t, x(t), u(t), s\right)\right)$$

attains its maximum for $s = 0$. Therefore

$$\left.\frac{\mathrm{d}M_t(s)}{\mathrm{d}s}\right|_{s=0} = 0\,,$$

that is,

$$\psi_0\frac{\partial L\left(t, x(t), u(t)\right)}{\partial u} \cdot U\left(t, x(t), u(t)\right) \\ + \psi(t) \cdot \frac{\partial \varphi\left(t, x(t), u(t)\right)}{\partial u} \cdot U\left(t, x(t), u(t)\right) = 0\,. \tag{9}$$

Using (9) for simplification of (8) one obtains

$$\psi_0\left(\frac{\partial L}{\partial t}T + \frac{\partial L}{\partial x} \cdot X + L\frac{\mathrm{d}}{\mathrm{d}t}T - \frac{\mathrm{d}}{\mathrm{d}t}F\right) \\ + \psi(t) \cdot \left(\frac{\partial \varphi}{\partial t}T + \frac{\partial \varphi}{\partial x} \cdot X + \varphi\frac{\mathrm{d}}{\mathrm{d}t}T - \frac{\mathrm{d}}{\mathrm{d}t}X\right) = 0\,,$$

and the proof is complete.

Remark 4. For $\psi \cdot X(t, x, u) - H(t, x, u, \psi_0, \psi)T(t, x, u)$ to be an abnormal conservation law under the family of transformations (6), it suffices that condition (ii) of Definition 5 holds.

4 Examples

It is not difficult to find examples for which Theorem 3 provides conservation laws which are not easily derived from the previously known results. Two such examples follow.

Example 1 $(n = r = 1)$.

$$J[x(\cdot), u(\cdot)] = \int_a^b \mathrm{e}^{tx(t)} u(t)\, \mathrm{d}t \longrightarrow \min\,,$$

$$\dot{x}(t) = tx(t)u(t)^2\,, \quad u \in \Omega\,.$$

We claim that

$$\psi_0 t \mathrm{e}^{tx(t)} u(t) + \psi(t)x(t)\left((tu(t))^2 + 1\right)$$

is constant along any Pontryagin extremal $(x(\cdot), u(\cdot), \psi_0, \psi(\cdot))$ of the problem. Indeed, setting

$$\mathcal{T} = \mathrm{e}^{-s} t\,,$$
$$\mathcal{X} = \mathrm{e}^{s} x\,,$$
$$\mathcal{U} = \mathrm{e}^{s} u\,,$$
$$\mathcal{F} \equiv 0\,,$$

it follows from Theorem 3 that

$$\psi x + Ht$$

is a conservation law.

The conservation law holds independently of the control set Ω, and can be viewed as a necessary optimality condition.

Example 2 $(n = 3,\ r = 2)$.

$$\int_a^b (u_1(t))^2 + (u_2(t))^2\ \mathrm{d}t \longrightarrow \min\,,$$

$$\begin{cases} \dot{x_1}(t) & = u_1(t) \\ \dot{x_2}(t) & = u_2(t) \\ \dot{x_3}(t) & = u_1(t)\,(x_2(t))^2\,. \end{cases}$$

Choosing $\mathcal{F} \equiv 0$ and

$$\mathcal{T} = \mathrm{e}^{-2s} t\,, \mathcal{X}_1 = \mathrm{e}^{-s} x_1\,, \mathcal{X}_2 = \mathrm{e}^{-s} x_2\,, \mathcal{X}_3 = \mathrm{e}^{-3s} x_3\,, \mathcal{U}_1 = \mathrm{e}^{s} u_1\,, \mathcal{U}_2 = \mathrm{e}^{s} u_2\,,$$

one concludes from Theorem 3 that

$$\psi_1 x_1 + \psi_2 x_2 + 3\psi_3 x_3 - 2Ht$$

is a conservation law for the Pontryagin extremals

$$(x_1(\cdot), x_2(\cdot), x_3(\cdot), u_1(\cdot), u_2(\cdot), \psi_0, \psi_1(\cdot), \psi_2(\cdot), \psi_3(\cdot))\,.$$

In previous results (cf. [16,25]) no time transformation appears in the autonomous case. Example 2 is autonomous but transformation of the time-variable is required in order to obtain the conservation law.

Acknowledgments

I would like to thank Emmanuel Trélat for the stimulating conversations during the workshop; Andrey Sarychev for the suggestions regarding improvement of the text.

References

1. Alekseev V. M., Tikhomirov V. M., Fomin S. V. (1987) Optimal Control. Plenum, New York
2. Arnold V. I. (1987) Métodos Matemáticos da Mecânica Clássica. Mir, Moscovo
3. Arnold V. I. (1988) Geometrical Methods in the Theory of Ordinary Differential Equations. Springer-Verlag, New York
4. Baez J. C., Gilliam J. W. (1994) An Algebraic Approach to Discrete Mechanics. Lett. Math. Phys. **31**, 205–212
5. Blankenstein G., van der Schaft A. J. (1999) Symmetry and Reduction in Implicit Generalized Hamiltonian Systems. Memorandum No. 1489, Faculty of Mathematical Sciences, University of Twente, The Netherlands
6. Cesari L. (1983) Optimization – Theory and Applications. Springer-Verlag, New York
7. Clarke F. H. (1993) An Indirect Method in the Calculus of Variations. Trans. Amer. Math. Soc. **336**, 655–673
8. Clarke F. H., Vinter R. B. (1985) Regularity Properties of Solutions to the Basic Problem in the Calculus of Variations. Trans. Amer. Math. Soc. **289**, 73–98
9. Erdmann G. (1877) Über Unstetige Lösungen in der Variationsrechnung. J. für reine und angew. Math. **82**, 21–30
10. Euler L. (1744) Methodus Inveniendi Lineas Curvas Maximi Minimive Proprietate Gaudentes, Sive Solutio Problematis Isoperimetrici Lattisimo Sensu Accepti. Bousquet, Lausannae et Genevae, E65A.O.O. Ser. I **24**
11. Gelfand I. M., Fomin S. V. (2000) Calculus of Variations. Dover, New York
12. Gilliam J. W. (1996) Lagrangian and Symplectic Techniques in Discrete Mechanics. Ph.D. dissertation, Univ. California Riverside
13. Grizzle J. W., Marcus S. I. (1985) The Structure of Nonlinear Control Systems Possessing Symmetries. IEEE Trans. Automat. Control **30**, 248–258
14. Heinricher A. C., Mizel V. J. (1988) The Lavrentiev Phenomenon for Invariant Variational Problems. Arch. Rational Mech. Anal. **102**, 57–93
15. Jost J., Li-Jost X. (1998) Calculus of Variations. Cambridge University Press, Cambridge
16. Jurdjevic V. (1997) Geometric Control Theory. Cambridge University Press, Cambridge
17. Jurdjevic V. (1999) Optimal Control, Geometry, and Mechanics. In: Baillieul J., Willems J. C. (Eds.) Mathematical Control Theory, Springer, New York, 227–267
18. Nijmeijer H., van der Schaft A. J. (1990) Nonlinear Dynamical Control Systems. Springer-Verlag, New York
19. Noether E. (1918) Invariante Variationsprobleme. Nachr. Akad. Wissen. Göttingen, Math.-Phys. Kl. II, 235–257 (see Transport Theory and Stat. Phys. **1** (1971), 186–207 for an English translation)

20. Olver P. J. (1993) Applications of Lie Groups to Differential Equations. Springer-Verlag, New York
21. Pontryagin L. S., Boltyanskii V. G., Gamkrelidze R. V., Mischenko E. F. (1962) The Mathematical Theory of Optimal Processes. John Wiley, New York
22. Respondek W. (1982) On Decomposition of Nonlinear Control Systems. Syst. Contr. Lett. **1**, 301–308
23. Rund H. (1966) The Hamilton–Jacobi Theory in the Calculus of Variations, Its Role in Mathematics and Physics. D. Van Nostrand Co., Ltd., London–Toronto, Ont.–New York
24. Sarychev A. V., Torres D. F. M. (2000) Lipschitzian Regularity of Minimizers for Optimal Control Problems with Control-Affine Dynamics. Appl. Math. Optim. **41**, 237–254
25. Sussmann H. J. (1995) Symmetries and Integrals of Motion in Optimal Control. In: Fryszkowski A., Jakubczyk B., Respondek W., Rzezuchowski T. (Eds.) Geometry in Nonlinear Control and Differential Inclusions, Banach Center Publications **32**, Mathematics Institute of the Polish Academy of Sciences, Warsaw, Poland, 379–393
26. Torres D. F. M. (2000) On the Noether Theorem for Optimal Control. Cadernos de Matemática CM00/I-13, Dep. Mat., Univ. Aveiro, Portugal. Accepted to European Control Conference ECC 2001. To appear in the proceedings.
27. van der Schaft A. (1981/82) Symmetries and Conservation Laws for Hamiltonian Systems With Inputs and Outputs: A Generalization of Noether's Theorem. Systems Control Lett. **1**, 108–115
28. van der Schaft A. (1987) Symmetries in Optimal Control. SIAM J. Control and Optimization **25**, 245–259.
29. Yemelyanova I. S. (1999) Some Properties of Hamiltonian Symmetries. European women in mathematics (Trieste, 1997), Hindawi Publ. Corp., Stony Brook, New York, 193–203

Participants of the 3rd NCN Workshop "Dynamics, Bifurcations and Control" Kloster Irsee, Germany, April 1–3, 2001

Prof. Abed, Eyad
Department of Electrical Engineering and Systems Research Center
University of Maryland, College Park, MD 20742, USA
abed@Glue.umd.edu

Dr. Atay, Fatihcan M.
Artesis A.S.
Tuzla, Istanbul 81719, Turkey
atay@member.ams.org

Prof. Baños, Alfonso
Facultad de Informática
Universidad de Murcia, 30100 Campus de Espinardo, Spain
abanos@dif.um.es

Prof. Bacciotti, Andrea
Dipartimento di Matematica
Politecnico di Torino, Corso Duca Degli Abruzzi 24, 10129 Torino, Italy
Bacciotti@Polito.lt

Dr. Barreiro, Antonio
Universidade de Vigo
Departamento de Ingenieria de Sistemas y Automática, Spain
abarreiro@uvigo.es

Dr. Chambrion, Thomas
Universite de Bourgogne
Laboratoire Analyse Numérique UFR Sciences et Techniques
9, Avenue Alain SAVARY BP 47870, 21078 DIJON Cedex, France
Thomas.Chambrion@u-bourgogne.fr

Dr. Clemente–Gallardo, Jesús
Faculty of Math. Sciences, Dept. of Signals, Systems and Control
University of Twente, P.O.Box 217, 7500 AE Enschede, The Netherlands
jesus@math.utwente.nl

Prof. Colonius, Fritz
Institut für Mathematik
Universität Augsburg, 86135 Augsburg, Germany
fritz.colonius@math.uni-augsburg.de

F. Colonius, L. Grüne (Eds.): Dynamics, Bifurcations, and Control, LNCIS 273, pp. 297–301, 2002.

Dr. Crauel, Hans
Institut für Mathematik
Technische Universität Ilmenau, Weimarer Straße 25,
98693 Ilmenau, Germany
crauel@mathematik.tu-ilmenau.de

Dipl.–Math. Dirr, Gunter
Fakultät für Mathematik und Informatik
Universität Würzburg, Am Hubland, 97074 Würzburg, Germany

Dr. Fabbri, Roberta
Dipartimento di Sistemi e Informatica
Università di Firenze, Via di S. Marta, 3, 50139 Firenze, Italy
fabbri@dsi.unifi.it

Dipl.–Ing. Feiler, Matthias
Lehrstuhl für elektrische Antriebssysteme
Technische Universität München, 80290 München, Germany

Dipl.–Math. Gayer, Tobias
Institut für Mathematik
Universität Augsburg, 86135 Augsburg, Germany
tobias.gayer@math.uni-augsburg.de

Prof. Giovanardi, Lorenzo
Dipartimento di Sistemi e Informatica
Università di Firenze, Via di S. Marta, 3, 50139 Firenze, Italy
giovanardi@dsi.unifi.it

Dr. Gordillo, Francisco
Escuela Superior de Ingenieros
Universidad de Sevilla, Paseo de los Descubrimientos s/n, 41092 Sevilla, Spain
gordillo@esi.us.es

Dr. Grüne, Lars
Fachbereich Mathematik
J.W. Goethe–Universität, Postfach 111932,
60054 Frankfurt am Main, Germany
gruene@math.uni-frankfurt.de

Mr. Hatonen, Jari
Department of Automatic Control and Systems Engineering
University of Sheffield, Mappin Street, Sheffield, S1 3JD, United Kingdom
J.Hatonen@sheffield.ac.uk

Mr. Hecker, Simon
Lehrstuhl für Steuerungs– und Regelungstechnik
Technische Universität München, 80290 München, Germany
simon.hecker@gmx.de

Prof. Helmke, Uwe
Fakultät für Mathematik und Informatik
Universität Würzburg, Am Hubland, 97074 Würzburg, Germany
helmke@mathematik.uni-wuerzburg.de

Prof. Ilchmann, Achim
Institut für Mathematik
Technische Universität Ilmenau, Weimarer Straße 25,
98693 Ilmenau, Germany
ilchmann@mathematik.tu-ilmenau.de

Mr. Jerouane, Mohamed
Laboratoire des Signaux et Systèmes
LSS, CNRS SUPELEC, 91190 Gif-sur-Yvette, France
jerouane@lss.supelec.fr

Prof. Johnson, Russell
Dipartimento di Sistemi e Informatica
Università di Firenze, Via di S. Marta 3, 50139 Firenze, Italy
johnson@ingfi1.ing.unifi.it

Prof. Kang, Wei
Deartment of Mathematics
Naval Postgraduate School, Monterey, CA 93943, USA
wkang@nps.navy.mil

Prof. Krener, Arthur J.
Department of Mathematics
University of California, One Shields Ave, Davis, CA 95616-8633, USA
ajkrener@ucdavis.edu

Mr. Lakehal–Ayat, Mohsen
Laboratoire des Signaux et Systèmes
LSS, CNRS SUPELEC, 91190 Gif-sur-Yvette, France

Dr. Mareczek, Jörg
Lehrstuhl für Steuerungs– und Regelungstechnik
Technische Universität München, 80290 München, Germany
mareczek@ei.tum.de

Dipl.–Math. Marquardt, Albert
Institut für Mathematik
Universität Augsburg, 86135 Augsburg, Germany

Prof. Moiola, Jorge Luis
Universidad Nacional del Sur, Bahia Blanca, Argentina
Currently at: Mathematisches Institut
Universität zu Köln Weyertal 86-90, 50931 Köln, Germany
jmoiola@MI.Uni-Koeln.DE

Prof. Piccardi, Carlo
Dipartimento di Elettronica e Informazione
Politecnico di Milano, Piazza Leonardo da Vinci 32, 20133 Milano, Italy
piccardi@elet.polimi.it

Dr. Pisano, Alessandro
Dipartimento di Ingegneria Elettrica ed Elettronica
Università di Cagliari, Piazza D'Armi, 09123 Cagliari, Italy
pisano@diee.unica.it

Dr. Raimúndez, José Cesáreo
Departamento de Ingenieria de Sistemas y Automática
Universidade de Vigo, CEP 08544, Spain
cesareo@uvigo.es

Prof. Respondek, Witold
Laboratoire de Mathématiques, BP 08
Institut National des Sciences Appliquées de Rouen
Place Emile Blondel, 76131 Mont Saint Aignan, Cedex, France
wresp@lmi.insa-rouen.fr

Mrs. Ricardo, Sandra
Fakultät für Mathematik und Informatik
Universität Würzburg, Am Hubland, 97074 Würzburg, Germany

Dr. Schneider, Klaus
Weierstraß–Institut für Angewandte Analysis und Stochastik
Mohrenstraße 39, 10117 Berlin, Germany
schneider@wias-berlin.de

Dr. Schweiger, Christian
Lehrstuhl für Steuerungs– und Regelungstechnik
Technische Universität München, 80290 München, Germany
schweiger@lsr.ei.tum.de

Prof. Sepulchre, Rodolphe
Institut Montefiore, B28
University of Liège, B-4000 Liège Sart-Tilman, Belgium
r.sepulchre@ulg.ac.be

Prof. Soravia, Pierpaolo
Dipartimento di Matematica Pura e Applicata
Università di Padova, Via Belzoni 7, 35131 Padova, Italy
soravia@math.unipd.it

Dr. Spadini, Marco
Universita di Firenze Dipartimento di Matematica Applicata "G. Sansone"
Via di S. Marta 3, 50139 Firenze, Italy
spadini@dma.unifi.it

Dr. Szolnoki, Dietmar
Institut für Mathematik
Universität Augsburg, 86135 Augsburg, Germany
szolnoki@math.uni-augsburg.de

Dr. Tall, Issa A.
Laboratoire de Mathématiques, BP 08
Institut National des Sciences Appliquées de Rouen
Place Emile Blondel, 76131 Mont Saint Aignan, Cedex, France

Prof. Tesi, Alberto
Dipartimento di Sistemi e Informatica
Università di Firenze, Via di S. Marta, 3, 50139 Firenze, Italy
atesi@dsi.unifi.it

Prof. Torres, Delfim Fernando Marado
Departamento de Matematica
Universidade de Aveiro,
Campus Universitario de Santiago 3810-193 Aveiro, Portugal
delfim@mat.ua.pt

Dr. Trélat, Emmanuel
Laboratoire de Topologie – UMR 5584 du CNRS,
UFR des Sciences et Techniques, Université de Bourgogne
9, avenue Alain Savary, B.P. 47870, 21078 Dijon Cedex, France
trelat@topolog.u-bourgogne.fr

Prof. van der Schaft, Arjan
Department of Applied Mathematics
University of Twente, P.O.Box 217, 7500 Enschede, Netherlands
twarjan@math.utwente.nl

Dr. Wirth, Fabian
Zentrum für Technomathematik
Universität Bremen, Postfach 330 440, 28334 Bremen, Germany
fabian@math.uni-bremen.de

Dr. Zhang, Qinghua
IRISA-INRIA, Rennes
IRISA Campus de Beaulieu, 35042 Rennes Cedex, France
zhang@irisa.fr

Lecture Notes in Control and Information Sciences

Edited by N. Thoma and M. Morari
1998–2001 Published Titles:

Vol. 231: Emel'yanov, S.V.; Burovoi, I.A.; Levada, F.Yu.
Control of Indefinite Nonlinear Dynamit Systems
196 pp. 1998 [3-540-76245-0]

Vol. 232: Casals, A.; de Almeida, A.T. (Eds)
Experimental Robotics V: The Fifth International Symposium Barcelona, Catalonia, June 15–18, 1997
190 pp. 1998 [3-540-76218-3]

Vol. 233: Chiacchio, P.; Chiaverini, S. (Eds)
Complex Robotic Systems
189 pp. 1998 [3-540-76265-5]

Vol. 234: Arena, P.; Fortuna, L.; Muscato, G.; Xibilia, M.G.
Neurat Networks in Multidimensional Domains: Fundamentals and New Trends in Modelling and Control
179 pp. 1998 [1-85233-006-6]

Vol. 235: Chen, B.M.
$H\infty$ Control and Its Applications
361 pp. 1998 [1-85233-026-0]

Vol. 236: de Almeida, A.T.; Khatib, O. (Eds)
Autonomous Robotic Systems
283 pp. 1998 [1-85233-036-8]

Vol. 237: Kreigman, D.J.; Hagar, G.D.; Morse, A.S. (Eds)
The Confluence of Vision and Control
304 pp. 1998 [1-85233-025-2]

Vol. 238: Elia, N.; Dahleh, M.A.
Computational Methods for Controller Design
200 pp. 1998 [1-85233-075-9]

Vol. 239: Wang, Q.G.; Lee, T.H.; Tan, K.K.
Finite Spectrum Assignment for Time-Delay Systems
200 pp, 1998 [1-85233-065-1]

Vol. 240: Lin, Z.
Low Gain Feedback
376 pp. 1999 [1-85233-081-3]

Vol. 241: Yamamoto, Y.; Hara S.
Learning, Control and Hybrid Systems
472 pp. 1999 [1-85233-076-7]

Vol. 242: Conte, G.; Moog, C.H.; Perdon A.M.
Nonlinear Control Systems
192 pp. 1999 [1-85233-151-8]

Vol. 243: Tzafestas, S.G.; Schmidt, G. (Eds)
Progress in Systems and Robot Analysis and Control Design
624 pp. 1999 [1-85233-123-2]

Vol. 244: Nijmeijer, H.; Fossen, T.I. (Eds)
New Directions in Nonlinear Observer Design
552 pp. 1999 [1-85233-134-8]

Vol. 245: Garulli, A.; Tesi, A.; Vicino, A. (Eds)
Robustness in Identification and Control
448 pp. 1999 [1-85233-179-8]

Vol. 246: Aeyels, D.; Lamnabhi-Lagarrigue, F.; van der Schaft, A. (Eds)
Stability and Stabilization of Nonlinear Systems
408 pp. 1999 [1-85233-638-2]

Vol. 247: Young, K-D.; Özgüner, Ü. (Eds)
Variable Structure Systems, Sliding Mode and Nonlinear Control
400 pp. 1999 [1-85233-197-6]

Vol. 248: Chen, Y.; Wen C.
Iterative Learning Control
216 pp. 1999 [1-85233-190-9]

Vol. 249: Cooperman, G.; Jessen, E.; Michler, G. (Eds)
Workshop on Wide Area Networks and High Performance Computing
352 pp. 1999 [1-85233-642-0]

Vol. 250: Corke, P.; Trevelyan, J. (Eds)
Experimental Robotics VI
552 pp. 2000 [1-85233-210-7]

Vol. 251: van der Schaft, A.; Schumacher, J.
An Introduction to Hybrid Dynamical Systems
192 pp. 2000 [1-85233-233-6]

Vol. 252: Salapaka, M.V.; Dahleh, M.
Multiple Objective Control Synthesis
192 pp. 2000 [1-85233-256-5]

Vol. 253: Elzer, P.F.; Kluwe, R.H.; Boussoffara, B.
Human Error and System Design and Management
240 pp. 2000 [1-85233-234-4]

Vol. 254: Hammer, B.
Learning with Recurrent Neural Networks
160 pp. 2000 [1-85233-343-X]

Vol. 255: Leonessa, A.; Haddad, W.H.; Chellaboina V.
Hierarchical Nonlinear Switching Control Design with Applications to Propulsion Systems
152 pp. 2000 [1-85233-335-9]

Vol. 256: Zerz, E.
Topics in Multidimensional Linear Systems Theory
176 pp. 2000 [1-85233-336-7]

Vol. 257: Moallem, M.; Patel, R.V.; Khorasani, K.
Flexible-link Robot Manipulators
176 pp. 2001 [1-85233-333-2]

Vol. 258: Isidori, A.; Lamnabhi-Lagarrigue, F.; Respondek, W. (Eds)
Nonlinear Control in the Year 2000 Volume 1
616 pp. 2001 [1-85233-363-4]

Vol. 259: Isidori, A.; Lamnabhi-Lagarrigue, F.; Respondek, W. (Eds)
Nonlinear Control in the Year 2000 Volume 2
640 pp. 2001 [1-85233-364-2]

Vol. 260: Kugi, A.
Non-linear Control Based on Physical Models
192 pp. 2001 [1-85233-329-4]

Vol. 261: Talebi, H.A.; Patel, R.V.; Khorasani, K.
Control of Flexible-link Manipulators Using Neural Networks
168 pp. 2001 [1-85233-409-6]

Vol. 262: Dixon, W.; Dawson, D.M.; Zergeroglu, E.; Behal, A.
Nonlinear Control of Wheeled Mobile Robots
216 pp. 2001 [1-85233-414-2]

Vol. 263: Galkowski, K.
State-space Realization of Linear 2-D Systems with Extensions to the General nD (n>2) Case
248 pp. 2001 [1-85233-410-X]

Vol. 264: Baños, A.; Lamnabhi-Lagarrigue, F.; Montoya, F.J
Advances in the Control of Nonlinear Systems
344 pp. 2001 [1-85233-378-2]

Vol. 265: Ichikawa, A.; Katayama, H.
Linear Time Varying Systems and Sampled-data Systems
376 pp. 2001 [1-85233-439-8]

Vol. 266: Stramigioli, S.
Modeling and IPC Control of Interactive Mechanical Systems – A Coordinate-free Approach
296 pp. 2001 [1-85233-395-2]

Vol. 267: Bacciotti, A.; Rosier, L.
Liapunov Functions and Stability in Control Theory
224 pp. 2001 [1-85233-419-3]

Vol. 268: Moheimani, S.O.R. (Ed)
Perspectives in Robust Control
390 pp. 2001 [1-85233-452-5]

Vol. 269: Niculescu, S.-I.
Delay Effects on Stability
400 pp. 2001 [1-85233-291-316]

Vol. 270: Nicosia, S. et al.
RAMSETE
294 pp. 2001 [3-540-42090-8]

Vol. 271: Rus, D.; Singh, S.
Experimental Robotics VII
585 pp. 2001 [3-540-42104-1]

Vol. 272: Yang, T.
Impulsive Control Theory
363 pp. 2001 [3-540-42296-X]

Vol. 252: Salapaka, M.V.; Dahleh, M.
Multiple Objective Control Synthesis
192 pp. 2000 [1-85233-256-5]

Vol. 253: Elzer, P.F.; Kluwe, R.H.; Boussoffara, B.
Human Error and System Design and Management
240 pp. 2000 [1-85233-234-4]

Vol. 254: Hammer, B.
Learning with Recurrent Neural Networks
160 pp. 2000 [1-85233-343-X]

Vol. 255: Leonessa, A.; Haddad, W.H.; Chellaboina V.
Hierarchical Nonlinear Switching Control Design with Applications to Propulsion Systems
152 pp. 2000 [1-85233-335-9]

Vol. 256: Zerz, E.
Topics in Multidimensional Linear Systems Theory
176 pp. 2000 [1-85233-336-7]

Vol. 257: Moallem, M.; Patel, R.V.; Khorasani, K.
Flexible-link Robot Manipulators
176 pp. 2001 [1-85233-333-2]

Vol. 258: Isidori, A.; Lamnabhi-Lagarrigue, F.; Respondek, W. (Eds)
Nonlinear Control in the Year 2000 Volume 1
616 pp. 2001 [1-85233-363-4]

Vol. 259: Isidori, A.; Lamnabhi-Lagarrigue, F.; Respondek, W. (Eds)
Nonlinear Control in the Year 2000 Volume 2
640 pp. 2001 [1-85233-364-2]

Vol. 260: Kugi, A.
Non-linear Control Based on Physical Models
192 pp. 2001 [1-85233-329-4]

Vol. 261: Talebi, H.A.; Patel, R.V.; Khorasani, K.
Control of Flexible-link Manipulators Using Neural Networks
168 pp. 2001 [1-85233-409-6]

Vol. 262: Dixon, W.; Dawson, D.M.; Zergeroglu, E.; Behal, A.
Nonlinear Control of Wheeled Mobile Robots
216 pp. 2001 [1-85233-414-2]

Vol. 263: Galkowski, K.
State-space Realization of Linear 2-D Systems with Extensions to the General nD (n>2) Case
248 pp. 2001 [1-85233-410-X]

Vol. 264: Baños, A.; Lamnabhi-Lagarrigue, F.; Montoya, F.J
Advances in the Control of Nonlinear Systems
344 pp. 2001 [1-85233-378-2]

Vol. 265: Ichikawa, A.; Katayama, H.
Linear Time Varying Systems and Sampled-data Systems
376 pp. 2001 [1-85233-439-8]

Vol. 266: Stramigioli, S.
Modeling and IPC Control of Interactive Mechanical Systems – A Coordinate-free Approach
296 pp. 2001 [1-85233-395-2]

Vol. 267: Bacciotti, A.; Rosier, L.
Liapunov Functions and Stability in Control Theory
224 pp. 2001 [1-85233-419-3]

Vol. 268: Moheimani, S.O.R. (Ed)
Perspectives in Robust Control
390 pp. 2001 [1-85233-452-5]

Vol. 269: Niculescu, S.-I.
Delay Effects on Stability
400 pp. 2001 [1-85233-291-316]

Vol. 270: Nicosia, S. et al.
RAMSETE
294 pp. 2001 [3-540-42090-8]

Vol. 271: Rus, D.; Singh, S.
Experimental Robotics VII
585 pp. 2001 [3-540-42104-1]

Vol. 272: Yang, T.
Impulsive Control Theory
363 pp. 2001 [3-540-42296-X]

Printing (Computer to Film): Saladruck Berlin
Binding: Stürtz AG, Würzburg